AF522368

Wheat Crop Cultivation Management

Wheat Crop Cultivation Management

Shashi Shekhar Prasad

RANDOM PUBLICATIONS
NEW DELHI (INDIA)

Wheat Crop Cultivation Management

ISBN 978-93-5111-781-0

Published in 2016 in India by
RANDOM PUBLICATIONS
4376-A/4B, Gali Murari Lal, Ansari Road
New Delhi-110 002
Phone : +9111-43580356, 011-23289044, 011-43142548
e-mail: sales@randompublications.com,
info@randompublications.com, randomexports@gmail.com

Reprint 2023

Type Setting by : Friends Media, Delhi-110089
Digitally Printed at: Replika Press Pvt. Ltd.

Preface

Wheat is the most widely grown crop in the world and is produced under a wide range of environments. The delineation of these production environments is largely conditioned by temperature and moisture availability. By far the most important agro-ecological zone for wheat production in the developing world is the temperate, irrigated condition characterized mainly by large areas in northwest India, Pakistan, Nepal and northern Bangladesh, south-central China, Afghanistan, parts of West Asia and the central Asian republics. Asia now produces over 90 percent of irrigated wheat in the developing world. There are smaller, yet important temperate, irrigated wheat production areas located in North Africa, Zimbabwe, South Africa, northwest Mexico and Chile. There are a multitude of cropping systems that occur where spring wheat is grown under temperate conditions with irrigation, and by far the largest such wheat production areas are found in South and East Asia. For example, India, Pakistan, Bangladesh and Nepal have a total wheat area of over 34 million ha.

In more and more irrigated production areas where wheat is important, there has been the trend in recent years for marked reductions in the diversification of crop rotations. The current prevalence of continuous rice-wheat over large areas in northwest India is a classic example. The economic realities that farmers must face have, unfortunately, brought this to bear. However, data are now being accumulated that strongly indicate that such narrow rotation patterns can exacerbate production declines, especially when continuous cereal crop rotations are followed as compared to more complex rotation systems, particularly those including legume or *Brassica* spp. crops.

– ***Author***

Contents

1

Introduction

Wheat is a cereal grain, originally from the Levant region of the Near East but now cultivated worldwide. In 2013, world production of wheat was 713 million tons, making it the third most-produced cereal after maize (1,016 million tons) and rice (745 million tons). Wheat was the second most-produced cereal in 2009; world production in that year was 682 million tons, after maize (817 million tons), and with rice as a close third (679 million tons).

This grain is grown on more land area than any other commercial food. World trade in wheat is greater than for all other crops combined. Globally, wheat is the leading source of vegetable protein in human food, having a higher protein content than other major cereals, maize (corn) or rice. In terms of total production tonnages used for food, it is currently second to rice as the main human food crop and ahead of maize, after allowing for maize's more extensive use in animal feeds.

Wheat was a key factor enabling the emergence of city-based societies at the start of civilization because it was one of the first crops that could be easily cultivated on a large scale, and had the additional advantage of yielding a harvest that provides long-term storage of food. Wheat contributed to the emergence of city-states in the Fertile Crescent, including the Babylonian and Assyrian empires. Wheat grain is a staple food used to make flour for leavened, flat and steamed breads, biscuits, cookies, cakes, breakfast cereal, pasta, noodles, couscous and for fermentation to make beer, other alcoholic beverages, and biofuel.

There are six wheat classifications:

- Hard red winter,
- Hard red spring,
- Soft red winter,
- Durum (hard),
- Hard white, and
- Soft white wheat.

The hard wheats have the most amount of gluten and are used for making bread, rolls and all-purpose flour. The soft wheats are used for making flat bread, cakes, pastries, crackers, muffins, and biscuits. A high percentage of wheat production in the EU is used as animal feed, often surplus to human requirements or low-quality wheat.

Wheat is planted to a limited extent as a forage crop for livestock, although the straw cannot be used as feed. Its straw can be used as a construction material for roofing thatch. The whole grain can be milled to leave just the endosperm for white flour. The by-products of this are bran and germ. The whole grain is a concentrated source of vitamins, minerals, and protein, while the refined grain is mostly starch.

Wheat is one of the first cereals known to have been domesticated, and wheat's ability to self-pollinate greatly facilitated the selection of many distinct domesticated varieties. The archaeological record suggests that this first occurred in the regions known as the Fertile Crescent. Recent findings estimate the first domestication of wheat down to a small region of southeastern Turkey, and domesticated Einkorn wheat at Wadi el Jilat in Jordan—has been dated to 7,500-7,300 BCE.

ORIGIN

Cultivation and repeated harvesting and sowing of the grains of wild grasses led to the creation of domestic strains, as mutant forms ('sports') of wheat were preferentially chosen by farmers. In domesticated wheat, grains are larger, and the seeds (inside the spikelets) remain attached to the ear by a toughened rachis during harvesting. In wild strains, a more fragile rachis allows the ear to easily shatter and disperse the spikelets. Selection for these traits by farmers might not have been deliberately intended, but simply have occurred because these traits made gathering the seeds easier; nevertheless such 'incidental' selection was an important part of crop domestication. As the traits that improve wheat as a food source *also* involve the loss of the plant's natural seed dispersal mechanisms, highly domesticated strains of wheat cannot survive in the wild.

Cultivation of wheat began to spread beyond the Fertile Crescent after about 8000 BCE. Jared Diamond traces the spread of cultivated emmer wheat starting in the Fertile Crescent sometime before 8800 BCE. Archaeological analysis of wild *emmer* indicates that it was first cultivated in the southern Levant with finds dating back as far as 9600 BCE. Genetic analysis of wild *einkorn* wheat suggests that it was first grown in the Karacadag Mountains in southeastern Turkey. Dated archeological remains of einkorn wheat in settlement sites near this region, including those at Abu Hureyra in Syria, suggest the domestication of einkorn near the Karacadag Mountain Range. With the anomalous exception of two grains from Iraq ed-Dubb, the earliest carbon-14 date for einkorn wheat remains at Abu Hureyra is 7800 to 7500 years BCE.

Remains of harvested emmer from several sites near the Karacadag Range have been dated to between 8600 (at Cayonu) and 8400 BCE (Abu Hureyra), that is, in the Neo-lithic period. With the exception of Iraq ed-Dubb, the earliest carbon-14 dated remains of domesticated emmer wheat were found in the earliest levels of Tell Aswad, in the Damascus basin, near Mount Hermon in Syria. These remains were dated by Willem van Zeist and his assistant Johanna Bakker-Heeres to 8800 BCE. They also concluded that the settlers of Tell Aswad did not develop this form of emmer themselves, but brought the domesticated grains with them from an as yet unidentified location elsewhere.

The cultivation of emmer reached Greece, Cyprus and India by 6500 BCE, Egypt shortly after 6000 BCE, and Germany and Spain by 5000 BCE. "The early Egyptians were developers of bread and the use of the oven and developed baking into one of the first large-scale food production industries." By 3000 BCE, wheat had reached England and Scandinavia. A millennium later it reached China. The first identifiable bread wheat (*Triticum aestivum*) with sufficient gluten for yeasted breads has been identified using DNA analysis in samples from a granary dating to approximately 1350 BCE at Assiros in Greek Macedonia. Wheat continued to spread throughout Europe. In England, wheat straw (thatch) was used for roofing in the Bronze Age, and was in common use until the late 19th century.

FARMING TECHNIQUES

Technological advances in soil preparation and seed placement at planting time, use of crop rotation and fertilizers to improve plant growth, and advances in harvesting methods have all combined to promote wheat as a viable crop. Agricultural cultivation using horse collar leveraged plows (at about 3000 BCE) was one of the first innovations that increased productivity. Much later, when the use of seed drills replaced broadcasting sowing of seed in the 18th century, another great increase in productivity occurred.

Yields of pure wheat per unit area increased as methods of crop rotation were applied to long cultivated land, and the use of fertilizers became widespread. Improved agricultural husbandry has more recently included threshing machines and reaping machines (the 'combine harvester'), tractor-drawn cultivators and planters, and better varieties. Great expansion of wheat production occurred as new arable land was farmed in the Americas and Australia in the 19th and 20th centuries.

GENETICS

Wheat genetics is more complicated than that of most other domesticated species. Some wheat species are diploid, with two sets of chromosomes, but many are stable polyploids, with four sets of chromosomes (tetraploid) or six (hexaploid).

- Einkorn wheat (*T. monococcum*) is diploid (AA, two complements of seven chromosomes, 2n=14).

- Most tetraploid wheats (*e.g.* emmer and durum wheat) are derived from wild emmer, *T. dicoccoides*. Wild emmer is itself the result of a hybridization between two diploid wild grasses, *T. urartu* and a wild goatgrass such as *Aegilops searsii* or *Ae. speltoides*. The unknown grass has never been identified among now surviving wild grasses, but the closest living relative is *Aegilops speltoides*. The hybridization that formed wild emmer (AABB) occurred in the wild, long before domestication, and was driven by natural selection.
- Hexaploid wheats evolved in farmers' fields. Either domesticated emmer or durum wheat hybridized with yet another wild diploid grass (*Aegilops tauschii*) to make the hexaploid wheats, spelt wheat and bread wheat. These have *three* sets of paired chromosomes, three times as many as in diploid wheat.

The presence of certain versions of wheat genes has been important for crop yields. Apart from mutant versions of genes selected in antiquity during domestication, there has been more recent deliberate selection of alleles that affect growth characteristics. Genes for the 'dwarfing' trait, first used by Japanese wheat breeders to produce short-stalked wheat, have had a huge effect on wheat yields world-wide, and were major factors in the success of the Green Revolution in Mexico and Asia, an initiative led by Norman Borlaug. Dwarfing genes enable the carbon that is fixed in the plant during photosynthesis to be diverted towards seed production, and they also help prevent the problem of lodging. 'Lodging' occurs when an ear stalk falls over in the wind and rots on the ground, and heavy nitrogenous fertilization of wheat makes the grass grow taller and become more susceptible to this problem. By 1997, 81 per cent of the developing world's wheat area was planted to semi-dwarf wheats, giving both increased yields and better response to nitrogenous fertilizer.

Wild grasses in the genus *Triticum* and related genera, and grasses such as rye have been a source of many disease-resistance traits for cultivated wheat breeding since the 1930s.

Heterosis, or hybrid vigour (as in the familiar F1 hybrids of maize), occurs in common (hexaploid) wheat, but it is difficult to produce seed of hybrid cultivars on a commercial scale (as is done with maize) because wheat flowers are perfect and normally self-pollinate. Commercial hybrid wheat seed has been produced using chemical hybridizing agents; these chemicals selectively interfere with pollen development, or naturally occurring cytoplasmic male sterility systems. Hybrid wheat has been a limited commercial success in Europe (particularly France), the USA and South Africa. F1 hybrid wheat cultivars should not be confused with the standard method of breeding inbred wheat cultivars by crossing two lines using hand emasculation, then selfing or inbreeding the progeny many (ten or more) generations before release selections are identified to be released as a variety or cultivar.

Synthetic hexaploids made by crossing the wild goatgrass wheat ancestor *Aegilops tauschii* and various durum wheats are now being deployed, and these increase the genetic diversity of cultivated wheats. Stomata (or leaf pores) are involved in both uptake of carbon dioxide gas from the atmosphere and water vapor losses from the leaf due to water transpiration. Basic physiological investigation of these gas exchange processes has yielded valuable carbon isotope based methods that are used for breeding wheat varieties with improved water-use efficiency. These varieties can improve crop productivity in rain-fed dry-land wheat farms. In 2010, a team of UK scientists funded by BBSRC announced they had decoded the wheat genome for the first time (95 per cent of the genome of a variety of wheat known as Chinese Spring line 42). This genome was released in a basic format for scientists and plant breeders to use but was not a fully annotated sequence which was reported in some of the media.

On 29 November 2012, an essentially complete gene set of bread wheat has been published. Random shotgun libraries of total DNA and cDNA from the *T. aestivum* cv. Chinese Spring (CS42) were sequenced in Roche 454 pyrosequencer using GS FLX Titanium and GS FLX+ platforms to generate 85 Gb of sequence (220 million reads), equivalent to 5X genome coverage and identified between 94,000 and 96,000 genes.

This sequence data provides direct access to about 96,000 genes, relying on orthologous gene sets from other cereals. and represents an essential step towards a systematic understanding of biology and engineering the cereal crop for valuable traits. Its implications in cereal genetics and breeding includes the examination of genome variation, association mapping using natural populations, performing wide crosses and alien introgression, studying the expression and nucleotide polymorphism in transcriptomes, analyzing population genetics and evolutionary biology, and studying the epigenetic modifications. Moreover, the availability of large-scale genetic markers generated through NGS technology will facilitate trait mapping and make marker-assisted breeding much feasible.

Moreover, the data not only facilitate in deciphering the complex phenomena such as heterosis and epigenetics, it may also enable breeders to predict which fragment of a chromosome is derived from which parent in the progeny line, thereby recognizing crossover events occurring in every progeny line and inserting markers on genetic and physical maps without ambiguity. In due course, this will assist in introducing specific chromosomal segments from one cultivar to another. Besides, the researchers had identified diverse classes of genes participating in energy production, metabolism and growth that were probably linked with crop yield, which can now be utilized for the development of transgenic wheat. Thus whole genome sequence of wheat and the availability of thousands of SNPs will inevitably permit the breeders to stride towards identifying novel traits, providing biological knowledge and empowering biodiversity-based breeding.

PLANT BREEDING

In traditional agricultural systems wheat populations often consist of landraces, informal farmer-maintained populations that often maintain high levels of morphological diversity. Although landraces of wheat are no longer grown in Europe and North America, they continue to be important elsewhere. The origins of formal wheat breeding lie in the nineteenth century, when single line varieties were created through selection of seed from a single plant noted to have desired properties. Modern wheat breeding developed in the first years of the twentieth century and was closely linked to the development of Mendelian genetics. The standard method of breeding inbred wheat cultivars is by crossing two lines using hand emasculation, then selfing or inbreeding the progeny. Selections are *identified* (shown to have the genes responsible for the varietal differences) ten or more generations before release as a variety or cultivar.

The major breeding objectives include high grain yield, good quality, disease and insect resistance and tolerance to abiotic stresses, including mineral, moisture and heat tolerance. The major diseases in temperate environments include the following, arranged in a rough order of their significance from cooler to warmer climates: eyespot, Stagonospora nodorum blotch (also known as glume blotch), yellow or stripe rust, powdery mildew, Septoria tritici blotch (sometimes known as leaf blotch), brown or leaf rust, Fusarium head blight, tan spot and stem rust. In tropical areas, spot blotch (also known as Helminthosporium leaf blight) is also important. Wheat has also been the subject of mutation breeding, with the use of gamma, x-rays, ultraviolet light, and sometimes harsh chemicals. The varieties of wheat created through this methods are in the hundreds (varieties being as far back as 1960), more of them being created in higher populated countries such as China.

HYBRID WHEAT

Because wheat self-pollinates, creating hybrid varieties is extremely labour-intensive; the high cost of hybrid wheat seed relative to its moderate benefits have kept farmers from adopting them widely despite nearly 90 years of effort. F1 hybrid wheat cultivars should not be confused with wheat cultivars deriving from standard plant breeding. Heterosis or hybrid vigour (as in the familiar F1 hybrids of maize) occurs in common (hexaploid) wheat, but it is difficult to produce seed of hybrid cultivars on a commercial scale as is done with maize because wheat flowers are perfect in the botanical sense, meaning they have both male and female parts, and normally self-pollinate. Commercial hybrid wheat seed has been produced using chemical hybridizing agents, plant growth regulators that selectively interfere with pollen development, or naturally occurring cytoplasmic male sterility systems. Hybrid wheat has been a limited commercial success in Europe (particularly France), the United States and South Africa.

HULLED VERSUS FREE-THRESHING WHEAT

The four wild species of wheat, along with the domesticated varieties einkorn, emmer and spelt, have hulls. This more primitive morphology (in evolutionary terms) consists of toughened glumes that tightly enclose the grains, and (in domesticated wheats) a semi-brittle rachis that breaks easily on threshing. The result is that when threshed, the wheat ear breaks up into spikelets. To obtain the grain, further processing, such as milling or pounding, is needed to remove the hulls or husks. In contrast, in free-threshing (or naked) forms such as durum wheat and common wheat, the glumes are fragile and the rachis tough. On threshing, the chaff breaks up, releasing the grains. Hulled wheats are often stored as spikelets because the toughened glumes give good protection against pests of stored grain.

NAMING

There are many botanical classification systems used for wheat species, discussed in a separate article on Wheat taxonomy. The name of a wheat species from one information source may not be the name of a wheat species in another.

Within a species, wheat cultivars are further classified by wheat breeders and farmers in terms of:

- Growing season, such as winter wheat vs. spring wheat.
- Protein content. Bread wheat protein content ranges from 10 per cent in some soft wheats with high starch contents, to 15 per cent in hard wheats.
- The quality of the wheat protein gluten. This protein can determine the suitability of a wheat to a particular dish. A strong and elastic gluten present in bread wheats enables dough to trap carbon dioxide during leavening, but elastic gluten interferes with the rolling of pasta into thin sheets. The gluten protein in durum wheats used for pasta is strong but not elastic.
- Grain colour (red, white or amber). Many wheat varieties are reddish-brown due to phenolic compounds present in the bran layer which are transformed to pigments by browning enzymes. White wheats have a lower content of phenolics and browning enzymes, and are generally less astringent in taste than red wheats. The yellowish colour of durum wheat and semolina flour made from it is due to a carotenoid pigment called lutein, which can be oxidized to a colorless form by enzymes present in the grain.

MAJOR CULTIVATED SPECIES OF WHEAT

Hexaploid Species:

- Common wheat or Bread wheat (*T. aestivum*) – A hexaploid species that is the most widely cultivated in the world.

- Spelt (*T. spelta*) – Another hexaploid species cultivated in limited quantities. Spelt is sometimes considered a subspecies of the closely related species common wheat (*T. aestivum*), in which case its botanical name is considered to be *Triticum aestivum* subsp. *spelta*.

Tetraploid Species:

- Durum (*T. durum*) – The only tetraploid form of wheat widely used today, and the second most widely cultivated wheat.
- Emmer (*T. dicoccon*) – A tetraploid species, cultivated in ancient times but no longer in widespread use.
- Khorasan (Triticum turgidum ssp. turanicum also called Triticum turanicum) is a tetraploid wheat species. It is an ancient grain type; Khorasan refers to a historical region in modern-day Afghanistan and the northeast of Iran. This grain is twice the size of modern-day wheat and is known for its rich nutty flavour.

Diploid Species:

- Einkorn (*T. monococcum*) – A diploid species with wild and cultivated variants. Domesticated at the same time as emmer wheat, but never reached the same importance.

Classes used in the United States:

- Durum – Very hard, translucent, light-colored grain used to make semolina flour for pasta and bulghur; high in protein, specifically, gluten protein.
- Hard Red Spring – Hard, brownish, high-protein wheat used for bread and hard baked goods. Bread Flour and high-gluten flours are commonly made from hard red spring wheat. It is primarily traded at the Minneapolis Grain Exchange.
- Hard Red Winter – Hard, brownish, mellow high-protein wheat used for bread, hard baked goods and as an adjunct in other flours to increase protein in pastry flour for pie crusts. Some brands of unbleached all-purpose flours are commonly made from hard red winter wheat alone. It is primarily traded on the Kansas City Board of Trade. One variety is known as "turkey red wheat", and was brought to Kansas by Mennonite immigrants from Russia.
- Soft Red Winter – Soft, low-protein wheat used for cakes, pie crusts, biscuits, and muffins. Cake flour, pastry flour, and some self-rising flours with baking powder and salt added, for example, are made from soft red winter wheat. It is primarily traded on the Chicago Board of Trade.
- Hard White – Hard, light-colored, opaque, chalky, medium-protein wheat planted in dry, temperate areas. Used for bread and brewing.
- Soft White – Soft, light-colored, very low protein wheat grown in temperate moist areas. Used for pie crusts and pastry. Pastry flour, for example, is sometimes made from soft white winter wheat.

Red wheats may need bleaching; therefore, white wheats usually command higher prices than red wheats on the commodities market.

AS A FOOD

Raw wheat can be ground into flour or, using hard durum wheat only, can be ground into semolina; germinated and dried creating malt; crushed or cut into cracked wheat; parboiled (or steamed), dried, crushed and de-branned into bulgur also known as groats. If the raw wheat is broken into parts at the mill, as is usually done, the outer husk or bran can be used several ways. Wheat is a major ingredient in such foods as bread, porridge, crackers, biscuits, Muesli, pancakes, pies, pastries, cakes, cookies, muffins, rolls, doughnuts, gravy, boza (a fermented beverage), and breakfast cereals (*e.g.*, Wheatena, Cream of Wheat, Shredded Wheat, and Wheaties).

NUTRITION

Ch. = Choline; Ca = Calcium; Fe = Iron; Mg = Magnesium; P = Phosphorus; K = Potassium; Na = Sodium; Zn = Zinc; Cu = Copper; Mn = Manganese; Se = Selenium; per centDV = per cent daily value *i.e.* per cent of DRI (Dietary Reference Intake) Note: All nutrient values including protein are in per centDV per 100 grams of the food item. Significant values are highlighted in light Gray colour and bold letters.

Cooking reduction = per cent Maximum typical reduction in nutrients due to boiling without draining for ovo-lacto-vegetables group 100 g (3.5 oz) of hard red winter wheat contain about 12.6 g (0.44 oz) of protein, 1.5 g (0.053 oz) of total fat, 71 g (2.5 oz) of carbohydrate (by difference), 12.2 g (0.43 oz) of dietary fibre, and 3.2 mg (0.00011 oz) of iron (17 per cent of the daily requirement); the same weight of hard red spring wheat contains about 15.4 g (0.54 oz) of protein, 1.9 g (0.067 oz) of total fat, 68 g (2.4 oz) of carbohydrate (by difference), 12.2 g (0.43 oz) of dietary fibre, and 3.6 mg (0.00013 oz) of iron (20 per cent of the daily requirement).

Much of the carbohydrate fraction of wheat is starch. Wheat starch is an important commercial product of wheat, but second in economic value to wheat gluten. The principal parts of wheat flour are gluten and starch. These can be separated in a kind of home experiment, by mixing flour and water to form a small ball of dough, and kneading it gently while rinsing it in a bowl of water. The starch falls out of the dough and sinks to the bottom of the bowl, leaving behind a ball of gluten.

In wheat, phenolic compounds are mainly found in the form of insoluble bound ferulic acid and are relevant to resistance to wheat fungal diseases. Alkylresorcinols are phenolic lipids present in high amounts in the bran layer (*e.g.* pericarp, testa and aleurone layers) of wheat and rye (0.1-0.3 per cent of dry weight).

WORLDWIDE CONSUMPTION

Wheat is grown on more than 218,000,000 hectares (540,000,000 acres), larger than for any other crop. World trade in wheat is greater than for all other crops combined. With rice, wheat is the world's most favored staple food. It is a major diet component because of the wheat plant's agronomic adaptability with the ability to grow from near arctic regions to equator, from sea level to plains of Tibet, approximately 4,000 m (13,000 ft) above sea level. In addition to agronomic adaptability, wheat offers ease of grain storage and ease of converting grain into flour for making edible, palatable, interesting and satisfying foods. Wheat is the most important source of carbohydrate in a majority of countries.

Wheat protein is easily digested by nearly 99 per cent of the human population (all but those with gluten-related disorders), as is its starch. Wheat also contains a diversity of minerals, vitamins and fats (lipids). With a small amount of animal or legume protein added, a wheat-based meal is highly nutritious.

The most common forms of wheat are white and red wheat. However, other natural forms of wheat exist. For example, in the highlands of Ethiopia grows purple wheat, a tetraploid species of wheat that is rich in anti-oxidants. Other commercially minor but nutritionally promising species of naturally evolved wheat species include black, yellow and blue wheat.

HEALTH CONCERNS

Several screening studies in Europe, South America, Australasia, and the USA suggest that approximately 0.5–1 per cent of these populations may have undetected coeliac disease. Coeliac (also written as celiac) disease is a condition that is caused by an adverse immune system reaction to gliadin, a gluten protein found in wheat (and similar grains of the tribe Triticeae which includes other species such as barley and rye). Upon exposure to gliadin, the enzyme tissue transglutaminase modifies the protein, and the immune system cross-reacts with the bowel tissue, causing an inflammatory reaction. That leads to flattening of the lining of the small intestine, which interferes with the absorption of nutrients. The only effective treatment is a lifelong gluten-free diet.

The estimate for celiac disease among people in the United States is between 0.5 and 1.0 per cent of the population. While gluten sensitivity is caused by a reaction to wheat proteins, it is not the same as a wheat allergy. Recently non-celiac gluten sensitivity has been identified as a further gluten sensitivity condition that differs from celiac disease and wheat allergy.

COMMERCIAL USE

Harvested wheat grain that enters trade is classified according to grain properties for the purposes of the commodity markets. Wheat buyers use these to decide which wheat to buy, as each class has special uses, and producers use them to decide which classes of wheat will be most profitable to cultivate.

Wheat is widely cultivated as a cash crop because it produces a good yield per unit area, grows well in a temperate climate even with a moderately short growing season, and yields a versatile, high-quality flour that is widely used in baking. Most breads are made with wheat flour, including many breads named for the other grains they contain, for example, most rye and oat breads. The popularity of foods made from wheat flour creates a large demand for the grain, even in economies with significant food surpluses.

In recent years, low international wheat prices have often encouraged farmers in the USA to change to more profitable crops. In 1998, the price at harvest was $2.68 per bushel. USDA report revealed that in 1998, average operating costs were $1.43 per bushel and total costs were $3.97 per bushel. In that study, farm wheat yields averaged 41.7 bushels per acre (2.2435 metric ton/hectare), and typical total wheat production value was $31,900 per farm, with total farm production value (including other crops) of $173,681 per farm, plus $17,402 in government payments. There were significant profitability differences between low- and high-cost farms, mainly due to crop yield differences, location, and farm size.

In 2007 there was a dramatic rise in the price of wheat due to freezes and flooding in the northern hemisphere and a drought in Australia. Wheat futures in September, 2007 for December and March delivery had risen above $9.00 a bushel, prices never seen before. There were complaints in Italy about the high price of pasta.

PRODUCTION AND CONSUMPTION

In 2011, global per capita wheat consumption was 65 kg (143 lb), with the highest per capita consumption of 210 kg (460 lb) found in Azerbaijan. In 1997, global wheat consumption was 101 kg (223 lb) per capita, with the highest consumption 623 kg (1,373 lb) per capita in Denmark, but most of this (81 per cent) was for animal feed. Wheat is the primary food staple in North Africa and the Middle East, and is growing in popularity in Asia. Unlike rice, wheat production is more widespread globally though China's share is almost one-sixth of the world.

"There is a little increase in yearly crop yield comparison to the year 1990. The reason for this is not in development of sowing area, but the slow and successive increasing of the average yield. Average 2.5 tons wheat was produced on one hectare crop land in the world in the first half of 1990s, however this value was about 3 tons in 2009. In the world per capita wheat producing area continuously decreased between 1990 and 2009 considering the change of world population. There was no significant change in wheat producing area in this period. However, due to the improvement of average yields there is some fluctuation in each year considering the per capita production, but there is no considerable decline. In 1990 per capita production was 111.98 kg/capita/year, while it was already 100.62 kg/capita/year in 2009. The decline is evident and the per capita production level of the year 1990 can not be feasible simultaneously with the growth of world population in spite of the increased average yields. In the whole period the lowest per capita production was in 2006."

In the 20th century, global wheat output expanded by about 5-fold, but until about 1955 most of this reflected increases in wheat crop area, with lesser (about 20 per cent) increases in crop yields per unit area. After 1955 however, there was a dramatic ten-fold increase in the rate of wheat yield improvement per year, and this became the major factor allowing global wheat production to increase. Thus technological innovation and scientific crop management with synthetic nitrogen fertilizer, irrigation and wheat breeding were the main drivers of wheat output growth in the second half of the century. There were some significant decreases in wheat crop area, for instance in North America.

Better seed storage and germination ability (and hence a smaller requirement to retain harvested crop for next year's seed) is another 20th century technological innovation. In Medieval England, farmers saved one-quarter of their wheat harvest as seed for the next crop, leaving only three-quarters for food and feed consumption. By 1999, the global average seed use of wheat was about 6 per cent of output.

Several factors are currently slowing the rate of global expansion of wheat production: population growth rates are falling while wheat yields continue to rise, and the better economic profitability of other crops such as soybeans and maize, linked with investment in modern genetic technologies, has promoted shifts to other crops.

FARMING SYSTEMS

In the Punjab region of India and Pakistan, as well as North China, irrigation has been a major contributor to increased grain output. More widely over the last 40 years, a massive increase in fertilizer use together with the increased availability of semi-dwarf varieties in developing countries, has greatly increased yields per hectare. In developing countries, use of (mainly nitrogenous) fertilizer increased 25-fold in this period. However, farming systems rely on much more than fertilizer and breeding to improve productivity. A good illustration of this is Australian wheat growing in the southern winter cropping zone, where, despite low rainfall (300 mm), wheat cropping is successful even with relatively little use of nitrogenous fertilizer. This is achieved by 'rotation cropping' (traditionally called the ley system) with leguminous pastures and, in the last decade, including a canola crop in the rotations has boosted wheat yields by a further 25 per cent. In these low rainfall areas, better use of available soil-water (and better control of soil erosion) is achieved by retaining the stubble after harvesting and by minimizing tillage.

In 2009, the most productive farms for wheat were in France producing 7.45 metric tonnes per hectare (although French production has low protein content and requires blending with higher protein wheat to meet the specifications required in some countries). The five largest producers of wheat in 2009 were China (115 million metric tonnes), India (81 MMT), Russian Federation (62 MMT), United States (60 MMT) and France (38 MMT). The wheat farm productivity in India and Russia were about 35 per cent of the wheat farm productivity in France. China's farm productivity for wheat, in 2009, was about double that of Russia.

In addition to gaps in farming system technology and knowledge, some large wheat grain producing countries have significant losses after harvest at the farm and because of poor roads, inadequate storage technologies, inefficient supply chains and farmers' inability to bring the produce into retail markets dominated by small shopkeepers. Various studies in India, for example, have concluded that about 10 per cent of total wheat production is lost at farm level, another 10 per cent is lost because of poor storage and road networks, and additional amounts lost at the retail level. One study claims that if these post-harvest wheat grain losses could be eliminated with better infrastructure and retail network, in India alone enough food would be saved every year to feed 70 to 100 million people over a year.

GEOGRAPHICAL VARIATION

There are substantial differences in wheat farming, trading, policy, sector growth, and wheat uses in different regions of the world. In the EU and Canada for instance, there is significant addition of wheat to animal feeds, but less so in the USA.

The biggest wheat producer in 2010 was EU-27, followed by China, India, USA and Russian Federation.

The largest exporters of wheat in 2009 were, in order of exported quantities: United States, EU-27, Canada, Russian Federation, Australia, Ukraine and Kazakhstan. Upon the results of 2011, Ukraine became the world's sixth wheat exporter as well. The largest importers of wheat in 2009 were, in order of imported quantities: Egypt, EU-27, Brazil, Indonesia, Algeria and Japan. EU-27 was on both export and import list, because EU countries such as Italy and Spain imported wheat, while other EU-27 countries exported their harvest. The Black Sea region – which includes Kazakhstan, the Russian Federation and Ukraine – is amongst the most promising area for grain exporters; it possess significant production potential in terms of both wheat yield and area increases. The Black Sea region is also located close to the traditional grain importers in the Middle East, North Africa and Central Asia.

In the rapidly developing countries of Asia, westernization of diets associated with increasing prosperity is leading to growth in *per capita* demand for wheat at the expense of the other food staples.

In the past, there has been significant governmental intervention in wheat markets, such as price supports in the USA and farm payments in the EU. In the EU these subsidies have encouraged heavy use of fertilizer inputs with resulting high crop yields. In Australia and Argentina direct government subsidies are much lower.

WORLD'S MOST PRODUCTIVE WHEAT FARMS AND FARMERS

The average annual world farm yield for wheat was 3.3 tonnes per hectare (330 grams per square meter), in 2013.

New Zealander wheat farms were the most productive in 2013, with a nationwide average of 9.1 tonnes per hectare. Ireland was a close second.

Various regions of the world hold wheat production yield contests every year. Yields above 12 tonnes per hectare are routinely achieved in many parts of the world. Chris Dennison of Oamaru, New Zealand, set a world record for wheat yield in 2003 at 15.015 tonnes per hectare (223 bushels/acre). In 2010, this record was surpassed by another New Zealand farmer, Michael Solari, with 15.636 tonnes per hectare (232.64 bushels/acre) at Otama, Gore.

AGRONOMY

CROP DEVELOPMENT

Wheat normally needs between 110 and 130 days between sowing and harvest, depending upon climate, seed type, and soil conditions (winter wheat lies dormant during a winter freeze). Optimal crop management requires that the farmer have a detailed understanding of each stage of development in the growing plants. In particular, spring fertilizers, herbicides, fungicides, and growth regulators are typically applied only at specific stages of plant development. For example, it is currently recommended that the second application of nitrogen is best done when the ear (not visible at this stage) is about 1 cm in size (Z31 on Zadoks scale). Knowledge of stages is also important to identify periods of higher risk from the climate. For example, pollen formation from the mother cell, and the stages between anthesis and maturity are susceptible to high temperatures, and this adverse effect is made worse by water stress. Farmers also benefit from knowing when the 'flag leaf' (last leaf) appears, as this leaf represents about 75 per cent of photosynthesis reactions during the grain filling period, and so should be preserved from disease or insect attacks to ensure a good yield.

Several systems exist to identify crop stages, with the Feekes and Zadoks scales being the most widely used. Each scale is a standard system which describes successive stages reached by the crop during the agricultural season.

DISEASES

There are many wheat diseases, mainly caused by fungi, bacteria, and viruses. Plant breeding to develop new disease-resistant varieties, and sound crop management practices are important for preventing disease. Fungicides, used to prevent the significant crop losses from fungal disease, can be a significant variable cost in wheat production. Estimates of the amount of wheat production lost owing to plant diseases vary between 10–25 per cent in Missouri. A wide range of organisms infect wheat, of which the most important are viruses and fungi.

The main wheat-disease categories are:

- *Seed-borne diseases:* These include seed-borne scab, seed-borne *Stagonospora* (previously known as *Septoria*), common bunt (stinking smut), and loose smut. These are managed with fungicides.
- *Leaf- and head- blight diseases:* Powdery mildew, leaf rust, *Septoria tritici* leaf blotch, *Stagonospora* (*Septoria*) nodorum leaf and glume blotch, and *Fusarium* head scab.
- *Crown and root rot diseases:* Two of the more important of these are 'take-all' and *Cephalosporium* stripe. Both of these diseases are soil borne.
- *Stem rust diseases:* Caused by basidiomycete fungi *e.g.* Ug99
- *Viral diseases:* Wheat spindle streak mosaic (yellow mosaic) and barley yellow dwarf are the two most common viral diseases. Control can be achieved by using resistant varieties.

PESTS

Wheat is used as a food plant by the larvae of some Lepidoptera (butterfly and moth) species including The Flame, Rustic Shoulder-knot, Setaceous Hebrew Character and Turnip Moth. Early in the season, many species of birds, including the Long-tailed Widowbird, and rodents feed upon wheat crops. These animals can cause significant damage to a crop by digging up and eating newly planted seeds or young plants.

They can also damage the crop late in the season by eating the grain from the mature spike. Recent post-harvest losses in cereals amount to billions of dollars per year in the USA alone, and damage to wheat by various borers, beetles and weevils is no exception. Rodents can also cause major losses during storage, and in major grain growing regions, field mice numbers can sometimes build up explosively to plague proportions because of the ready availability of food. To reduce the amount of wheat lost to post-harvest pests, Agricultural Research Service scientists have developed an "insect-o-graph," which can detect insects in wheat that are not visible to the naked eye. The device uses electrical signals to detect the insects as the wheat is being milled. The new technology is so precise that it can detect 5-10 infested seeds out of 300,000 good ones. Tracking insect infestations in stored grain is critical for food safety as well as for the marketing value of the crop.

ECONOMIC IMPORTANCE AND TAXONOMY OF THE WHEAT

Wheat (*Triticum aestivum* L. em Thell.) is the first important and strategic cereal crop for the majority of world's populations. It is the most important staple food of about two billion people (36 per cent of the world population). Worldwide, wheat provides nearly 55 per cent of the carbohydrates and 20 per cent of the food calories consumed globally (Breiman and Graur, 1995). It exceeds in acreage and production every other grain crop (including rice, maize, etc.) and is therefore, the most important cereal grain crop of the world, which is cultivated over a wide range of climatic conditions and the understanding of genetics and genome organization using molecular markers is of great value for genetic and plant breeding purposes. The grass family *Poaceae* (*Gramineae*) includes major crop plants such as wheat (*Triticum aestivum* L.), barley (*Hordeum vulgare* L.), oat (*Avena sativa* L.), rye (*Secale cereale* L.), maize (*Zea mays* L.) and rice (*Oryza sativa* L.). *Triticeae* is one of the tribes containing more than 15 genera and 300 species including wheat and barley. Wheat belongs to the tribe *Triticeae* (= *Hordeae*) in the grass family *Poaceae* (*Gramineae*) (Briggle and Reitz, 1963) in which the one to several flowered spikelets are sessile and alternate on opposite sides of the rachis forming a true spike. Wheats (*Triticum*) and ryes (*Secale*) together with *Aegilops, Agropyron, Eremopyron* and *Haynalidia* form the subtribe *Triticineae* (Simmonds, 1976). Linnaeus in 1753 first classified wheat. In 1918, Sakamura reported the chromosome number sets (genomes) for each commonly recognized type. This was a turning point in *Triticum* classification. It separated wheat into three groups. Diploids had 14 (n=7), tetraploids had 28 (n=14) and the hexaploids had 42 (n=21) chromosomes. Bread wheat is *Triticum aestivum*. *T. durum* and *T. compactum* are the other major species. All three are products of natural hybridization among ancestrals no longer grown commercially (Briggle, 1967).

SIGNIFICANCE OF WHEAT

World economy role of wheat production is significance both in terms of cultivated land and food supply, feeding and commerce. In my opinion this significance has to be tried to make clear as well as it is worth to examine the role and importance of Hungary in it. Hungary has 4.5 million hectares of arable land. Shares of cereals' sowing area from Hungarian arable land fluctuated between 68.4 per cent and 69.9 per cent in period of 2004 and 2008. Difference between different years is negligible. Significance of wheat and corn is nearly the same within cereals. Both plants meet with approximately rate of 28 per cent in comparison to entire arable land. In connection with international trade of Hungarian wheat production our country has to cope with considerable competitive disadvantages. Since we have to cope with logistics handicap, the reason of it is that due to geographical situation of Hungary in many cases we are not able to transit our agricultural products at a competitive price. Other countries, such as the United States, with maritime transport, by much more logistics costs, stay easier competitive at international markets. In spite of this, Hungary as regards exported wheat quantity is placed as 11. of the world in 2008.

PRODUCTION

World has plant production area of around 1500 million hectares. Figure shows the sowing structure of the world in 2009. In this figure it can be properly seen that growers are producing cereals on 48 per cent of sowing area in the world. Corn, rice and wheat have to be pointed out within cereals. Share of these three plants is 36 per cent from entire sowing area, which is equivalent to 546 hectares. Wheat has the largest proportion within cereals. Owing to its sowing area of 225 million hectares it occupies 15 per cent of area being under plant production in the world. Proportions of three main cereals are stable on average of number of years.As perme in the future significant displacement are not expected in sowing structure.

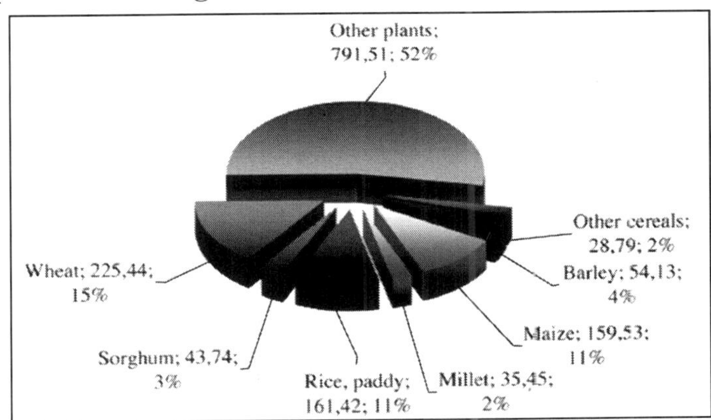

Fig. Division of Sowing Area in the World in 2009 (in Million Hectares and percentage)

Beside wheat, proportion of the previously mentioned rice and corn is also significant. Proportion of remaining cereals in comparison to entire sowing area of the world is under 5 per cent. From Hungary's point of view wheat and corn production have highlighted role. These two plants occupy approximately 58–60 per cent of 4.5 million hectares of domestic arable land. Considering the territorial proportion of rice production in Hungary its significance is negligible, however in terms of the utilization of areas having poor productivity its role is deemed to be significant. Further in this article I specifically intend to concentrate on wheat production.

Table. Wheat Production is between 1990 and 2009.

Years	Area harvested *(million hectars)*	Production *(million tonnes)*	Average yield *(t/ha)*	Area/capita *(hectar/ capita)*	Production/ capita *(kg/capita/ year)*
1990	231.26	592.31	2.56	0.0437	111.9880
1991	223.35	546.88	2.45	0.0416	101.8027
1992	222.49	565.29	2.54	0.0408	103.5951
1993	222.95	564.47	2.53	0.0403	101.9122
1994	215.12	527.04	2.45	0.0383	93.7892
1995	216.32	542.60	2.51	0.0379	95.1881
1996	226.85	585.20	2.58	0.0392	101.2357
1997	226.25	613.36	2.71	0.0386	104.6855
1998	220.11	593.53	2.70	0.0371	99.9768
1999	213.34	587.62	2.75	0.0355	97.7170
2000	215.44	585.69	2.72	0.0354	96.1781
2001	214.60	589.82	2.75	0.0348	95.6557
2002	213.81	574.75	2.69	0.0343	92.0721
2003	207.66	560.13	2.70	0.0329	88.6559
2004	216.88	632.67	2.92	0.0339	98.9515
2005	219.74	626.84	2.85	0.0340	96.8895
2006	211.82	602.89	2.85	0.0324	92.0959
2007	216.65	612.61	2.83	0.0327	92.4841
2008	222.76	683.41	3.07	0.0332	101.9860
2009	225.44	681.92	3.02	0.0333	100.6256

Table shows the tendency of world's wheat production from 1990 to 2009. There are minimal differences in case of sowing area regarding each year. There is a little increase in yearly crop yield comparison to the year 1990. The reason for this is not in development of sowing area, but the slow and successive increasing of the average yield. Average 2.5 tons wheat was produced on one hectare crop land in the world in the first half of 1990s, however this value was about 3 tons in 2009.

In the world per capita wheat producing area continuously decreased between 1990 and 2009 considering the change of world population. There was no significant change in wheat producing area in this period. However due to the improvement of average yields there is some fluctuation in each year considering the per capita production, but there is no considerable decline. In 1990 per capita production was 111.98 kg/capita/year, while it was already 100.62 kg/capita/year in 2009.

The decline is evident and the per capita production level of the year 1990 can not be feasible simultaneously with the growth of world population in spite of the increased average yields. In the whole period the lowest per capita production was in 2006. *Figure* shows the hierarchy based on produced amount of wheat by EU 27 countries in 2009. The European Union produced 138.7 million tons wheat in 2009 and the first 10 countries produced the 86 per cent of this amount. France is the biggest wheat-producer in the EU. The French produced 38 million tons wheat in 2009. Germany stays on the second place with its 25 million tons. The UK is the third with its 14 million tons yield. In the hierarchy based on produced amount of wheat France and Germany take place within the top 10 list of the world's wheat-producer countries.

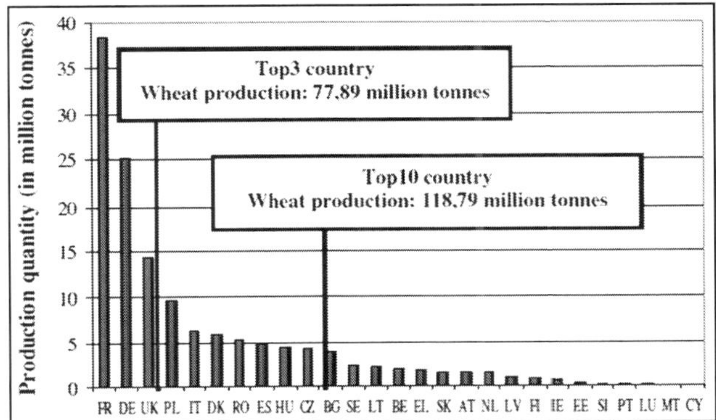

Fig. The EU's 27 Countries Ranked in Terms of Wheat Production in 2009.

The harvested production of 4th placed Poland was 9.7 million tons in the examined year. There are not so many differences between the productions of the countries in the 5th to 10th places of the list, than in case of the top 3 countries. Italy produced 6.3 million tons wheat in 2009, while Czech Republic made 4.3 million tons. Hungary stays in the 9th place in the hierarchy in the examined year. Czech Republic has been overtaken by our country with minimal difference.

Our country produced 4.4 million tons in 2009. *Figure* demonstrates the production quantity produced by the ten largest wheat producing country in the world in comparison to Hungarian volume in production year 2009. In the given year 681 million tons were produced, from which share of top 10 countries is 69.6 per cent. 659.8 million tons were consumed from the wheat produced in the given year. 69.8 per cent of the total consumed quantity used for food supply, 18.5 per cent used for feeding, the remaining 11.5 per cent were used for other purposes. In ranks of leading wheat growers in the world the participants are the same apart from minimum deviation between the period of 2000 and 2009. In relation to different years there are not significant differences in sequences. In the given period there was an example that two large wheat producing countries changed their place in the rank certain years. Considering an exact case is for example the USA and Russia. Between 2003 and 2008 the USA were the third largest wheat producing country of the world while in 2009 Russia gained the third place of the imaginary stage. The position reached in the rank of the above-mentioned two countries will also be interesting in 2010, thus significant yield decrease is experienced in both countries due to unusual weather.

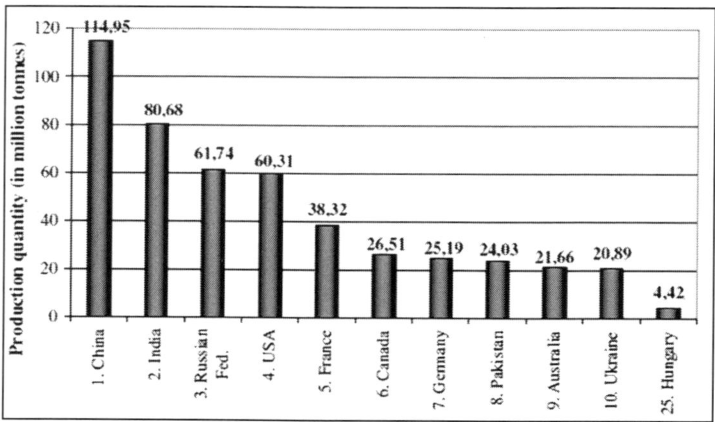

Fig. Production Quantity of the Top 10 Wheat Producing Countries in the World and Situation of Hungary in Comparison to them in 2009.

In connection to the given year Hungary was placed as 25th considering the wheat production of the world. This position of Hungary is comparatively stable between the period of 2000 and 2009, in the given period Hungary was between 20th and 25th in the quantity rank of the wheat production in the world in various years.

Between 2000 and 2009 the domestic produced wheat quantity was around 2.9 and 6 million tons each year. The reason of this fluctuation is the different amount of precipitation fallen in different production year. It is clear that wheat quantity of 2.9 million tons is in droughty year while quantity of 6 million tons is the result of a good year.

Figure shows the rank of top ten 10 countries occupied largest harvested area in the world. There are differences in consideration of rank between produced quantity and cultivated area. Not the country with the largest production produces wheat on the largest territory. The reason of this can be searched in various natural endowments of each country and various technological level of its production. In the given year wheat were produced on 225 million hectares in the world, from which the share of 10 countries occupied the largest harvested area is 71.3 per cent that is equivalent to 160.6 million hectares.

In the given year wheat were produced on 225 million hectares in the world, from which the share of 10 countries occupied the largest harvested area is 71.3 per cent that is equivalent to 160.6 million hectares. From both the produced quantity's and cultivated area's viewpoint it can be stated that the production is considerable concentrated at international level. In 2009, Hungary with 1.15 million hectares of wheat producing area was placed at 29th in the world rank.Minimum differences are experienced regarding area being under wheat production in Hungary. Sowing area of the wheat decreased in Hungary compared to 2009 both in 2010 and 2011.

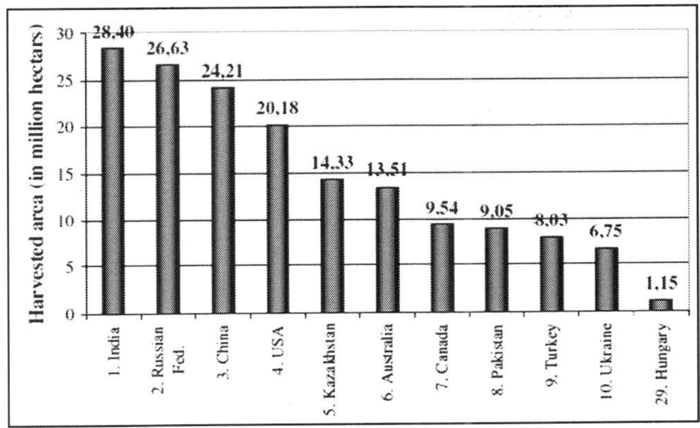

Fig. The Rank of top 10 Countries Occupied Largest Harvested Area in the world and Situation of Hungary in Comparison to them in 2009.

This had various reasons on behalf of growers. Profitability of wheat production was low in 2009, due to it a number of growers decided for the change of sowing structure, causing disadvantages of the wheat. Sowing period in autumn, 2010 was exposed to various meteorological conditions and varied by internal water troubles; due to these a number of growers could not sow the previously planned wheat areas. To sum up it has to be stated that compared to 2009 both in 2010 and 2011 wheat production area in Hungary decreased in Hungary, however there were completely different reasons of decrease in case of two years. In 2010 the profitability of the wheat, in case of honest average yield due to increased buying-in prices, considered to be satisfied.

CONCLUSION

Considering the world wheat production there was no significant change in wheat producing area of the last 20 years, but average yields improved over the years. World market position of Hungary is not favourable due to its specific geographical situation. In case of transportation we have to calculate with high shipping expenses towards Rotterdam and Constanta as well. In spite of this fact our wheat export is respectable. Hungary has negligible market share in world trade of wheat in spite of its favourable place among wheat exporter countries. In the future we might have to calculate with the decline of wheat producing area because wheat production was not able to compete with maize production in income-generating capacity. The year of 2010 can be an exception; however the major part of the increased prices on the product market will not be realized by the producers but by the corn-merchants at the end of 2010 and in the beginning of 2011.

2

Crop Plants and Agriculture: Origins and Diversity

CENTERS OF ORIGIN AND DIVERSITY. IN THIS HE RECOGNIZES EIGHT PRIMARY CENTERS, AS FOLLOWS

I. The Chinese Center In which he recognizes 138 distinct species of which probably the earlier and most important were cereals, buckwheats and legumes.

II. The Indian Center (including the entire subcontinent) - Based originally on rice, millets and legumes, with a total of 117 species.

IIa. The Indo-Malayan Center (including Indonesia, Philippines, etc.) - With root crops (*Dioscorea* spp., *Tacca,* etc.) preponderant, also with fruit crops, sugarcane, spices, etc., some 55 species.

III. The Inner Asiatic Center (Tadjikistan, Uzbekistan, etc.) - With wheats, rye and many herbaceous legumes, as well as seed-sown root crops and fruits, some 42 species.

IV. Asia Minor (including Transcaucasia, Iran and Turkmenistan) - With more wheats, rye, oats, seed and forage legumes, fruits, etc., some 83 species.

V. The Mediterranean Center Of more limited importance than the others to the east, but including wheats, barleys, forage plants, vegetables and fruits - especially also spices and ethereal oil plants, some 84 species.

VI. The Abyssinian (now Ethiopian) Center - Of lesser importance, mostly a refuge of crops from other regions, especially wheats and barleys, local grains, spices, etc., some 38 species.

VII. The South Mexican and Central American Center - important for maize, *Phaseolus* and Cucurbitaceous species, with spices, fruits and fibre plants, some 49 species.

VIII. South America Andes region (Bolivia, Peru, Ecuador) - Important for potatoes, other root crops, grain crops of the Andes, vegetables, spices and fruits, as well as drugs (cocaine, quinine, tobacco, etc.), some 45 species.

VIIIa. The Chilean Center - Only four species - Outside the main area of crop domestication, and one of these (*Solarium tuberosum*) derived from the

Andean center in any case. This could hardly be compared with the eight main centers.

VIIIb. Brazilian-Paraguayan Center - Again outside the main centers with only 13 species, though *Manihot* (cassava) and *Arachis* (peanut) are of considerable importance; others such as pineapple, *Hevea* rubber, *Theobroma cacao* were probably domesticated much later. the very wide range of plant diversity in the tropical and warm temperate regions of the world our major food crops have come mainly from high mountain valleys, isolated from each other to a large extent and with a very great habitat range. Here people made selections of wheat, barley, oats, rye, potatoes and maize which were eventually cultivated.

These plants were weeds or possessed the syndrome of not being able to compete well with climax vegetation. Hence they grew in areas where nature or humans had reduced competition from other species, were noticed, eaten, resown by chance and eventually became domesticated. Several other weedy plants were never or only temporarily domesticated, remaining as weeds but often hybridizing by chance with the cultivated ones and thus enhancing their diversity. the restricted access of the mountain valleys and the wide range of altitudes helped to produce and select the diversity needed for domestication. Similar selection pressures even in unrelated crops produced similar types of adaptation, a process developed by Vavilov into his Law of Homologous Series. Because such adaptation in only partially related crops must surely have been due to mutations on distinct loci in each crop, this writer feels that a more correct title might have been the Law of Analogous Series. However, the phrase has persisted as Homologous Series and we must retain it as just one of the extraordinarily innovative ideas put forward by the great genius,

Weedy relatives of the crops were also present, adding to the genetic diversity by random mating with the primitive crops themselves. In the crop I know best -potatoes - one can find not only a range of ploidy, from diploid to pentaploid, but clear evidence for crosses between weed species and cultivated ones as well as the natural introduction of frost resistance from wild species into the cultivated ones. that evolution of domesticates is very much promoted by the factors of varied microclimate, aspect, altitude, restricted habitats and human selection which are all present in the intertropical mountain zones of the Old and New Worlds. Such a wide range of natural and human selection pressures is not available to the same degree of intensity in other world regions. Vavilov was the first investigator to study and understand these systems and to use them as a basis for present and future plant breeding. At the same time, since Vavilov was not only a geneticist and plant breeder, but also a man of wide interests and intelligence, he was able to provide a theoretical background to interpret crop genetic diversity in all its aspects and thus to make it available for the whole world.

Introduction

the origin and geography of cultivated plants, have served as a systematic collection of such work with plants and its presentation to the world in general. *Centers of origin of cultivated plants,* was dedicated to Alphonse de Candolle

The Centers of type-formation (centers of origin) of Cultivated Plants

The centers of type-formation or the centers of diversity the 'differential phyto-geographical method'

1. A strict differentiation of the plants Linnaean species and intraspecific groups by all available means of various disciplines beginning with morphology, agrobotany, phytopathology, cytology and recently by molecular methods.
2. Delimitation of the distribution areas of these plants and, if possible, also of the distribution areas in the remote past when communication and seed exchange were more difficult than at present.
3. A detailed determination of the composition of the varieties and races of each species, and a general system of the genetic variability within the different species.
4. Establishment of the distribution of the genetic variability of the forms of a given species as far as regions and areas are concerned, and the establishment of the geographical centers where these varieties are now accumulated. Regions of maximum diversity, usually also including a number of endemic types and characteristics, can also be centers of type-formation.
5. For a more exact definition of the center of origin and type-formation it is necessary to establish the geographical centers of concentrations of species that are botanically closely related as well.
6. Finally, the establishment of the areas of diversity of wild subspecies and species that are closely related to the cultivated species in question should be used for amendment and addition to the area defined as area of origin, when the differential method for studying races is applied to them.

In 1940 Vavilov stated that the method of differential taxonomy offers an opportunity to trace the dispersal of many cultivated plants. It demonstrates their stages of evolution with respect to both the initial origin and their introduction into cultivation within different areas. It shows their relation to wild subspecies and species, but also demonstrates the subsequent evolution under domestication of these plants when dispersed from the basic centers and undergoing changes under new conditions and the further effects of natural and artificial selection. the origin of different cultivated plants led Vavilov to the establishment of new concepts, *i.e.* primary and more ancient crops in contrast to secondary ones, allowing him to characterize with good precision the centers where agriculture originated and the pathways along which it was dispersed. The geographical distribution of plant resources on earth and the establishment of the enormous infraspecific diversity of the majority of crops allowed not only a determination of their localization but also offered an opportunity to ascertain the period of origin of the plants most important for cultivation.

"The history and origin of human civilizations and agriculture are, no doubt, much older than what any ancient documentation in the form of objects, inscriptions and bas-reliefs reveals to us. A more intimate knowledge of cultivated plants and their differentiation into geographical groups helps us attribute their origin to very remote epochs, where 5000 to 10,000 years represent but a short moment" The number of centers listed in the terms 'center', 'focus' and also 'area' of origin. Their definition is important: the geographical centers are basic and independent foci where agricultural crops originated but are also geographic areas where cultivated plants are grown the centers of origin of cultivated plants but they amount only to a correction of details.

"The basic composition of cultivated plants, typical of this or that center, remains stable" As far as the historical character of Vavilov's works towards the establishment of the centers of agricultural crops is concerned, Sinskaya calls our attention to the prevalent use of the expressions "historical-geographical area" or "geographical areas of the historic development of a cultivated flora" Sinskaya elaborated a more detailed approach to the analysis of the cultivated plants in their centers of origin. This approach is based on a differentiated characterization of the endemism the various taxa have in a given area which are divided into the following categories:

- Genera originating from the area
- Genera having one of their centers of origin or their most important secondary center of origin in the given area
- Species strictly endemic for the given area
- Species endemic to the given area, but having their first origin in another area
- Species having one of their centres of origin or their most important secondary centre of origin in the given area.

The main categories and proposed to differentiate five basic geographical areas of historical development of cultivated plants, each having its subareas. The basic geographical areas of origin of cultivated plants after the respective cultivated plants together with their characterization by the above-named categories. The development of important genera (such as *Triticum, Secale, Hordeum, Beta, Brassica, Daucus, Lens, Linum, Olea, Mandragora, Pisum, Melilotus* and many others), which include the major proportion of all crops, forms the basis for agricultural production in countries around the Mediterranean, but most of them are intensively grown in Asia as well as in

Southwest Asia. Phytogeographical studies have revealed that compared with the areas of Africa south of the Mediterranean, there is a characteristic cultivated flora that is not less rich than those in other centers where agriculture arose. Many cultivated plants have undergone very old but secondary development there, *e.g.* in Ethiopia.

While further developing Vavilov's ideas about the centers of origin, singled out the African region for the historic development of cultivated flora. Ancient Mediterranean elements (actually both Mediterranean and Southwest Asiatic ones) predominated in the composition of the cultivated flora of Ethiopia but are not sharply delimited from those of other African areas. Elements from South Asia also occur there. This area is rather an area of introduction than of distribution of cultivated plants to other places. calls such territories "dependent areas", to which belong not only Ethiopia, but also North America, where agriculture developed on the basis of Mexican and Central American crops and, later on, that of crops from the Old World. In central and northern Europe, on the Russian steppes and in Siberia, agriculture is based primarily also on cultivated plants, introduced from the subareas Southwest Asia and Mediterranean, etc. The agriculture of the "dependent areas" underwent a certain period of development and, therefore, is not limited, judging by the large quantity of plants introduced into cultivation from less rich, wild flora of these territories.

Table. Geographical areas of historical development of cultivated flora

Basic areas of origin	Subareas
I. Ancient Mediterranean	Southwest Asia1a. Anterior Asia (Transcaucasus, Asia Minor, Near East, West Iran)1b. Middle Asia (Turkistan, Afghanistan, East Iran, North West India, Pakistan)Mediterranean
II. East Asia	Northeast Asia (Japan, Manchuria) Southeast and Central China
III. South Asia	South China, India and Sri LankaMalesian
IV. Africa	
V. New World	Central AmericaSouth America

Vavilov referred emphatically to the division of the globe into floristic regions and subregions such as those accepted by conventional phytogeography for elaborating the geographic origin of cultivated plants. The origin of cultivated wheat is located in the Ancient Mediterranean (syn.=Old Mediterranean) which includes, the Mediterranean region and Southwest Asia. The latter was divided by Vavilov into Asia Minor in a broader sense and Inner Asia. Sinskaya considers these territories as subareas of the Ancient Mediterranean Many species, genera and families which were responsible for the development of cultivated plants originated from these subareas. Nevertheless, every subarea has its characteristics due to the ecological conditions and the richness of the flora of wild and cultivated plants as well as to the ancient history of agriculture. The area of origin of wheat is Southwest Asia. The greatest amount of endemic species and a huge amount of different intraspecific taxa is found there. The greater the distance from this primary area of origin, the less diversity of the species is observed.

From 26 species of the genus *Triticum* the following species are found in Anterior Asia and are endemic wild plants: *T. urartu, T. araraticum* and *T. dicoccoides.* Endemic cultivated plants are: *T. timopheevii, T. zhukovskyi, T. carthlicum, T. karamyschevii, T. ispahanicum, T. macha, T. vavilovii* and *T. sinskajae. Triticum turanicum* and the wild einkorn *T. boeoticum* mainly occur there. Sinskaya considers *T. sphaerococcum* a further species, which occurs in northwest India, as an endemic species

Subarea			**Dependent area**
Mediterranean †(147)	**Southwest Asia**		**Ethiopia (250)‡**
	Anterior Asia (412)	**Middle Asia (260)**	
T. boeoticum (16)§	*T. boeoticum* (57) *T. urartu* (6) *T. araraticum* (13) *T. dicoccoides* (25)		
T. monococcum (13)	*T. monococcum* (14) *T. sinskajae* (1)		
T. dicoccum (7)	*T. dicoccum* (15) *T. ispahanicum* (2) *T. karamyschevii* (3) *T. timopheevii* (4) *T. militinae* (2) *T. zhukovskyi* (1) *T. macha* (14) *T. vavilovii* (7)		*T. dicoccum* (8)§
T. spelta (14)	*T. spelta* (14)	*T. spelta* (19)	
T. durum (80)	*T.* durum (75)	*T. durum* (8)	
T. turanicum (4)	*T. turanicum* (34)	*T. turanicum* (7)	
T. turgidum (34)	*T. turgidum* (54) *T. carthlicum* (18)	*T. turgidum* (3) *T. jakubzineri* (1)	
T. polonicum (11)	*T. polonicum* (14) *T. sphaerococcum* (17)	*T. polonicum* (3)	*T. polonicum* (6)
T. compactum (13)	*T. compactum* (40)	*T. compactum* (64)	
T. aestivum (25)	*T. aestivum* (59)	*T. aestivum* (142) *T. petropavlovskyi* (4)	*T. aestivum* (33) *T. aethiopicum* (203)

A second group of Southwest Asian wheat originated in the Near East subarea but then spread to other areas. They were later replaced, at the beginning of the century, by higher-yielding species and therefore can only be found as relics isolated from each other. They are: *Triticum monococcum, T. dicoccum, T. aethiopicum* and *T. spelta* (Sinskaya 1969; Padulosi *et al.* 1996). These species, which are at present to be found in areas far away from each other, were more intensively developed in other subareas of the Ancient

Mediterranean. *Triticum aestivum* and *T. compactum* were intensively developed in Middle Asia and subsequently spread all over the world. *Triticum durum* and *T. turgidum* developed in the more central and western parts of the Ancient Mediterranean, particularly close to the sea coast. East of this area of origin, on the other hand, the process of formation of the further wheat varieties is not observed. A.M. Gorskyi during his expedition to Sinkiang (Western China) found a new endemic wheat named *T. petropavlovskyi* (Dorofeev *et al.* 1979). Vavilov had completed the investigation of Sinkiang in 1929 and regarded this western part of China as one of the geographically Distribution of the species of wheat in the Ancient Mediterranean area of origin of cultivated plants Number of taxa which occur in the given subarea: † number of botanical varieties per area;§ number of botanical varieties per species. Species of wheat, *e.g.* Ancient Mediterranean elements of northeast African flora. Distribution of taxa of some genera in the Ancient Mediterranean area of origin of cultivated plants.

	Subarea			
	Mediterranean	Southwest Asia		Dependent area
		Anterior Asia	Middle Asia	Ethiopia‡
***Hordeum vulgare*†**				
subsp. *vulgare*	14/3	7	21/3	38/8
subsp. *distichon*	10/1	18/2	7	38/20
***Pisum*§**				
P. formosum	-	1	-	-
P. fulvum	1	1	-	-
P. sativum	8	7	9	-
subsp. *abyssinicum*	-	-	-	1
Beta vulgaris	9	5	5	-
***Lens*††**	63/3	-	54/9	2/2

Note:

Species of different crops, *e.g.* Ancient Mediterranean elements of northeast African flora. Number of varieties: † Lukjanova *et al*total/endemic varieties Makasheva Krasochkin Barulina Such peripheral areas were only reached by a few infraspecific varieties of cultivated plants. Nevertheless, Vavilov acknowledged the possibility of finding endemic forms of wheat in these areas. Yue Dahue) also reports findings of *T. petropavlovskyi* by Chinese expeditions to Tibet. *Triticum spelta* was formerly considered a European crop. But *T. spelta* was found in Iran (Kuckuck and Sin the Transcaucasus and in Middle Asia This supports the existence of a common Southwest Asia and Mediterranean area of origin.

Other crops

The diversity of barley (*Hordeum vulgare*) is more evenly spread across the Ancient Mediterranean area and several infraspecific taxa are endemic to Ethiopia The botanical varieties of the genus *Pisum* are also more or less evenly spread in the Ancient Mediterranean area of origin Several monographs dealing with many different crops have been published by the VIR. Such work is important not only for studies on centers of origin of cultivated plants but also for theoretical and practical agronomy. Such work is the basis for searching for initial material for plant breeding. Taxonomic studies in the tradition of Vavilov, based on the investigation of variation within a species, have also been carried out in Gatersleben Recently information on coriander has been provided by Diederichsen The detailed investigation of the variation of the species *Coriandrum sativum* by several characters caused the author to divide this species into several groups (so far called ecological types), which are connected with the geographical origin in the Ancient Mediterranean area.

The infraspecific classification helps to indicate areas where different types of a given species are to be found. It also is an excellent method to single out and preserve rare accessions in a collection. However in the case of wheat a very interesting group - *T. durum* convar. *villosum* - collected by Vavilov in Syria, Jordan and Lebanon is untraceable in the collection at VIR. It was an extremely xerophytic type with a very hairy ear and similar leaves. A herbarium specimen of it exists at VIR but even that is threatened. The extinction of durum wheat without ligula (*T. durum* convar. *aglossicon* Flaksb.) from Cyprus was prevented. In its area of origin on Cyprus this type is already extinct. In bulk populations of wheat accessions single plants of this type occurred. Such plants were singled out and received their own number in the VIR catalogue.

The Caucasus comprised more than ten botanical varieties. Of these only one or two are still part of the VIR collection. The VIR collection of Ethiopian wheats has also lost many varieties. Every botanical variety has to be preserved as a single accession, if the original landrace, which contained several varieties, is not reproduced under the conditions which are similar to its natural area of origin The taxonomical category 'varietas' was introduced by Fr. Alefeld and Fr. Körnicke for crop plants and is based on the differentiation by distinct characters.

This category makes it possible to orientate quickly and properly in the diversity of a given area. At the same time such classification delivers clear information for the given species with respect to the Law of Homologous Series in variation The basic taxonomical category, however, is the species. For geobotanical investigations in wild plants the use of more detailed categories, *i.e.* infraspecific taxa, is not essential.

In the early days researchers concentrated on the level of the genera; later the species level was elaborated. At present taxonomists of cultivated plants should focus on the infraspecific level. The taxonomical classification of cultivated plants depends on the methods on which it is based, and the attention which was paid to a given species. In general the economic relevance of a species favors scientific interest.

The classification of wheats

As early as 1935 Vavilov stated that "a basic handicap of all genetic investigation in wheat, as well as in other plants, is the accidental choice of the material... and the neglect of the wide range of geographical variation." For crops, which have a short history of domestication, the characterization of a cultivar can be as general as for the botanical species. Wheat has been the basic element of food for humans for 10,000 years, and is to be found nearly all over the world. The resulting range of variation of wheat,, is astonishing. The formation of new infraspecific varieties is a continuous process during cultivation of the species. The modern techniques used in plant breeding of today never led to formation of such varieties. The latest systematical overview for wheat was and it differs from the system established by Flaksberger In particular, the latest system was cleared of contradictions and inconsistencies. The first successful approach to such a classification of wheat was done by Flaksberger in 1915. The system proposed by Percival does not differ very much from the latter. The ideas about the infraspecific differentiation of the species *T. aestivum* were very much changed owing to the expeditions of Vavilov to Central Asia, Iran, Afghanistan and India.

Methods

In southwestern Asia from the crucial period 20,000 to 8500 BP (non-calibrated). The quantity and quality of the archaeobotanical Because biological decomposition is rapid in the aerobic archaeological sediments of this area, archaeobotanists rely on plant materials which have been rendered stable through charring in hearths or other fires. These remains are recovered by flotation and sieving. Under the best circumstances large-scale flotation has obtained thousands of charred seeds, fruits and fragments of charcoal within the chronological At worst no sampling was carried out or only a few chance finds were collected. Concerning the archaeobotanical criteria for morphological domestication and cultivation, not all archaeobotanists agree on the best criteria. The solid rachides in barley and naked wheats are clear indicators for archaeobotanists. However, the more primitive hulled wheats such as einkorn and emmer are more problematical. For the archaeobotanist the distinction between domestic and wild hulled wheats is based on the disarticulation scar left by the abscission layer. The break occurs in the same place on the rachis, and on domestic modern material it is rough or torn and on wild material it is smooth. However, with ancient carbonized remains the surface is very often too poorly preserved to allow this distinction.

Grain size is another criterion used, because domestic grains are generally more plump Plump barley grains, unknown in the wild, occur with fragile rachis fragments These are not considered to be evidence of domestication. However there is some reason to reconsider these finds in the light of modern semi-solid rachis barley which occurs in Syria. The author has collected specimens of semi-solid rachis, two-row 'black' barley near Bosra in southern Syria. The disarticulation scar is similar to that of wild barley, and the rachis fragments would be difficult to distinguish from wild types in carbonized material. This morphological type could explain why domestic-type grains occur with apparent wild-type rachis fragments. As for evidence for cultivation, one might expect digging tools to provide the answer. However, for the moment this is not the case and it is possible that these tools were wooden and have not survived. Archaeobotanical research has concentrated on the presence of weed assemblages The most common taxa in these assemblages include the following: *Adonis, Aegilops, Astragalus, Avena, Bromus, Bupleurum, Camelina, Centaurea, Centranthus, Coronilla, Fumaria, Galium, Glaucium, Hordeum, Lathyrus, Lithospernum, Lolium, Malva, Papaver, Polygonum, Reseda, Silene, Valerianella* and *Vicia.* It is difficult to be sure that these taxa really represent a weed assemblage because these plants make up part of the original steppe flora, and identification at the species level is rarely possible. On the other hand, what one might expect is an increase in the frequency of these taxa at the expense of other steppe plants which were not preadapted to become part of the weed flora.

Food grains have the immense advantage that they can be stored. The wild grasses become ripe in late spring and promptly fall to the ground. They need to be harvested just before maturity. Given the dry climatic conditions, grains can be conserved or stored relatively easily. This facility for storage was one of the main advantages of seed-gathering which led to a more secure subsistence base and prepared the way for a sedentary way of life. Increasingly favorable climatic conditions resulted in a rich environment for Early Natufian inhabitants of the western Mediterranean (12,000-11,000 BP). The climate appears to have been more favorable than at present or at any time since the Natufian.

Archaeobotanical evidence indicates that the Syrian and Jordanian steppes had a much richer vegetation. This is indicated by the presence of forest steppe species, for example *Pistacia* and *Amygdalus.* Cultural factors such as an increased reliance on stored grain (although there is little archaeological evidence for this) permitted a sedentary existence, such as Mureybit and Abu Hureyra on the Euphrates in Syria, Hayonim in Israel, Wadi Hammeh in Jordan, and a little later Qermez Dere, Nemrik 9 and M'lefaat in northern Iraq have round architecture, large hearths and groundstone equipment. that wild cereals were exploited together with a number of edible fruits and pulses Charcoal and fruit remains indicate the forest/steppe with *Pistacia* and almond; in favorable conditions, deciduous oak was present. Archaeobotanical evidence clearly indicates that this vegetation penetrated further east into what is now arid steppe. This habitat provided wild cereals, pulses and an abundance of game the use of grasses comes from glossed flint tools indicating the harvesting of plants with high silica content Hatoula and Kebara in Israel and Beidha in Jordan It is clear that morphologically wild progenitors of Old World cereals and legumes were exploited for several millennia, before the appearance of their domestic counterparts. The geographical extent is impressive, stretching from northern Iraq to the southern Levant, Anatolia and even southeast Europe. During this period Einkorn is dominant at Mureybit and Abu Hureyra, barley and some emmer are present at Ohalo II. Rye is also present at a number of the Late Natufian there is wide evidence for climatic deterioration from about 11,000, usually referred to as the Younger Dryas (which appears to have adversely affected settlements in the more arid zones of the Jordanian and Syrian steppe and the Negev highlands. With few exceptions, near permanent water continued to be occupied into the next period. During the following period, the Pre-Pottery Neo-lithic A, the climate became more favorable again. Charcoal evidence indicates that the Syrian steppe was at least partly wooded near reliable water sources. The architecture of small round houses is more substantialFor the very earliest levels at Aswad IA and Jericho, remains are numerically too meagre to be certain of domestication, Emmer is dominant at Aswad einkorn at Mureybit, barley at Jerf al Ahmar in northern Syria suggests that the inhabitants of these were still gathering local cereals but this does not exclude small-scale cultivation as described by Harris using locally available wild cereals as seed stock.. The controversial plump domestic-type barley grains

Established during this period or earlier show no clear-cut domestication in the lowest levels, but morphological domestication does appear at higher levels. over a very long period, more than a millennium. This would have led to degradation of the local vegetation Thus resource depletion could have been a contributing factor for the adoption of agriculture. During the next chronological period architecture becomes rectangular and the transition. near. At Aswad, between 9730 and 8560 BP, 26 per cent of the barley rachis fragments are solid domestic types, but it is not clear whether they occur in the earliest levels. At preliminary studies indicate that the cereals are not yet domesticated, but indirect evidence of weed associations strongly suggests the presence of cultivation, and similar assemblages are seen at Aswad, Cayönü and Cafer Höyük.

More frequent and cover a wider geographical area with expansion of agricultural communities into central Anatolia (Asikli Höyük) and Cyprus (Shillourokambos). Crop evolution and morphological domestication are clearly shown by the appearance of a solid rachis in barley and naked wheat, for example at Aswad West phase II In the Euphrates Valley there is no evidence for *in situ* domestication unless we consider the plump barley Instead domesticates were introduced. Emmer absent in the area during earlier periods appears to have been introduced from elsewhere at Abu Hureyra and Halula together with a naked wheat wild types remain at significant frequencies. We can see this at Cayönü, Cafer Höyük (wild wheats), Aswad, Ganj Dareh (wild barley), Halula and Azraq (wild wheats and barley) These mixed finds can be interpreted in three ways:

- As evidence of the exploitation of wild stands,
- As unwanted weeds, and
- As an integral part of the crop consisting of a mixture of wild and domestic cereals.

The relatively high proportion of wild types and the lack of pure finds of domesticates suggests that the wild plants may have been considered as a useful part of the crop, as opposed to unwanted weeds. This suggests cultivation of wild and domestic types together but does not exclude gathering from wild stands in a kind of mixed economy. Even during later periods for example at Ramad between 8210 and 7880 BP, domestic barley rachis fragments are only at 52 per cent. A similar situation was noted at Magzalia such as Bouqras and Ras Shamra (phase Vc) wild types are rare or absent. wheat. During the Late PPNB, einkorn becomes a minor component and could be interpreted as a weed for most of the Near East. It reappears as a major component

DOMESTICATION OF PLANTS

Introduction

Lentil (*Lens culinaris* Medikus) is a short-statured, annual, self-pollinating, food legume mainly grown in the Indian subcontinent, the Mediterranean region and North America. The crop is grown in dryland cereal-based rotations because of its nitrogen-fixing ability, its high-protein seeds for human consumption, and its straw, which is a valued livestock feed. ICARDA has a global mandate for research on lentil improvement and is situated in the Near East arc, where the crop was domesticated. The temporal dimension given by the history of the spread of the lentil following domestication and the consequent selective forces give a perspective of applied evolution to plant breeding which is highly relevant to ICARDA's current role in any new 'spread' of the crop.

Lentil Origins

The putative progenitor of the cultivated lentil is *Lens culinaris* subsp. *orientalis* (Boiss.) Ponert which is distributed from Greece in the west to Uzbekistan in the east, and from the Crimean Peninsula in the north to Jordan in the south The oldest carbonized remains of lentil are from Franchthi cave in Greece dated to 11,000 BC and from Tell Mureybit in Syria dated 8500-7500 BC But as it is not possible to differentiate wild from cultivated small-seeded lentil, the state of domestication of these and other carbonized remains in the aceramic farming villages in the 7th millennium BC in the Near East arc is unknown. The finding of a large hoard of lentil at Yiftah-el dated to 6800 BC is, however, suggestive of domestication The oldest find of lentil seeds that are larger than wild seeds, and therefore unequivocally domesticated was at Tepe Sabz, Iran; they have been dated to 5500-5000 BC. The overlap in the distribution of wild lentil and the early archaeological record indicates that lentil was domesticated in the Near East arc.

Spread

From these beginnings the crop spread to the Nile, and to Central Europe via the Danube. As lentils are repeatedly found in the early agricultural settlements of the 5th millennium BC in Europe, situated outside the distribution of *L. culinaris* subsp. *orientalis,* this is indicative of earlier domestication. Lentils were definitely associated with the start of the 'agricultural revolution' in the Old World, which was initiated by the domestication of einkorn and emmer wheats, barley, pea, flax and lentil The crop was part of the assemblage of Near Eastern grain crops introduced to Ethiopia by the invaders of the Hamites. From the Bronze Age onward, maintained itself as an important companion of wheat and barley throughout the expanding realm of Mediterranean-type agriculture. The dissemination eastward of the Near Eastern grain crops, including lentil, reached Georgia in the 5th and early 4th millennia BC. The crop appears in the archaeological record in India around 2500 BC as part of the Harappan crop assemblage. Alphonse de Candolle wrote that on linguistic grounds, "It may be supposed that the lentil was not in this country (India) before the invasion of the Sanskrit-speaking race." The invasion occurred before 2000 BC. The crop probably reached its current Old World range about 3000 years ago. It was carried to the New World after Columbus.

Selective forces Operating During Spread

the selective forces operating during the spread of the culture of lentil derives from an analysis of the variation found in current landrace populations from different geographic regions. Following widespread collecting and evaluation classified the assembled variation into six groups (grex varietatum), each of which was geographically differentiated and also characterized by a complex of morphological characters, mainly qualitative, common within a group but differing in other groups. This type of geographic association is mirrored by variation in quantitative morphological traits Such morphological characters are readily observable and consequently often the subject of human selection. Geographic differentiation between landraces also has been found for other more cryptic, ecophysiological factors, such as those resulting from selection for soil conditions and for climatic conditions, as the following examples illustrate. Iron-deficiency symptoms are observable on some accessions of lentil grown in calcareous soil. In a germplasm collection of 3512 accessions originating from

18 countries, landraces from those Mediterranean countries where lentil originated (Syria and Turkey) exhibited Fe-deficiency symptoms only at very low frequencies (<1.5 per cent) In these regions the soils are generally highly calcareous with pH >8.0, conditions known to reduce Fe availability. Those landraces exhibiting symptoms of Fe deficiency mostly originate from relatively warm climates, such as India (37.5 per cent accessions showing Fe deficiency) and Ethiopia (30 per cent), where the pH is generally 6.5-7.0 This arose from either the chance introduction to these regions of Fe-deficient founder populations, or selection pressure in favour of either Fe-deficient types or genes linked to Fe-deficiency.

In a survey of patterns of morphological variation in the world germplasm collection of lentil, phenology was found as the key to the adaptation of the crop on a macrogeographic scale To further understand phenology, the flowering responses to temperature and photoperiod of a world collection of 369 accessions from 13 major lentil-producing countries (25 randomly selected accessions per country) together with lines from the ICARDA breeding programme were studied The distribution of country means for sensitivity to temperature and photoperiod illustrates the responses to selection for adaptation to new ecological environments following the spread of the crop from its origin Dissemination to lower latitudes such as into Egypt, Ethiopia and India was accompanied by a reduction in photoperiodic response. Obligate photoperiodic control of the onset of flowering ensures that flowering starts annually in the same calendar period, irrespective of fluctuations in temperature. Consequently, selection against photoperiodic control in a long-day plant such as lentil implies an adaptation to relatively short days, which occur at low latitudes and which would otherwise delay flowering to an unacceptable extent. Under these conditions, the crop relies rather more on temperature than photoperiod to ensure that flowering occurs at an ecologically and agronomically appropriate time. There was also evidence that flowering in the subtropical group is more temperature-sensitive than in West Asian germplasm, on average.

Movement from West Asia to higher latitudes, for example to Russia, resulted in a modest reduction in photoperiod sensitivity and an increase in temperature sensitivity, probably reflecting the change in sowing date from winter to spring. Temperature sensitivity to processes other than flowering, such as the base temperature for rate of germination (T_b), also has been altered during dissemination from the Near East to environments with higher temperature regimes. The mean T_b of 15 randomly selected accessions from Ethiopia and India was higher than that of similar samples from Lebanon and Turkey

The degree of winterhardiness also has been affected in the spread of the lentil from its origins as a winter-sown crop in the Near East. A world collection of 3910 accessions was screened for winterhardiness near Ankara, Turkey, in the winter of 1979/80, when temperatures dropped to a low of -26.8°C and there were 47 days of snow cover In total, 238 accessions were undamaged by the cold winter. Their origins were mostly from Chile (frequency of winterhardy accessions was 28 per cent), Greece (33 per cent), Syria (33 per cent) and Turkey (15 per cent), where selection for winterhardiness had occurred through winter sowing although lentil production is found in countries farther north, where the winter is colder, such as Russia (5 per cent), and Hungary (4 per cent), where sowing is exclusively done in spring. The frequency of winterhardy accessions from countries with warmer winter environments, such as Egypt, Ethiopia, India and Pakistan, was <1 per cent.

The locus *aat-p* from the world collection show differences among countries and regions in the frequency of alleles and the extent of polymorphism, reflecting adaptation to local conditions There is evidence that a low frequency of $Aat\text{-}p^F$ is associated with the spread into South Asia, whereas a higher frequency of $Aat\text{-}p^F$ is characteristic of northern European material. Such regional differences result from the spread of the crop to new physical environments with the consequent natural and artificial selection for local adaptation. To summarize with an extreme example, the spread of lentil from the Near East into India resulted in a loss of winterhardiness and the ability to extract iron from calcareous, high-pH soils, a reduction in photoperiodic sensitivity for flowering, increased intrinsic earliness in flowering, an increase in temperature sensitivity for flowering, and in the base temperature for rate of germination. This list of factors affecting adaptation is not exhaustive. Each factor individually may be of minor importance, but collectively they illustrate the complex of interacting ecophysiological factors that determine adaptation. Superimposed on these natural selective forces is the effect of human selection on observable morphological traits, particularly the seed, for culinary and agronomic

The *pilosae* group of the Indian subcontinent is also characterized by two endemic qualitative morphological traits: precocity in flowering and maturity, and a low biomass. Furthermore, lentil germplasm from India is among the least variable among lentil-producing countries, despite India being the largest lentil-producing country in the world. Recent evidence from isozyme (7 loci) and random amplified polymorphic DNA (RAPD) (22 loci) analysis indicates a similar striking difference between germplasm from South Asia and that from the rest of the world; additionally South Asian germplasm was found low in diversity germplasm. The simplest explanation has the introduction around 2000 BC of a founder population adapted to the new environment that was probably both small in number and low in variability which resulted in a genetic bottleneck in South Asia (Erskine *et al.,* unpublished). This still limits breeders' progress today. Indeed, a reconstruction of the phenological problems associated with the initial spread of the crop into the Indo-Gangetic Plain was inadvertently made with the introduction into that region of lentil selected in West Asia through ICARDA international nurseries in its early years Lentil selected in West Asia, when sown in India and Pakistan, mostly came into flower as the indigenous lentils were maturing.

Use of Information in Breeding

Armed with an understanding of the specific adaptation of lentil, the local constraints to production and the various consumer requirements of different geographic areas for seed, the breeding programme at ICARDA aims to produce genetic material suitable for national cooperators. The programme has been designed as a series of separate, but finely targeted streams linked closely to national breeding many of the selective forces important in the adaptation of the crop are environment-specific, such as temperature and photoperiodic sensitivities. Clearly, selection undertaken in conditions very different from those in the target environment will have a lower response in the target environment than selection conducted directly within the target environment. Harnessing the specific comparative research advantages of ICARDA and its national programme partners, crosses (previously agreed with cooperators) are made and segregating generations advanced at ICARDA. The selection of bulk segregating populations is undertaken in the target environment by national programmes for national programmes, for example, Algeria, Bangladesh, Egypt, India, Jordan, Morocco, Nepal, Syria and Turkey. Details may be found in ICARDA.

A major effort to widen the genetic base in the Indian subcontinent using three approaches, namely plant introduction, hybridization and mutation breeding. As mentioned above, plant introductions from West Asia flower as indigenous material matures in the Indian subcontinent. The asynchrony in flowering has isolated *pilosae* lentils reproductively. However, the introduction of ILL 4605, an early flowering, large-seeded line, has resulted in its release as 'Manserha 89' for wetter areas of Pakistan and its widespread use as a parent in breeding programmes in the region. Hybridization between *pilosae* and exotic germplasm, primarily at ICARDA, followed by selection in the Indian subcontinent has resulted in cultivars with improved disease resistance and yield in Bangladesh and Pakistan. Mutation breeding has given new morphological markers and several promising lines.

Target agro-ecological regions of production of lentil and key breeding aims.

Region **Mediterranean low to medium elevation**	**Key traits for recombination**
300-400 mm annual rainfall	Biomass (seed + straw), attributes for mechanical harvest and wilt resistance
<300 mm annual rainfall	Biomass, drought escape through earliness
Morocco	Biomass, attributes for mechanical harvest and rust resistance

Egypt	Seed yield, response to irrigation, earliness and wilt resistance
High elevation	
Anatolian highlands	Biomass and winterhardiness
N. African highlands	Seed yield and low level of winter hardiness
South Asia and E. Africa	
India, Pakistan, Nepal, Ethiopia	Seed yield, early maturity and resistance to rust, ascochyta and wilt
Bangladesh	Seed yield, extra earliness and resistance to rust and *Stemphylium*

The example of the widening of the genetic base of the lentil in South Asia illustrates the value of the historical perspective in plant improvement. Although ICARDA and its national programme partners are undertaking the first systematic lentil improvement programme with access to a wide range of genetic diversity and modern biotechnological tools, the efficiency of this research has been increased by an awareness that we belong to the approximately 200th generation of cultivators/selectors who have grown the crop since its domestication.

Introduction

Alphonse de Candolle, the founder of the study of crop evolution, was primarily interested in biogeography, the study of distribution of plants in relation to their environment. In his opinion, among the factors influencing this distribution are 'historic' factors such as glaciation events that are not related to the intrinsic adaptation characteristics of the species. When he considered crop plants, he posited that one particular historic event - domestication - had had a paramount importance in determining their adaptation and, hence, distribution, as well as their genetic characteristics Four types of evidence could be utilized to determine the origin of domestication of crop plants These were archaeological (or more specifically archaeobotany), botanical (*i.e.* the distribution of the wild, ancestral relative), historical (or the existence of a written record documenting the existence or importance of the crop) and linguistic evidence (*i.e.* the existence of words designating the crop or objects or concepts related to the crop in native languages). Of these four types, the first two provide the most abundant and reliable evidence. Since the time of de Candolle, however, additional scientific methods have increased considerably our power to determine centers of crop domestication based on archaeological and botanical arguments

Although common bean is not a species originating in the Near East, the focus of the present symposium, it presents nevertheless an interesting domestication history, which is only recently being elucidated. The first botanically accurate description of common bean in European herbalsWhereas barely 50 years had passed after the conquest of the Americas, the actual origin of the crop had already been lost to the writer of the herbal and later European botanists. Linnaeus assigned the origin of common bean to India; de Candolle himself expressed doubts that common bean had been domesticated in the Americas. It was only during and after the Second World War that it became generally admitted that common bean originated in the Americas. The conclusion was based on the discovery of wild common bean in Argentina and Guatemala and archaeological remains in the Americas

Based on the extensive distribution of the wild relative speculated that common bean originated through multiple domestications as did other crops originating in the Americas, such as cotton, pepper and amaranth. observed that there was a Mesoamerican cultivar group with smaller seeds and an Andean group with larger seeds. None of these observations, however, provided substantial experimental evidence and hence a clearer picture of the actual process of domestication for this species. For example, one could ask whether the two cultivated groups identified resulted from distinct domestications or a single domestication event followed by divergence into a small-seeded and a large-seeded group. One could also ask where these single or multiple domestications took place. The archaeological record of the origins of agriculture in the Americas is rather limited. In Mesoamerica, This important type of information therefore pales in comparison with what is available in the Near East In part, this may be due to the relatively humid environment required by many crop relatives in the Americas, which render long-term conservation of crop remains less likely. Therefore, in relative terms compared with archaeological information, there is a relatively higher emphasis on botanical evidence to decipher crop evolution in the Americas than in the Near East.

The use of molecular marker information has increased our understanding of the domestication process in common bean a crop originating in the Americas. In addition, I hope to demonstrate how molecular analyses can bridge the gap that has separated botanical and archaeological evidence, the twin pillars on which rests the study of crop evolution. The power of molecular biology to change scientific paradigms is by no means unique to this field but extends to many other fields as well.

Once upon a time there was a bean... General Methodological Principles

Our current understanding of the origins of domestication in common bean is based on a two-pronged approach relying on the study of wild-growing *P. vulgaris* beans, on one hand, and the use of biochemical or molecular markers, on the other. The use of wild beans is predicated on the fact that their dispersal occurs over much smaller distances (at most a few meters after each vegetation cycle, if at all) than their domesticated descendants. The latter are subject to long-distance exchange, trade, gift or human migration, all of which tend to obscure the original geographic distribution patterns of genetic diversity. Recent explorations have provided a more complete representation of the actual distribution of wild *P.* beans are distributed from approximately 30°N Lat. to 30°S Lat. at mid-elevations from 1000-2500 m asl in regions with moderate rainfall. A common characteristic of the environments in which wild beans are found is the existence of a marked dry season, an important feature needed for seed dispersal.

Genetic relatedness is traditionally assessed with morphological phenotypic traits. Plants with a similar phenotype are inferred to be genetically related either through common ancestry or through geneflow (hybridization). Occasionally, however, inferences based on phenotypic similarity are erroneous. Plants can arise from different selection episodes leading to the same phenotype (convergent evolution). In addition, similar phenotypes can have a different genetic basis (genetic heterogeneity). In both cases, an inference of genetic relatedness would be unwarranted. In contrast, biochemical, and even more so, molecular markers provide information on genetic relatedness that is generally (although not always) devoid of selective influences.

All biochemical and molecular markers were created equal! Differences have been observed with regard to the ease of analysis, reproducibility, level of polymorphism, number and genome distribution of loci Of particular note is the molecular basis of polymorphisms. The more complex the basis, the more desirable the marker, because complexity will reduce the probability of repeat mutations, especially in conspecific materials, which are, by definition, closely related. For example, polymorphisms based on DNA nucleotide substitutions are simple changes that could presumably occur repeatedly. Analyses based on single or few RAPD or RFLP markers are therefore unreliable in determining genetic relatedness that sample a large number of changes at the molecular level, because they include either a large number of markers or complex markers such as actual DNA sequences or multigene families, are more robust with regard to repeated changes. For example, banding patterns of seed proteins such as phaseolin in common bean, when analyzed in denaturing SDS-polyacrylamide gel electrophoresis, represent a compomy image resulting from changes at the DNA nucleotide level (duplication of genes, within-gene duplications, nucleotide substitutions) and at the translation level (co - and post-translational modifications). Therefore each pattern is unlikely to have arisen more than once. As a consequence, genotypes that exhibit the same protein pattern are likely to share a common ancestor

One of the major experimental difficulties in crop evolutionary studies is the distinction one has to make between ancestry (domestication) and geneflow as a cause of genetic relatedness. This is particularly the case given that wild ancestors and domesticated descendants belong to the same biological species and can therefore freely hybridize. Hence, similarity between ancestors and descendants even at the molecular level cannot be assumed to be due to domestication For example, DNA sequence data and the utilization of an outgroup (*e.g.* closely related species) can shed additional light on this problem because they can provide information on the nature of the polymorphism and the direction of the change, *i.e.* the ancestral vs. descendant states of the molecular polymorphism. This will be illustrated below with the case of phaseolin.

Domestication Pattern

The origin of the *Phaseolus vulgaris* progenitor in its wild state can be traced to the Pacific slope of Ecuador and northern Peru, based on the distribution pattern of direct repeats in phaseolin seed protein genes Phaseolin is coded by a small multigene family inherited as a single Mendelian unit and located on linkage group B7 The phaseolin locus is at most 190 Kb long and contains 6 to 9 genes Differences among these phaseolin sequences include nucleotide substitutions and the presence or absence of tandem direct repeats. Three repeats have been identified so far: a 21 bp repeat in the third intron, a 15 bp repeat in the fourth exon and a 27 bp repeat in the sixth exon. The existence of these tandem repeats provides an interesting landmark to follow the evolution of phaseolin genes. Indeed, the generation of a repeat is a more likely event than the precise loss of such a repeat, which would be needed to restore a sequence without repeats.

Hence, phaseolin genes without repeats represent probably a more ancestral state than phaseolin genes with repeats. This was verified by analyzing the presence of tandem repeats with a polymerase chain reaction (PCR) test of phaseolin sequences in two subspecies of a related species, *P. coccineus* subsp. *polyanthus* and *P. coccineus* subsp. *coccineus*. In these two taxa, none of the three repeats could be identified, suggesting that the repeatless state is indeed the ancestral state. A survey was then conducted among wild *P. vulgaris* to determine if any phaseolin haplotypes lacked genes with tandem repeats. Only wild bean populations from Ecuador and northern Peru on the Pacific slope of the Andes appeared to be devoid of phaseolin genes with tandem repeats It was inferred that these wild populations therefore represent the presumed ancestor of the species.

This conclusion is somewhat unexpected as most wild *Phaseolus* species are distributed in Mexico. One could have thought that *P. vulgaris* would have originated in that region as well and that other wild *P. vulgaris* populations distributed elsewhere were ultimately derived from Mexican ancestors. It may be, however, that some wild *Phaseolus* ancestors ('proto-vulgaris' and other species) were dispersed from Mexico to Central America and the Andes to give rise to locally distributed wild *Phaseolus* species, such as *P. costaricensis* From such a local distribution, a species such as *P. vulgaris* could have acquired a secondary, broader distribution by dispersal towards the north and south (see below). It is worth remembering that lima bean (*P. lunatus*) also shows an extensive distribution of its wild relative from Mexico to Argentina and that the species consists of two genepools, an Andean genepool originating on the Pacific side of the Andes in Ecuador and northern Peru (although at lower altitudes than wild *P. vulgaris*) and a 'Mesoamerican' genepool distributed from Mexico to the southern Andes. Although the domestication region of the 'Mesoamerican' genepool is unknown so far, it is likely to be located somewhere in Mesoamerica, given that most Mesoamerican cultivars are located there

From the small nuclear area in Ecuador and northern Peru, wild *P. vulgaris* were dispersed towards both the north (Colombia and Venezuela, Central America and Mexico) and the south (southern Peru, Bolivia and Argentina) to achieve their current distribution In a subsequent step, multiple domestications took place in the Mesoamerican and Andean regions. At least one domestication may have taken place in west-central Mexico (Jalisco), a conclusion based on phaseolin data. Most Mesoamerican cultivars exhibit the same phaseolin type ('S'-type). Wild beans with the same phaseolin type and without morphological signs of past hybridizations to domesticated beans are concentrated in the Mexican state of Jalisco, Interestingly, Doebley *et al.* suggested a similar domestication area for maize.

A second major domestication took place in the southern Andes although a more specific region awaits confirmatory. Phaseolin suggest possible multiple domestications within the Andean region because at least four phaseolin types have been identified among Andean beans. Alternatively, the original populations that were domesticated were polymorphic for phaseolin. Further are needed to distinguish between these possibilities. A third, likely minor domestication may have taken place in Colombia, although regions in Central America deserve further study because samples of wild *P. vulgaris* from that area were unavailable.

Post-domestication divergence

Common bean is perhaps best known for the high level of phenotypic diversity, particularly for seed type (colour, colour pattern, size and shape). This broad diversity is a consequence of the wide range of cultural environments under which common bean is grown, both in the Americas and elsewhere. This diversity has made it difficult to recognize patterns of genetic diversity in the domesticated genepool because of the possibility of convergent and reticulated evolution.

Analyses of biochemical traits such as isozymes and phaseolin seed protein have provided additional information on this problem. Koenig and Gepts and Singh *et al.* analyzed isozyme diversity in wild and domesticated germplasm, respectively. From these studies, it was possible to identify clusters of cultivars within the Andean and Mesoamerican genepools that shared a common isozyme allele. Further analyses of phenotypic diversity were conducted with multivariate statistical analyses (canonical and discriminate analyses) by morphological, agronomic and ecological traits. For example, in the Mesoamerican genepool, three subdivisions or races can be distinguished Race Mesoamerica represents cultivars originating predominantly from the warmer, more humid areas of Mexico, Central America and Colombia. Possibly, this race could be split further to account for an additional, minor domestication area in Central America or Colombia

Race Durango includes cultivars from the northern highlands of Mexico, which have a cooler, more arid environment. Race Jalisco consists of the most traditional common bean cultivars, originating predominantly in the humid central and southern highlands of Mexico. A similar subdivision in three ecogeographical races has been established for the current status of the domestication scenario in common bean. process The arrows indicate rare cases of geneflow between the two major genepools.

What does it Take to get a Domesticated Bean

Wild relatives and their crop descendants show marked phenotypic differences, collectively called the domestication syndrome As a consequence, they have been classified in different taxonomic species, which is unjustified given that often they can be freely crossed. Based on the traditional neo-Darwinian theory of evolution, one would expect that these marked phenotypic differences would be controlled by a large number of genes, each with a small phenotypic effect. This view reflects a view of evolution involving very gradual changes in environment and adaptation.

This may not have been the case during domestication (nor in natural environments for that matter) where the domesticated and natural environments are quite different. Hence, adaptation to cultivation represented selection for a radically different environment over a time scale that would have been very short in an evolutionary context (even a few 100 generations should be considered a very short period in evolution). This suggested that major genes might have played a role during domestication because only such genes would have provided a fast response to the rapidly changing environment imposed by cultivation.

To test this view of evolution during domestication, an experiment was conducted to determine the inheritance of the domestication syndrome in common bean A recombinant inbred population was established from a cross between 'Midas' (a snapbean cultivar representing the 'ultimate' in bean domestication) and 'G12873' (a wild bean from Mexico). In this RI population, a linkage map was constructed with RFLP markers regularly spaced throughout the genome. In turn, this map was used to conduct a quantitative trait locus (QTL) analysis for quantitative traits and to map major genes involved in the domestication syndrome. The following traits were studied.

Pod dehiscence

The onset of the dry season coincides with the maturation phase of the life cycle of wild beans. During the progressive desiccation of the plant, fibres located in the pod walls and sutures shorten and, hence, cause the explosive dehiscence of the pods. (Without desiccation during the dry season, pods rarely open and seeds germinate in the pods, causing a marked reduction in fitness.) Pods of domesticated beans contain fewer or no fibres and therefore do not open at maturity.

This difference in the capability of seed dispersal is one of the most important characteristics distinguishing wild and domesticated wild beans. It can be easily recognized in dry pod remains: dried wild pods are strongly twisted (owing to the oblique orientation of fibres in the pod walls), whereas dried domesticated pods are not or only slightly twisted. Presence or absence of fibres is controlled by the *St* gene.

Seed dormancy

Because the onset of the growing (wet) season is irregular, plants have adopted seed-dormancy mechanisms to insure against premature germination during the first rains. In common bean, dormancy is essentially impermeability of the seed coat, which prevents immediate imbibition of the seed. Prior to this study, no studies had ever been undertaken about the inheritance of dormancy in common bean.

Growth habit

Wild beans are viny plants that use shrubs or trees as support. During common bean evolution, there has been a selection for a bush growth habit. Growth habit consists of several subtraits, including the number of nodes and pods on the main stem, the length of the internodes, and type of stem termination (vegetative or indeterminate vs. reproductive or determinate). The latter traits are conditioned by the *fin* gene.

Photoperiod sensitivity

Photoperiod sensitivity plays a crucial role in scheduling of flowering, such that maturation will take place at the beginning of the dry season. Earlier maturation will result in reduced or lack of seed dispersal. Later maturation will prevent sufficient development of seeds before the dry season sets in.

Common bean, having originated in the tropics, is a short-day (long-night) plant, which will flower only under daylengths of 12 hours or less. During dispersal from centers of domestication, dayleingth neutrality was selected, whereby common bean plants will flower under any daylength. Photoperiod sensitivity is controlled by the *Ppd* gene.

Seed size and colour

Selection by farmers and consumers has led to larger pods and seeds with different colours, among them white (absence of pigmentation) conditioned by the *p* gene. Four conclusions were drawn from the results. First, for many traits it was possible to identify major genes. For example, at least four genes controlled seed dormancy, one of which accounted for at least 50 per cent of the phenotypic variation. Second, for most of the traits, more than 50 per cent of the phenotypic variation could be accounted for in genetic terms, suggesting that heritability of most traits was high. Third, distribution of genes involved in the domestication syndrome appeared to be non-random, as there was a relatively higher concentration of genes on linkage groups 1, 2 and 7. Linkage group 1 appeared to be particularly important because it included genes for photoperiod (*Ppd*), determinacy (*fin*) and internode length.

Introduction

The crucial difference between cultivated cereal crops and their wild relatives is in ear fragility at maturity. Easy shattering of spikes into spikelets upon maturity is essential for seed dispersal and survival in the wild, whereas forms with non-brittle ears survive only under cultivation. It is generally assumed that most Triticeae crops have been domesticated from their wild relatives by selection of non-shattering individuals which sporadically appear in wild populations as rare mutants However, there are no documented cases of the appearance of non-brittle mutants in wild populations, except observations of introduction from cultivars into weedy forms. The appearance of non-brittle mutants seems to be a rare event which may be induced under specific conditions. Interspecific hybridization may activate transposition of mobile genetic elements which may be one potential source of increased mutation rates. A hypothesis was proposed that rye and wheat forms with varying ear fragility may have arisen as a result of interspecific hybridization processes between different wild species (Jaaska 1975). Taking into account that non-brittle mutants will persist in the wild only for a limited number of generations after their appearance, it follows that rye and barley should have been domesticated *in statu nascendi* from the ancient hybrid populations of their wild relatives.

Rye

Despite considerable disagreement among taxonomists about the delimitation of rye species and their intraspecific taxa the rye genus *Secale* L. may be treated as consisting of three valid biological species

- The Outcrossing annual *Secale cereale* L. s.l, including the European cultivated rye (subsp. *cereale*), the Transcaucasian cultivated and weedy rye with a tough rachis (subsp. *segetale* Zhuk.), and various weedy and wild forms of the annual self-incompatible rye with a brittle rachis such as subsp. *ancestrale* Zhuk. s.l, including weedy subsp, *ancestrale* Zhuk. s.str, weedy subsp. *afghanicum* (Vav.) Hammer, weedy subsp. *dighoricum* Vav., and wild subsp. *vavilovii* (Grossh.) Kobyl. s.str., non s.l. of Kobyljanskij

- The Outcrossing perennial *Secale strictum* Presl., syn. *S. montanum* Guss., with several subspecies: subsp. *strictum* (= *S. dalmaticum* Vis.), subsp. *anatolicum* (Boiss.) Hammer, subsp. *kuprijanovii* (Grossh.) Hammer, subsp. *ciliatoglume* (Boiss.) Hammer, and subsp. *africanum* (Stapf) Hammer.
- *Secale Sylvestre* Host. - a cleistogamous annual of sandy dunes and coasts.

A cleistogamous annual, originally collected by Kuckuck in Iran and attributed to *S. vavilovii* may deserve a specific rank as *S. iranicum* Kobyl., syn. *S. vavilovii* Grossh. sensu Khush, non- sensu Grossheim or sensu lato of Gandilyan owing to its reproductive isolation as a result of its cleistogamous breeding system. Likewise, South African endemic perennial rye may be recognized as *S. africanum* Stapf because of its autogamous breeding system.

Grossheim described *S. vavilovii* as a short-statured (stems 20-35 cm) wild annual with short (4-8 cm) fragile spikes which he collected in Nakhitshevan from a dry habitat on volcanic ash soil. Populations of short wild annual rye were later found to occur in different dry regions of Armenia by Gandilyan who attributed the short stature to a phenotypic adaptation to a dry and nutrient-poor habitat. On those grounds he proposed to treat all wild and weedy annual ryes with a brittle rachis as *S. vavilovii* s.l. separately from the cultivated *S. cereale* s.str.

To give these forms a specific rank is questionable because they are not reproductively isolated from non-brittle weedy *S. cereale* subsp. *segetale.* Their inclusion under *S. cereale* subsp. *vavilovii* s.l. as proposed by Kobyljanskij contradicts the botanical nomenclature rules because of the priority of subsp. *ancestrale* Zhuk. at the subspecies level Vavilov expressed a view that cultivated rye evolved from a wild perennial rye through the appearance of annual weedy forms which in turn were domesticated by unconscious selection of non-brittle forms under cultivation in barley and wheat fields. Later he reported finding a weedy rye in Afghanistan with fragile ears which he attributed to *S. cereale* var. *afghanicum* and proposed that it might be an initial form from which semi-brittle and non-brittle forms of weedy rye were spontaneously selected under cultivation in wheat and barley fields in northern areas and higher altitudes He also further argued that *S. cereale* may have evolved from a perennial *S. montanum* with wild *S. vavilovii* Grossh. acting as a link between them. rye was domesticated in the Near East region as a secondary weedy crop (subsp. *segetale*) and only thereafter spread to European countries and was used as a distinct crop.

Alternatively, the present-day weedy ryes with brittle and semi-brittle rachis are products of introgressive hybridization of the cultivated non-brittle subsp. *segetale* with wild rye subsp. *vavilovii* s.str. Segregants from this hybridization acquired a weedy trait from subsp. *segetale* and became field weeds. Artificial hybrids between wild, weedy and cultivated forms of *S. cereale* are completely interfertile and show regular meiosis (that they comprise a single biological species. It is generally accepted that all annual rye species have evolved from the perennial *S. strictum*. *Secale cereale* and *S. strictum* differ by two large translocations between three of the seven chromosomes, but intraspecific taxa have the same chromosome arrangement in both species Polymorphism for reciprocal translocations has been described in populations of *S. cereale* The evolution of *S. cereale* from *S. strictum* should thus involve fixation of two reciprocal translocations which would be possible only by the involvement of an autogamous intermediate. Stutz (suggested that cleistogamous annuals *S. sylvestre* and *S. vavilovii* sensu Khush (= *S. iranicum*) might represent intermediate stages in the evolution of *S. cereale* from *S. strictum.*

Secale sylvestre has the same chromosomal arrangement as *S. strictum,* but a low crossability with it, or differs in only a small chromosomal translocation, whereas *S. iranicum* was found to have the same chromosomal arrangement as *S. cereale* (Nürnberg-Krüger Stutz suggested that cleistogamous annual *S. sylvestre* has been derived from *S. strictum* and has then given rise to a similarly cleistogamous annual *S. iranicum* by fixation of two chromosomal translocations, while *S. cereale* was assumed to emerge from *S. iranicum* by acquisition of the outcrossing breeding system by introgression from *S. strictum.*

Variability among the rye species established that *S. cereale* and *S. strictum* display partially homologous polymorphism of many isozymes with most allozymes shared between them. that *S. iranicum* is fixed for one of the allozymes found in *S. cereale* and *S. strictum,* whereas *S. sylvestre* has unique allozymes of acid phosphatase ACP-B, anodal peroxidase PRX-C, and aromatic alcohol dehydrogenases AAD-A and AAD-E In the it has been concluded that *S. sylvestre* is a lateral branch of evolutionary divergence form S. *strictum* and not an intermediate leading to *S. iranicum* and *S. cereale.*

Morever, *S. strictum* and *S. sylvestre* are well isolated in nature by growing in different ecological habitats and do not hybridize freely. Instead, I suggest that some unknown inbreeding form of subsp. *vavilovii* related to *S. iranicum* might be an intermediate from which both brittle- and tough-rachis forms of *S. cereale* evolved by introgression of self-incompatibility from *S. strictum* It was also assumed that non-brittle rye *S. cereale* subsp. *segetale* might be domesticated *in statu nascendi,* directly from some such hybrid populations currently extinct, while fragile subsp. *vavilovii* s.str. has persisted in the wild. Later, *S. sylvestre* was found to differ from the other rye species also in allozymes of cathodal peroxidases CPX-4 and CPX-5 and chloroplast DNA whereas *S. cereale* and *S. strictum* revealed homologous polymorphism and could not be definitely distinguished by allozymes and chloroplast DNA

Zhukovsky (1pointed out that there have been no reports of spontaneous appearance of non-brittle forms for more than 100 years of cultivation of wild and weedy ryes in different botanical gardens and collections. The reason may be in the recessive nature of the non-brittle mutation combined with the outcrossing breeding system of rye The rare non-brittle mutants will remain phenotypically unexpressed in outcrossing rye populations in heterozygous genotypes until two parents with recessive non-brittle alleles cross and segregate into homozygotes in a progeny. The probability for this depends on the frequency of non-brittle mutations which may be extremely low. Therefore, the involvement of an inbreeding intermediate would be an important premise for the appearance of recessive non-brittle mutants phenotypically expressed in homozygotes.

Zhukovsky also has proposed a hypothesis that weedy subsp. *segetale* might have arisen from interspecific hybridization between wild perennial and annual ryes with the appearance of non-brittle forms in the hybrid progeny. He assumed that interspecific hybridization might be the crucial mutagenic factor which has caused the appearance of semi- and non-brittle forms of rye. Stutz "on the slopes of Mt. Ararat, *S. vavilovii* makes contact and hybridizes rather freely with *S. montanum.*"

The seed obtained from crosses between *S. cereale* and *S. strictum* were found to germinate only when *S. cereale* was the mother parent indicating that the introgression is possible from the perennial rye. Sympatric populations of wild perennial and annual ryes in the Near East region should be examined for the fragility, chromosome rearrangements and molecular characters in order to have more information about the origin of wild, weedy and cultivated forms of *S. cereale.* Electrophoretic variants of aliphatic and aromatic alcohol dehydrogenase (ADH and AAD), aspartate aminotransferase (AAT), acid phosphatase (ACP), esterase (EST) and anodal peroxidase (APX) isoenzymes in rye species: major electrophoretic variants (allozymes) are numbered in the order of their decreasing mobilities and listed in the order of decreasing occurrence; rare allozymes are labelled by letter 'r'; unique allozymes are in italics.

Species	**ADH-A**	**AAD-A**	**AAD-E**	**AAT-B**	**AAT-C**	**ACP-B**	**EST-A**	**EST-B**	**APX-C**
S. strictum	2;1r	1	2;1;3r	1	2;1	4;4f;5;3	2f;2;3	3;1;2	1
S. cereale									
subsp. *cereale*	2	1	2;1	1	2;3	4;1;3;2	2	1	1
subsp. *segetale*	2	1	2,1	1	2;3	4;1;3;2	2	1	1
subsp. *ancestral*	2	1	2;1	1;2r	2;3	4;3;1;2	2;2f;3r	1;3;2	1
S. iranicum	2	1	1	1	2	4	2	1	1
S. sylvestre	2	2	*1f*	1	2	*4f*	2;0	1	2

Barley

The cultivated barley *Hordeum vulgare* L., incl. *H. distichon* L. and *H. hexastichon* L., and its closest wild relative *H. spontaneum* C. Koch are autogamous annuals which with the allogamous perennial *H. bulbosum* L. share basic genome a number of allozymes Their hybrids are completely interfertile and have normal chromosome pairing in meiosis. Cultivated barley and its closest wild relative, *H. spontaneum,* are now commonly recognized as subspecies of *H. vulgare* s.l. However, taking into account their autogamous breeding system, which itself provides a reproductive barrier between them, and the fact that the biological species concept is not strictly applicable to uniparentals, they could formally be accepted at a species level as separate phylogenetic lineages. There is a general agreement among investigators that cultivated barley has originated from its closest wild relative *H. spontaneum* through the latter's domestication Different views, however, have been expressed with respect to the place, time and mechanisms of the origin of different forms of cultivated barley, particularly six-row forms

Monophyletic origin of the cultivated barley as a two-row form by domestication of *H. spontaneum* and the secondary origin of the six-row barley from a cultivated two-row form is now generally accepted thanks to experimental evidence on the secondary origin of the Tibetan semi-brittle six-row barley by hybridization between the cultivated six-row barley and wild *H. spontaneum* (Therefore, this aspect of barley evolution is not discussed here. to focus attention on the appearance of non-brittle mutants in pure, wild populations of *H. spontaneum* as a critical issue in the domestication of barley which needs to be documented. The semi-brittle forms in the present-day mixed populations of weedy and cultivated barley are evidently derived by introgression from the cultivated barley.

Genetic studies of hybrid progeny between *H. vulgare* and *H. spontaneum* have shown that ear fragility is controlled by two closely linked genes (*Bt* and *Bt2*) with a recessive mutation at either loci giving rise to non-brittle ears In the case of a simple *in statu nascendi* domestication of barley by selecting out non-brittle mutants, the cultivated barley has evidently been domesticated repeatedly from different genotypes of wild barley, *i.e.* it has a polytopic origin. This, however, does not explain a strikingly wider morphological variability among cultivated barley than in other cereals. Indeed, morphological variation within *H. spontaneum* is limited to only three botanical varieties which differ in the apex shape and awn length of lateral spikelets. At the same time, the cultivated barley is remarkably more variable with over 100 botanical varieties described among primitive barley landraces Although a remarkable morphometric variation was observed among a set of 77 accessions of wild barley under cultivation around Moscow (Russia) under unusual climatic conditions, it still remained relatively limited and no non-brittle cultivar-like forms were observed Characters of the wild parent dominate in hybrids with cultivated barley and segregation appears in later generations. No morphologically new types were found among the progeny of hybrid generations between various morphological types of *H. spontaneum.* Therefore, the question as to why cultivated barley is so morphologically variable still remains unanswered.

Comparative studies of isoenzyme variation among a set of morphologically different accessions of *Hordeum vulgare* s.str. and *H. spontaneum* showed that both-had the same allozymes of essentially monomorphic isoenzymes and displayed variation of 10 polymorphic esterase isozymes This shows that the isozyme loci which were monomorphic in wild barley have remained also in the cultivated barley, *i.e.* cultivation has not induced new variation at the isozyme loci in contrast to those controlling variable morphological characters. The same also has been demonstrated with respect to wild and cultivated wheats and The cultivated barley and its wild progenitor could not definitely be distinguished by any of the isoenzymes studied except by a frequency of allozymes of some isoesterases.

Isozyme	*vulgare*	*spontaneum*	*bulbosum*	*vulgare*	*spontaneum*
ADH-A	1	1	1;2r		
ADH-B	1	1	1		
ADH-D	3	3	3;2;1		
AAD-A	1	3	3;4;2;1		
AAD-B	2;0	2;0	2;2s;1;0		
AAD-E	2	2	2;1;3		
AAT-A	1	1;2	1		
AAT-B	1	1	1;2;3;0		
AAT-C	2	2	2;3;1		
EST-A				1	1
EST-B				1	1
EST-D				3;2;4;0	2;3;4;1;0
EST-E				2	2;1r;3r
EST-F				1;2;0	2;1;3;4
EST-H				2;0	2;1;0;3
EST-I				2;0;1;3	2;0;3;1
EST-J				1;0	1;0
EST-K				3;0;2	0;1

Hordeum bulbosum displays extensive intrapopulation polymorphism of several isozymes which were largely monomorphic in *H. vulgare* and *H. spontaneum.* The allozyme genepool of the two annuals was found to be a subset of most frequent allozymes of *H. bulbosum* This result is consistent with a view that *H. spontaneum* might have evolved from *H. bulbosum* by fixation of genes controlling self-compatibility and annual habit, polytopic origin of *H. spontaneum,* as well as against later, repeated introgression from the bulbous barley. Moreover, crossing experiments between *H. vulgare* s.l. and *H. bulbosum* have shown that hybrids could be obtained only by the embryo rescue technique, had disturbed meiosis, were sterile and frequently haploid owing to spontaneous elimination of *bulbosum* chromosomes

Hordeum spontaneum and *H. bulbosum* are reproductively isolated by strong sterility barriers. Observations that crossability and chromosome elimination varies depending on the genotype of parental species suggested that introgression from the bulbous barley in some genotype combinations may be possible (Origins and Domestication of Mediterranean Olive Determined through RAPD Marker Analyses - G. Besnard, A. Moukhli, H. Sommerlatte, H. Hosseinpour, M. Tersac, P. Villemur, F. Dosba and A. Bervillé

MATERIAL AND METHODS

Plant materials

Sixty-four cultivars from different countries around the Mediterranean basin were analyzed These cultivars were selected from a germplasm collection maintained at Montpellier (France) and provided by Dr Rallo (Cordoba) and Dr Baldoni (Perugia) from the collections of Cordoba (Spain) and Perugia (Italy). Oleasters (wild olive trees) were studied from three sources: from Corsica Morocco (9 individuals: 'Morocco 1' to '9'), and from southern France We have no criteria to distinguish wild from feral forms. Four individuals of *O. maroccana* from southern Morocco and four individuals of *O. europaea* subsp. *cuspidata* from four different areas [Kenya (*O. africana*), Iran, India, China] were also studied.

CONCLUSION

Biologists, ethnologists and geneticists pay tribute to the genius of Nicolay Ivanovich Vavilov - his enormous width of understanding and innovative vision. We may well consider him to be the 'Darwin of the 20th century' with just as enquiring a mind and capacity to recognize the basic similarities between several apparently quite distinct phenomena as Darwin had done in the 19th century. Whereas Darwin studied the diversity of all living organisms known to science, Vavilov directed his vision towards domesticated plants just as had Alphonse de Candolle before him. Clearly, Vavilov's depth of enquiry was much greater than that of de Candolle, to whom he pays tribute. Vavilov was basically a geneticist and plant breeder, approaching the problems of cultivated plant species in terms of the diversity within and between them that might be put to practical ends. For this reason Vavilov was not particularly interested in diversity as such, but only in the diversity that could be put to practical advantage. The world flora, even a quick survey that there are many areas of plant diversity which have little to do with cultivated plant origins. Thus, the 'fynbos' plant formation on the very southernmost tip of South Africa is extremely diverse. In quite a small area one can find hundreds of species in quite distinct genera and even plant families. These are very attractive and colorful but were never brought into cultivation by the indigenous people. They were not at all edible and are known today as beautiful and extraordinary examples of plant evolution.

A rather similar flora exists in southern Australia, and again, none of its species was brought into cultivation. the vast tropical rainforests of South America, Africa and Asia. These differ from the fynbos in that many plants have been used, perhaps for millennia, for food, medicine, clothing and in some instances building materials. But were any of these domesticated? Perhaps a few, such as cassava, pineapple, peanuts, etc., but these were plants from the drier forest margins, and occurred only in certain regions. Several nuts and fruits were gathered for food. They were gathered, eaten and used in a variety of ways. Again, the vastly rich fruit-tree floras of South East Asia were of immense value to the people of such areas but their actual cultivation is comparatively recent.

The wild spring bulb flora of the central and southern European mountains. These plant communities are very diverse and extremely beautiful, with their bulbs, annuals, shrubs and trees. Few were eaten by humans - in fact the plants themselves had developed mechanisms to prevent their being eaten, such as poisonous or bitter roots, bulbs, and spiny leaves and branches. None of these was domesticated in ancient times. Much further north in northern Europe and the steppes of Asia, as well as the northern part of North America and the southern part of South America, the species diversity diminishes very greatly. There are only a few species able to survive, grazed by wild and domesticated cattle, but these are not of direct use to humans as food, only through their horses and cattle.

There is considerable wild plant diversity up to about 45-50°N latitude in Europe, Asia and North America, and southward to 35-40°S latitude in the southern continents, apart from desert and semi-desert areas. However, only in certain areas between 45°N and 30°S latitudes were potential crop plants actually domesticated. the centers of origin of cultivated plants occurred mostly imountainous regions between the Tropic of Capricorn (23°28') south of the equator and about 45°N of the equator in the Old World. In the New World crop domestication occurred between the two tropics (Cancer and Capricorn) approximately. In all cases agricultural origins and primitive diversity occurred in high and complex mountain regions. Why only these?

Domesticated despite their evolutionary processes to protect them from being eaten by the development of various defensivemechanisms such as poisons, bitter substancesOne strategy of survival adopted by many plants is to produce so many seeds that even if most of them are eaten enough will remain to provide for the next generation. And the plants that do this superbly well are grasses. To a lesser extent those that do this quite well are herbaceous legumes. This, then, may be the solution to the problem of how humans began the process of domestication. Hunter-gatherers took and ate the natural surplus and left enough seeds (probably by chance) to provide for the next generation. Later, when non-shattering mutants occurred by chance, these were adopted automatically by the primitive farmers. From this point evolution under domestication began to take place.Vavilov considered that "as a rule the primary foci of crop origins were in mountainous regions, characterized by the presence of dominant alleles

3

Relationship between Environmental Impacts on Crops

THE HYDROSPHERE

The activity of water at or below the ground surface is part of the hydrological cycle, in which water arrives at the surface via precipitation and is eventually returned to the atmosphere via evapotranspiration, having passed along a number of possible pathways in between.

Some precipitation may be prevented from reaching the surface directly by obstructions such as vegetation or buildings, and some may not reach the ground at all, being intercepted and returned directly to the atmosphere by evaporation. That water which does reach the surface will either remain there as *surface water,* flowing over the surface or being stored in depressions, or it will infiltrate the ground to become *subsurface water;* in this form it can be stored or moved through the ground, and may eventually become surface water.

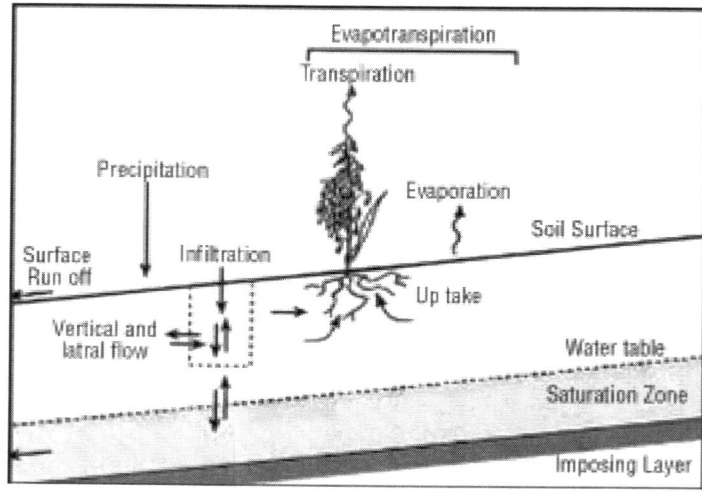

Fig. Components of the Hydrological Cycle in Relation to the Soil Layer.

Subsurface water can be classified into four major zones. The *soil zone* lies nearest the ground surface and therefore controls the infiltration of water into the ground. The underlying *intermediate zone* is one in which percolation of water is the dominant process and this can vary enormously in thickness depending on the relief and rock type.

This overlies the *capillary fringe* in which most of the pores are filled with water, and beneath this lies the *saturation zone;* the interface of these two zones is marked by the *water table*. While these zones may be separated on interfluves and valley sides, they usually converge downslope and may overlap on valley floors.

In this section we will examine the influence of soils on the hydrosphere under three main headings—the forces controlling water movement, soil water additions and losses in terms of infiltration, percolation and evaporation, and water storage and flow. It is important to recognise that soils can also influence the chemical characteristics of water, but this will be considered in the context of the geosphere.

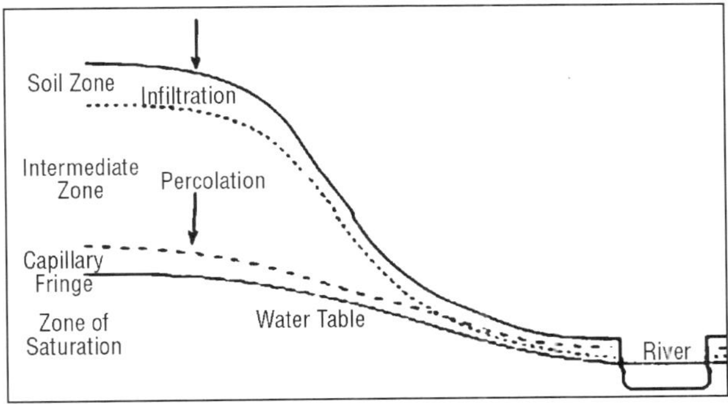

Fig. Classification of Subsurface Water

Forces Controlling Water Movement

The movement of water into, through and out of soils is controlled to a large extent by gravity, and also by three types of force determined by soil properties which can either encourage or restrict water movement—adsorption, capillarity and osmosis, the combined effects of adsorptive and capillary forces being known as *matric suction*. The total suction in a soil will depend to a large extent on its texture and pore size.

Higher clay or organic matter contents will have stronger adsorptive forces, and finer textures are also usually associated with smaller pore sizes, in which capillary forces will be greater. These factors therefore influence the retention and drainage of water. Coarse-textured soils generally drain more quickly and retain less water than finer-textured soils which, for any given water content, will have a greater suction. Water content will, however, also influence suction. Dry soils will have high values, but these will rapidly decrease as the pores become filled with water and the suctional effects are therefore lost. The speed and direction of water movement in a soil will depend on the magnitude of these suctional forces compared with the gravitational force. The rate at which water can move through a soil is measured by *hydraulic conductivity,* which is determined to a large extent by soil water content and also by texture and its associated pore size. Water transfer is more effective in wetter soils than in drier ones because drier soils have a greater volume of air in their pores which inhibits the conduction of water from one location to another.

Therefore at high moisture contents conductivity increases as texture coarsens; values can be less than 1 mm per day in the case of clays, 10 cm to 10 m per day in silty sand and over 100 m per day in gravel. However, at low moisture contents conductivity increases as texture becomes finer, because of higher suction forces.

The direction of movement under gravity is downwards, but soil water can move in other directions depending on the other forces present. For example, it can move laterally due to osmosis if there is a lateral variation in the concentration of salts dissolved in the soil water due to variations in parent material, or it can move upwards due to capillary forces operating as a soil dries out.

Infiltration, Percolation and Evaporation

When rainfall reaches the ground surface, some or all of the water will infiltrate the soil. Initially the matric suction gradient will be relatively high and infiltration will therefore be rapid, but as wetting increases the suction gradient lowers and the infiltration rate decreases as gravitational forces become more important and eventually a steady flow will be attained, approximating to the *saturated hydraulic conductivity*.

The *infiltration capacity* is a measure of the ease with which water can penetrate the surface. For example, because of their larger pores, coarse-textured soils will allow more rapid infiltration than finer materials. Infiltration will also be reduced if a crust has formed, for example by aggregates breaking down under raindrop impact, or if fines are washed into the surface pores.

The type and thickness of the litter layer can also affect infiltration; a thick cover of broadleaf litter with the leaves lying horizontally can reduce infiltration appreciably. Soil freezing will also impede infiltration as the pores become filled with ice, infiltration becoming zero in wet soils where the pores become completely ice-filled. Conversely, soils which develop cracks at the surface will have high infiltration rates, as in the case of Vertisols.

The percolation of water through the soil zone and intermediate zone towards the water table will continue while rainfall is infiltrating the surface, but once rainfall ceases the soil will start to dry out as water drains downwards and is also lost back to the atmosphere via evapotranspiration.

Drainage under gravity is usually more or less complete after two or three days and the soil is then said to have reached *field capacity*. The movement of water through a soil is enhanced by the presence of macropores; these are relatively large pores which can result from shrinkage of a soil on drying or from the development of burrowing or root channels.

They are particularly important in infiltration and percolation during intense rainfall, although the pores must be interconnected otherwise the continuity of flow is interrupted.

The loss of water from a soil by evaporation will be controlled by a number of factors, principally climate, moisture content and texture. High temperatures and windspeeds will enhance evaporation, as will high vapour pressure gradients between the soil and above-ground atmosphere. The moisture content of the top few centimetres of soil is important in controlling evaporation, which will decrease as the soil dries out, becoming zero once the soil is completely dry.

In contrast the moisture content of the subsoil is considered to have little effect on evaporation because of the slow rate of soil moisture movement. Eventually a point is reached when little water is available for movement in the liquid form, and any remaining movement occurs by the diffusion of water vapour. At this stage the soil is said to be at *wilting point*.

Vapour diffusion is controlled by matric and osmotic pressure to a minor extent, but the most important control under most soil conditions is temperature, with vapour diffusion occurring from relatively warm to cooler areas of a soil in response to the vapour pressure gradient. However, the movement of water vapour from the subsoil to the surface constitutes only a minor proportion of total evaporative losses. Hence, for any given volume of rainfall, soils which are regularly wetted at their surface will have greater evaporation values than those which are wetted more thoroughly but less frequently. Texture affects evaporation in that upward capillary movement of water is generally greater in fine-textured soils because of the greater suction forces. In some cases water can move upwards through a vertical distance of several metres, although in coarse-textured soils the distance is unlikely to exceed several centimetres. However, because the speed of capillary water movement is slow, this source of water does not usually contribute greatly to total evaporation except where the water table lies within a metre of the surface.

Other factors which will affect evaporation include soil colour and vegetation cover. Dark soils will generally have higher surface temperatures due to their lower reflectivities and therefore experience higher rates of evaporation than lighter coloured soils. A vegetation cover can shade the soil surface and thus decrease surface temperatures and evaporation.

It can also increase the relative humidity of the air near the surface, which will lower evaporation. However, these reduced moisture losses may well be offset by the loss of water via transpiration from the vegetation itself.

Water Storage and Flow

Water can be stored in or on a soil, or can flow over its surface or within it. Storage of surface water requires a soil of low permeability and a surface relief which will prevent lateral drainage. The storage may be either temporary, as in the case of puddles formed in microtopographic depressions, or it may be permanent if the supply of water is sufficiently great to exceed losses via evaporation and infiltration; in such cases ponds or lakes may form.

The supply of water for surface storage can be provided directly by precipitation, water running downslope over the surface, or by ground water reaching the surface. Surface storage is therefore associated with fine-textured, compacted soils, and is favoured in cooler climates, where evaporative losses are limited. Water storage within the soil can occur where the regional water table approaches the surface, for example in low-lying areas or enclosed depressions.

Storage can also occur where local conditions cause waterlogging near the surface, as in the case where a clay-rich Bt horizon or an iron pan impede the downward movement of water; this is known as a *perched water table*. However, where soils are located on a slope, water flow will usually occur. Three main types of flow can be recognised—*overland flow, throughflow* and *ground-water flow*.

Overland flow occurs when the rate of infiltration is exceeded by the rate at which water is arriving at the surface. This can occur, for example, in the case of frozen soil conditions, or where aggregate breakdown and surface sealing are caused by raindrop impact where soils lack a protective vegetation cover.

Throughflow occurs when water flows laterally through the soil and is confined near the surface, as opposed to ground-water flow which involves water movement through the saturated zone beneath the water table, often at depths below the soil layer; this usually occurs more slowly than throughflow.

Throughflow can constitute a large proportion of total run-off from a catchment in cases where soils encourage infiltration and lateral water movement but restrict deeper percolation. Such conditions will occur, for example, if a soil has vertical cracks, extensive burrowing networks or other forms of macropore which allow rapid infiltration and also a humified organic layer, Bt horizon or some form of pan which impedes vertical drainage.

In extreme cases these conditions can lead to the development of piping, whose networks can provide major pathways for water movement through soils. Soil thickness and slope angle are also important in that thin soils over impermeable parent materials will encourage rapid throughflow, as will steep slopes.

Even in the absence of such favourable factors for throughflow, soil hydraulic conductivity will usually be greater in the lateral than the vertical direction due to the development of structure and also to lower compaction than at depth, therefore on slopes, lateral flow will generally predominate over vertical drainage.

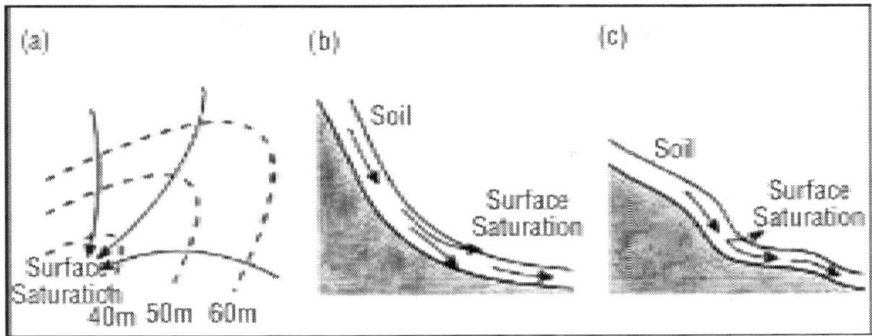

Fig. Principal Locations of Flow Convergence to Produce Surface Saturation: (a) Hillslope Hollow (Dotted Lines Represent Contours), (b) Footslope, (c) Soil thinning

Throughflow will be converted to surface run-off if the near-surface saturated zone reaches the surface. This can occur if the flow is concentrated in a hillslope hollow or towards the foot of a slope. Lateral thinning of the throughflow zone can also cause saturation to occur at the surface and therefore produce surface run-off.

The relative proportions of overland flow, throughflow and ground-water flow will determine the shape of the hydrograph of a catchment. A hydrograph is a plot of stream discharge against time, comprising one or more peaks, each of which has a rising limb and a recession limb. The rising limb represents an increase in discharge in response to an input of water to the catchment, usually by rainfall or snowmelt.

The discharge will reach a maximum value beyond which it gradually decreases through time, and this represents the recession limb of the hydrograph. The steepness and height of the rising limb of the hydrograph will therefore depend on the rate at which water can reach the stream channels. Where overland flow occurs, water will reach the channels quickly, whereas throughflow and ground-water flow will delay its passage.

The extent of the delay will be determined by slope angle and the way in which soil conditions influence vertical and lateral drainage. Therefore, soils on steep slopes and which impede infiltration will produce the most peaked hydrographs; soils which allow some infiltration but which have subsurface compaction, pans, Bt horizons or frozen subsurface layers, will tend to produce less peaked hydrographs, and soils which do not possess these features and therefore allow greater vertical infiltration will have the least steeply sloping and lowest hydrograph peaks.

It is important to recognise, however, that these trends can be complicated by antecedent moisture conditions. For example, dry soils will often produce a delayed discharge response compared to wet soils, regardless of their effects on the type of flow, although under very dry conditions a rapid response may be encouraged by hydrophobic coatings of soil particles, such as clay or organic coatings of mineral grains or peds.

THE ATMOSPHERE

Soils have their own climate and can also influence conditions in the lowest part of the above-ground atmosphere. The effect is, however, only on a small scale, and the study of soil-atmosphere interactions therefore forms part of the discipline known as *microclimatology*. The atmospheric components involved in this study will be considered under three headings—solar radiation and temperature, atmospheric moisture and air movement.

Solar Radiation and Temperature

Radiation reaches the ground surface from the sun, and leaves the ground surface, as electromagnetic energy of differing wavelengths. The wavelengths of the incoming radiation are predominantly shorter than those of the outgoing radiation, and these two types are therefore known respectively as *short-wave radiation* and *long-wave radiation*.

Some of the incoming short-wave radiation will be reflected off the ground surface back into the atmosphere; the amount reflected will depend on the reflectivity, or *albedo,* of the surface. The net all-wavelength radiation (Rn) can be shown as follows:

$$R_n = [S(1-a)] + L_n$$

Where S=incoming short-wave radiation, a=fractional albedo of the surface and Ln =net outgoing long-wave radiation. On unvegetated ground the albedo is determined by the nature of the soil. For example, dark soils, such as those rich in organic matter or derived from a dark-coloured parent material, can have an albedo of as little as 0.05 (or 5 per cent) and will therefore reflect little incoming radiation, whereas soils which are low in organic matter or derived from a light-coloured parent material can have values as high as 0.60 (60 per cent) and will therefore reflect much greater quantities of incoming radiation. Albedo also decreases with increasing water content due to internal reflection at the surfaces of water menisci in soil pores; a 20 per cent water content can decrease soil reflectivity to a quarter of its dry value.

The angle of incidence of incoming radiation can also have a marked effect on reflectivity, particularly in the case of a soil with standing water at its surface, as will microrelief. If the angle of incoming radiation is low, a water surface will be much more reflective than for higher incidence angles; the albedo of water is only around 0.05 for incidence angles greater than 45°, but rises rapidly for decreasing angles.

Where soils have a marked microrelief, some of the incoming radiation will be absorbed by other parts of the surface, following the initial reflection, and hence the overall albedo of the surface will be reduced.

The exchange of energy between the atmosphere and a surface can be shown by the surface energy budget equation:

$$R_n = G + H + LE$$

Where Rn =net all-wavelength radiation, G=ground heat flux, H= turbulent sensible heat flux to the atmosphere and LE=turbulent latent heat flux to the atmosphere. The *ground heat flux* represents the transfer of heat into the ground, while *turbulent sensible heat flux* to the atmosphere represents the transfer of heat by fluid (air) flow, and *latent heat* is heat absorbed during evaporation or released during condensation.

During the day the available net radiation has a positive value and is balanced by turbulent fluxes of sensible and latent heat into the atmosphere and by conductive heat flux into the soil. Soils with a low albedo will attain higher temperatures at and close to their surfaces than those which reflect a higher proportion of incoming radiation, as seen in extreme cases for artificially whitened and blackened soils.

Higher surface temperatures will usually give higher above-ground temperatures in the first metre or so of air due to sensible heat transfer. Some of the heat will also be transferred by evaporation if the soil contains moisture, and some will be transferred deeper into the soil by conduction.

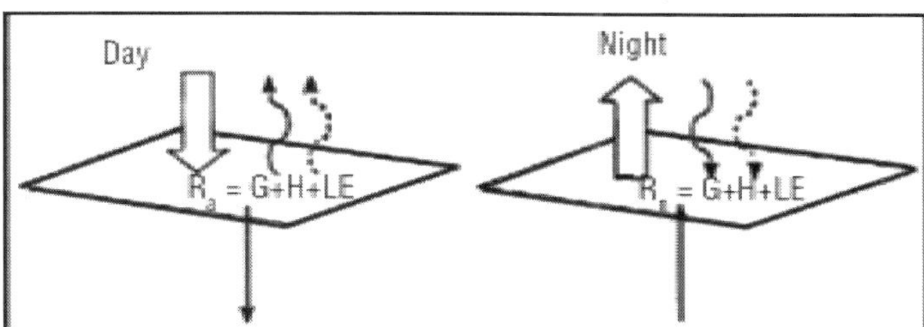

Fig. Energy Flows Involved in the Energy Balance of a Simple Surface

The rate at which heat transfer occurs through a material is expressed by its *thermal diffusivity (k)*. This is a function of its ability to conduct heat, known as its *thermal conductivity (k),* and the amount of heat necessary to cause a temperature change, expressed by its *heat capacity* (C).

The relationship of these factors can be shown as: k =*k*/C, indicating that the rate of heat transfer is directly proportional to the ability to conduct heat, but is inversely proportional to the heat capacity. When saturated, a sandy soil therefore has a higher thermal diffusivity than a clay soil, which in turn has a higher value than a peat soil, but these values decrease when the materials are dry.

However, the greatest diffusivity will occur at intermediate moisture contents, because at high contents it is reduced due to the high heat capacity of water. Although a dry peat soil will have a low albedo because of its dark surface, little heat will therefore be transmitted deeper into it and the surface will become very hot. Sensible heat transfer will also cause the overlying air to heat up. Conversely, a moist sandy soil is likely to have a higher albedo, reflecting more incoming radiation, and will also transmit more heat into the ground, therefore its surface temperature will be lower than that of the peat, as will that of the overlying air, but its subsurface temperature will be higher.

At night, the net radiation becomes negative due to the loss of outgoing long-wave radiation, and this is balanced by conductive heat supplied from the soil plus turbulent heat from the air. Under these conditions the surface of the peat will become colder than that of the sand because the peat is less able to replenish surface heat loss from below due to its lower diffusivity; this can cause a strong temperature inversion in the overlying air.

Most soils show greater diurnal and annual fluctuations in surface and overlying air temperatures than at depth, but soils with low diffusivity will have more extreme fluctuations in temperature at and near the surface than those with a higher diffusivity, and less extreme fluctuations at depth. For wet soils the depth over which diurnal and annual temperature fluctuations occur is around 0.5 m and 9.0 m respectively, whereas in dry soils the values are around 0.2 m and 3.0 m. Soil and near-ground temperatures will also be determined by vegetation cover. This may have a different albedo from the soil and therefore reflect different amounts of incoming short-wave radiation. For example, grassland, heathland and scrub have albedos of around 0.15-0.25, while forests have values of around 0.05-0.20.

Vegetation also allows heat loss via transpiration in addition to evaporation. The effect of vegetation cover is to lower the diurnal range of soil surface temperatures; these are suppressed during the day, due to radiation reflection and heat absorption by the overlying vegetation, but enhanced at night due to long-wave emission and limited transpiration from the vegetation.

Similarly, the effect of soil on air temperatures is reduced by the presence of a vegetation cover. Snow cover will also affect soil radiation and temperature conditions. Because of its high albedo (0.95, decreasing to 0.40 with age), large amounts of incoming radiation will be reflected, and because of its low thermal diffusivity, similar to that of dry peat soil, heat exchange occurs at its surface with little heat being transferred to or from the soil. Consequently, the soil is insulated by a snow cover from extreme temperature changes in the air, and the thicker the snow cover the greater this effect. Similarly, the effect of the soil on air temperatures will be suppressed under snow covers. Soil structure also influences thermal characteristics of the soil, with structured soils showing greater heat conduction than unstructured soil.

Atmospheric Moisture

Water occurring in a liquid form in the soil can influence the moisture of both the soil and above-ground atmosphere in terms of its water vapour characteristics. The amount of moisture which can be held as vapour in the air increases with temperature, and when the air is saturated with water vapour, condensation will therefore occur if cooling takes place.

The temperature at which this occurs is known as the *dew point*. Because air in pores of moist soils is in close contact with the water, it is usually close to saturation. However, vapour gradients will develop due to variations in soil temperature with depth. During the day soils are usually warmer towards the surface, therefore more water vapour can be held in the pores in the upper part of the soil than in those lower down. The resulting vapour concentration gradient produces a net flow of vapour down into the soil. Conversely, soils cool towards the surface at night which produces a net vapour flow upwards. If the vapour is cooled to its dew point on reaching the soil surface, condensation will occur; this process is known as *distillation*.

Moisture can also be added from the soil to the above-ground atmosphere via evaporation. This is greatest where temperatures are high, and is also enhanced by the desiccating effect of air moving across the soil surface, because fresh air is usually associated with lower humidities.

Most evaporation occurs from the top few centimetres of soil, and once this water has been lost, further evaporation occurs by water being moved upwards by capillary suction, although this constitutes a relatively minor proportion of soil evaporative losses. The extent of upward capillary movement depends on pore size, small-pored fine-textured soils being more conducive to this process than larger-pored coarser-textured soils.

Moisture lost from the soil via evaporation will generally increase the vapour pressure of the above-ground atmosphere, the increase being greatest close to the surface and decreasing with increasing altitude. The increase will be greatest during the day when temperatures, and therefore evaporation, are at a maximum.

If at night the temperature of the soil surface falls to the dew point, condensation occurs on the surface; this process is known as *dewfall*. The exchange of moisture from the soil to the above-ground atmosphere will also be influenced by vegetation cover. In addition to direct evaporative losses, moisture will also be lost to the atmosphere via transpiration, and within the vegetation layer itself the reduction in turbulent transfer allows vapour to accumulate close to the soil surface.

Air Movement

Soils can influence the movement of air above the ground in two main ways. Air movement occurs when it is heated which causes it to rise, a process known as *free convection,* and variations in soil surface types will cause different amounts of heating and therefore different extents of thermal uplift. For example, low albedo soils usually have higher surface temperatures during the day than adjacent higher albedo types, and areas of dry soil attain higher surface temperatures than adjacent wetter areas because of differences in thermal diffusivity. Because of such lateral variations in heating, the most favourable areas will experience greatest thermal uplift, which causes convectional circulation cells to develop, thus mixing the air and sometimes leading to the formation of cloud by condensation of the upward moving air. This mixing reduces vertical differences in temperature, humidity and wind speed, and if surface heating and convectional uplift are strong, the mixing effect allows faster-moving upper air layers to be brought down nearer the surface, thus increasing wind velocities close to the surface.

During the night a decrease in temperatures is accompanied by a decline in convectional activity, and therefore wind velocities near the surface also decrease. The second way in which soil surfaces can affect air movement is by them causing horizontally moving air to be set in turbulent motion, a process known as *forced convection*. This is a smaller-scale process than the previous one, and the extent to which it occurs will depend on the aerodynamic roughness of the surface.

This can be expressed as *roughness length* and will be determined by factors such as the size and shape of aggregates if the soil surface is unvegetated, or by the type of vegetation cover. For example, a sandy desert soil surface has a roughness length of 0.0003 m, while soils more typically have values of 0.001-0.01 m; in contrast, short grassland has values of 0.003-0.01 m, while taller grassland has values of 0.04-0.10 m and forests have values of 1.0-6.0 m. Surface roughness will increase turbulence, although vegetation will reduce air movement within it; for example, forests can reduce wind speeds by over 90 per cent beneath the canopy.

Air movement within a soil occurs by *diffusion* whereby oxygen diffuses into the soil and carbon dioxide diffuses out. It can also occur by *mass flow,* in which gas molecules move from an area of higher to lower pressure. This can occur for a number of reasons. For example,

Turbulence can produce pressure differences which are equalised by air movements, some from above the ground and some from within the soil. Pressure differences can also occur within the soil due to diurnal temperature variations with depth. During daytime surface heating, warm air will rise from the upper part of the soil and this will be replaced by cooler air from below.

Conversely, cooling of the surface during the night will cause relatively dense air to sink into the soil. In contrast, diurnal variations in vapour concentration will cause downward movement of air during the day and upward movement at night. Plant roots can also extract water from pores, causing it to be replaced by air.

Air movement can be restricted where the topsoil is wet or frozen, causing increased levels of carbon dioxide. Slower gaseous exchange with the above-ground atmosphere at depth can also lead to higher carbon dioxide concentrations than near the surface. Carbon dioxide concentrations also vary seasonally, with greater quantities being evolved during higher biotic activity at higher temperatures.

THE GEOSPHERE

A variety of processes operate in the formation of a landscape, as shown in the hypothetical 9-unit landsurface model, in which a slope profile is divided into 9 units, each characterised by a set of pedological and geomorphological processes. The interfluve is dominated by vertical movement within the soil, while the upper slope shows both vertical and lateral soil movement.

The free face is characterised by weathering and rapid mass movement, and the midslope represents an area of downslope transportation, involving both mass movement and water flow. Unit 6 marks the zone of net deposition on the footslope while unit 7 is an area of deposition of material from both upslope and upvalley, via the river channel.

Units 8 and 9 are dominated by river channel processes operating down-valley. In this model, soils are therefore important in three principal ways—they influence the nature and intensity of weathering, surface and subsurface sediment and solute transport, and mass movement, and each of these aspects will now be examined in turn.

Weathering

Soils are an important factor in the supply of moisture which is required for many weathering processes. For example, thick soils, or those of a fine texture or on gentle slopes, generally hold more moisture and are often associated with greater extents of bedrock weathering. However, during its development a soil may reach a critical thickness beyond which little bedrock weathering occurs because the soil acts as a protective mantle rather than as a catalyst to weathering. The variation in water availability can also be important in the case of weathering processes such as wetting and drying and salt crystal growth. Soils with a fluctuating moisture regime are therefore likely to allow more effective weathering of the underlying bedrock in these cases than continuously dry or wet soils.

Soil thickness can also be important in terms of protecting the bedrock from fluctuating temperatures which can cause freeze-thaw weathering. Although climate will obviously play an important role in determining soil moisture and temperature characteristics, properties of the soil itself can also have a marked effect. The chemical composition of water can influence its weathering effectiveness, and this can be determined by the soil through which the water has passed. Of particular importance in this respect is the pH of the soil and the speed at which water moves through it.

For example, water percolating through a soil of very high or low pH will generally have a greater weathering capability than that passing through a more neutral soil. Also, a fine-textured soil will hold the water for a longer period than a coarse-textured soil, and therefore allow a longer reaction time over which weathering can occur.

Sediment and Solute Transport

The transport of sediment as part of the denudation process can occur via the soil surface or subsurface, while solute transport tends to be confined within the soil. Surface sediment transport operates by aeolian processes, or by surface water movement; these processes have already been considered in the context of losses during soil formation, but soils can also influence the way in which these processes operate.

For example, soils with a high silt or fine sand content are most susceptible to wind erosion, which is also enhanced by low moisture contents and sparse vegetation covers as occur in arid and semi-arid regions. Soils themselves can also provide a source of material for aeolian abrasion of exposed bedrock.

The relative hardness of these materials will determine the effectiveness of abrasion, and since soils often contain many minerals with hardness values in the range 5-7 on the Mohs scale, these can be effective abrasives of relatively soft rock types such as shales and limestones.

Soils influence the erosion of sediment by water via their control on run-off. Low permeability soils will encourage rapid run-off, which can take the form of overland flow or throughflow. Overland flow is encouraged by soils with low permeability surfaces, particularly where vegetation cover is low so that interception of precipitation is limited and infiltration rates are therefore more likely to be exceeded by precipitation rates.

A limited vegetation cover will also expose aggregates at the soil surface to the direct impact of raindrops, which may cause disaggregation, resulting in blocking of the surface pores and therefore a decrease in permeability. Shallow soils will also encourage overland flow because of their limited water storage capacity. Unconcentrated flow or *sheetflow* will only occur on smooth surfaces, but on most surfaces their roughness causes the flow to be concentrated in the form of small ephemeral channels or *rills,* particularly on silty or clayey soils.

If these channels become sufficiently large to stabilise, gullies may develop which will then act as locations for concentrated erosion. Silt and fine sand will generally be most easily transported by surface wash, because clays and colloidal organic matter will resist movement due to their cohesiveness, while sand and gravel particles will be more difficult to move because of their greater weight.

However, in soils with a high sodium salt content, clays often become easily dispersed on wetting and drying, and this can encourage erosion and gullying, for example in southern Africa.

In the case of subsurface sediment and solute movement, the hydraulic conductivity of a soil will exert an important control on the ease with which movement can occur. The potential for sediment movement will also depend on the size of material available for transport, the extent of its aggregation and the size of the soil pores, while solute movement will also be determined by the ease of dissolution of the soil components and the acidity of the drainage waters; in some cases, soils may inhibit solute loss by buffering the acidity, or by providing exchange sites which can remove dissolved ions from the drainage waters.

As in the case of surface run-off, material may be removed from a slope by subsurface processes in either an unconfined or a concentrated manner. Unconfined movement can occur throughout the soil as a whole, whereas concentrated movement operates in zones which develop in soils often in response to topographic conditions, forming patterns similar to those associated with surface run-off, although usually with more diffuse boundaries.

The smaller features are sometimes known as *percolines,* as distinct from the larger and more easily recognisable *seepage lines*. These zones can be important in the movement of solutes and fine particulate material. The removal of larger material can lead to the development of subsurface pipes, particularly in soils which are susceptible to vertical cracking, which allows water to penetrate rapidly into the subsoil; organic soils and those rich in expanding lattice clays are particularly prone to this condition.

Piping will also be encouraged in soils which have an abrupt decrease in porosity at some point below the surface, as this will inhibit downward water percolation and promote lateral subsurface flow.

Pipes can range from a few centimetres to several metres in diameter, and can therefore be responsible for transporting large quantities of water down hillslopes, along with sediments and solutes. If a pipe develops to a size which exceeds that capable of being supported by the overlying soil, the roof will collapse, exposing the channel to the surface. A similar situation may result from the extension of vertical cracking down to the level at which the pipes occur. These can be important mechanisms in the formation of gullies, especially in arid and semi-arid environments and also in some peat soils.

Soils can influence landforms by their resistance to sediment and solute transport. This is particularly the case in low latitude regions where soils have become indurated by the formation of pans. In these instances they can behave like resistant rock types and a number of geomorphological features can result from erosion.

For example, indurated layers, or *duricrusts,* can form cappings over softer materials, producing flat-topped interfluves, benches along valley sides or pavements on valley floors; they can also form escarpment features if inclined at an angle to the horizontal, or irregular relief due to differential erosion.

Studies of sediment and solute transport can be made at a variety of scales. For example, within a river catchment in mid-Wales, it was reported that physical denudation of the catchment was occurring at a rate of 3.5 mm per 1,000 years, based on measurements of river suspended sediment and bedload, while input-output budgets for solutes indicated that chemical denudation was operating at a rate of 2.9 mm per 1,000 years.

At a smaller scale, rates of sediment and solute transport can be measured by hillslope plot studies; for example, material eroded from a known area can be collected as it moves downslope, or erosion rates can be calculated from the redistribution patterns of radioisotopes within the soil.

Mass Movement

The influence of soils on mass movement depends largely on its mechanical and hydrological properties. Of particular importance are cohesion and friction, which can be described by the Mohr-Coulomb equation:

$$\tau = C + \sigma_{\Pi} \tan \phi$$

where *ô*=shear stress at failure, c=cohesion, *ó n*=normal stress on the shear plane and *ø*=angle of internal friction. The extent of cohesiveness and friction depends mainly on soil texture and moisture content.

In sandy soils there is little cohesion, so the shear stress at failure will depend on the normal stress and angle of internal friction. Conversely, in clayey soils cohesion will be high but friction low, therefore the shear stress at failure will depend on cohesion, while most soils have a combination of both friction and cohesion. However, moisture content will also affect stability because of the effect of the normal stress on pore water pressure. By taking this into consideration, the previous equation can be rewritten as:

$$\tau = C + (\sigma_{\Pi} - u) \tan \phi$$

or:

$$\tau = C' + \sigma'_{\Pi} \tan \phi'$$

Where *u*=pore water pressure, and c^2 and *ø'* are the cohesion and friction angle with respect to the effective stress. If pore water is therefore compressed by a soil on a slope, it will come under pressure and the value of will decrease, thus reducing the shear strength. Conversely, negative pore water pressure, or suction, can increase soil strength; for example, damp sandy soil in which capillary forces are acting will be more stable than dry sandy soil in which there will be no such effect.

Mass movement comprises three main sets of processes—heave, flow and slide—and various types of movement can be identified according to the relative combination of these processes. Transfer by rivers, rockslide and talus creep are not strictly relevant in the context of soil influences, but soils can have important effects on the remaining processes.

Soil creep can occur in one of two forms—*continuous creep* and *seasonal creep*. Continuous creep is mainly confined to materials with low shear stress values such as clay-rich soils. Movement is usually most rapid near the surface and decreases progressively with depth. Experts report movement to depths of 10 m and rates of movement up to 22 cm a^{-1} in the upper layers. Seasonal creep is more common and results from forces which cause variations in shear stress such as expansion and contraction caused by wetting and drying or freeze-thaw, plus disturbances due to bioturbation.

Movement is often confined to the top metre of soil in which these processes are most marked, and can occur in any direction. Rates of downslope movement are therefore difficult to establish, although values in the order of 0.5-2.0 mm a^{-1} have been suggested for humid temperate regions and up to 15 mm a^{-1} for temperate continental climates.

Soil creep can produce small step-like features known as *terracettes,* which run roughly parallel with the contour. The origin of these features has been the subject of some debate, and suggested alternative mechanisms to soil creep have included micro-scale slumping and trampling by animals.

Gelifluction (solifluction) occurs by movement of material over a frozen subsoil. When ice lenses in the upper part of a soil melt, pore water pressures are increased if the water is unable to drain away because of the frozen subsoil, and this can reduce the shear stress at failure of the material, causing it to become unstable and move downslope.

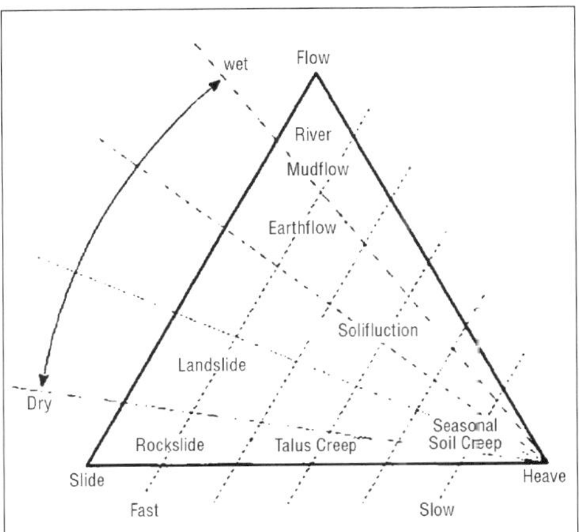

Fig. Classification of Mass Movement Processes

Expansion and contraction by freeze-thaw may form an additional component of the movement, as in the case of soil creep, as may the process of *frost creep*. The latter has been suggested to occur when particles on a sloping soil surface are displaced upwards at right angles to the slope when needle ice forms, and then on melting are returned to the surface, under gravity, via a more vertical path. Repeated freezing and thawing therefore produces a zigzag movement of particles down the slope in the vertical plane. However, the use of the term *creep* to describe this process has been criticised on the grounds that it has nothing to do with deformation of material in relation to shear stress, and its operation within a soil as a whole, as opposed to simply the surface particles, has also been questioned. Silty soils are often most susceptible to gelifluction as these are prone to large quantities of segregation ice development so that large volumes of water will be released into the soil on melting.

Geliflucted material can occur as a featureless sheet if moving at a similar rate over an entire slope, or as terraces or lobes if moving at different rates or confined to particular parts of a slope. In the latter case gelifluction may be confined to areas downslope of late-lying snow patches which can increase the water content of the soil and therefore contribute to its instability.

Rates of gelifluction vary widely, but are generally greater than those of seasonal soil creep, being typically in the order of 0.5-10 cm $a^{"1}$, while depths of operation will depend on the extent of thawing.

Landslides, earthflows and mudflows are generally more rapid movements than gelifluction and soil creep, and usually result from instability caused by the reduction of shear stress at failure by high pore water pressures. They are therefore often associated with clay-rich soils with low angles of internal friction, or with soils which receive sudden influxes of large volumes of water via heavy rainfall; if removal of the water is impeded by an impermeable bedrock layer beneath the soil this will enhance instability.

Soils with pans or other forms of low-permeability illuvial horizon can also act in a similar way, as for example in the case of podzol Bs horizons. Rates of rapid mass movement vary enormously from 1 cm $day^{"1}$ to over 1 m $s^{"1}$. Well-developed root networks in soils can increase the stability of slopes which are prone to slippage. In addition to the forms, mechanical transfer processes operating within the soil itself can produce small-scale microrelief and patterned features. For example, frost-susceptible soils can produce earth hummocks and sorted patterns in cold environments, while Vertisols can also produce mounds and hollows.

THE BIOSPHERE

Soils are influenced to a great extent by biota in terms of pedogenic processes. However, this is by no means a one-way relationship—soils can have an important effect on biota in terms of the extent to which they encourage or constrain habitation.

The principal influences of soil on biota are via nutrients, moisture and aeration, although a number of other controls, both chemical and physical, are also important. For ease of discussion these factors will be examined individually, although it should of course be recognised that in reality many of the factors operate in combination.

Additional Factors

Although nutrients, moisture and aeration are important influences on most biota, a number of other factors, both chemical and physical, can exert an influence on these properties and also act as additional influences in their own right. The chemical factors relate principally to soil pH and toxins, while the physical factors include temperature, soil depth and ground stability, and soil pore size, compaction and induration. Soil pH can affect biota both directly and indirectly via its influence on nutrient availability and toxicity.

The direct influence is often not great because most plants can tolerate a wide range of pH values as long as nutrients occur in sufficient quantity, although at very low pH, hydrogen ions become toxic to plants. However, the effect of pH on nutrient availability and toxicity can be major. For example, iron, manganese and zinc are less readily available at alkaline than acid pH values, while calcium and molybdenum availability is greater at higher pH values.

Phosphorus is often in short supply because it occurs in forms which are not readily soluble, but it is most easily extracted by plants at around pH 6.5. At pH values below about 5.0, aluminium, iron and manganese can become soluble to an extent which can make them toxic to certain plants.

For example, aluminium toxicity inhibits cell division in root tips, causing them to become stunted, while manganese toxicity produces leaf distortion and yellowing. Low pH is also associated with calcium deficiency, which can cause leaves to wilt and collapse. Conversely, at very high pH values, bicarbonate ions can occur in concentrations which reduce nutrient uptake. Carbonates can also accumulate as precipitates around roots, which inhibits water and nutrient uptake.

The type of organisms inhabiting a soil can also be influenced by pH. For example, earthworms, bacteria and actinomycetes prefer neutral to slightly alkaline soil conditions, whereas fungi generally require more acid conditions. There are exceptions, however; iron- and sulphur-oxidising bacteria can occur in extremely acid conditions, with pH values as low as 1 or 2.

Some substances can also occur at toxic levels independently of pH conditions. Perhaps the most important of these are sodium salts, which increase the osmotic gradient that plant roots have to overcome in order to absorb water. If this gradient cannot be overcome, *plasmolysis* will occur in which movement of water out of plant cells towards the salt solution will cause the cells to collapse.

Soils with high sodium concentrations, such as those occurring in arid or coastal areas, therefore support plant communities which are specially adapted to high osmotic pressures. Other naturally occurring toxins include copper, chromium, arsenic and mercury, although these are normally found at toxic concentrations only in soils developed from metalliferous parent materials, or close to sources of industrial contamination.

Toxins derived from vegetation can lead to lower populations of soil fauna, as in the case of earthworms. Micro-organisms are also susceptible to natural toxins, although they are probably more tolerant of salinity than most plants.

Soil temperature will have an important influence on vegetation in terms of seed germination and growth, and the growth of roots, with both water and nutrient uptake increasing with increasing temperature up to an optimum value. For temperate species, optimum growth occurs at soil temperatures of around 20°C, while for tropical species it occurs at 30°C or more.

Soils which warm up rapidly will therefore encourage growth to commence early. Rapid warming will be favoured by high thermal diffusivity; low diffusivities will encourage surface heating but inhibit heat penetration of the subsurface layers. If soil temperatures fall below freezing point this may be injurious to certain types of plant, because the expansion which accompanies freezing can damage organic tissue.

This problem is reduced if a soil is protected from sub-zero air temperatures by a cover of snow, although very thick snow covers can both delay soil warming and cause physical damage to vegetation by compression. Soil temperature can also affect mesofauna and micro-organisms.

Many mesofauna are susceptible to frost and also to drought conditions often associated with high soil temperatures and evapotranspiration rates. Many micro-organisms have an optimum temperature range of around 20-35°C, but some specialised groups have much lower (5-10°C) or higher (45-50°C) optimum temperatures. Soil temperature will therefore affect not only the populations of micro-organism groups, but also the rate at which processes such as organic matter decomposition and nitrogen fixation occur.

At extreme temperatures caused by fire, combustion of vegetation and organic accumulations in the soil can produce a marked increase in nutrient content, and therefore in soil fertility, although this can quickly decline due to subsequent rapid leaching and erosion.

Shallow soils or those on steep slopes can limit moisture availability, but can also limit vegetation growth in extreme cases due to restricted root penetration or instability of the ground. Plants adapted to such conditions are therefore usually not only drought-resistant but also shallow-rooted and fast-growing.

However, shallow soils do not always limit root depth; in some cases roots can penetrate great distances into the underlying bedrock. Shallow soils or those on steep slopes may also discourage habitation by burrowing macrofauna because of the potential instability of burrowing passages. Instability on more gentle surfaces can also limit plant growth, as in the case of the removal of topsoil by wind or water erosion, or mechanical disturbances such as cryoturbation. Soil pore size and interconnectivity, and soil compaction and induration, in addition to their effects on soil moisture and aeration, will influence plant growth in terms of the ease of root penetration through the soil. Many roots exceed 60 ìm in diameter, but in fine-textured soils many of the pores are much smaller than this, therefore these will be impenetrable; root penetration is therefore confined to the larger spaces between individual particles and aggregates.

Soils with compacted or indurated layers will similarly restrict root extension. This can be seen for example in the case of iron pans, which can cause shallow tree root systems to develop above them, leading to instability in strong winds.

Nutrients

The nutrients required for plant growth fall into two types—the *macronutrients,* which are used in relatively large quantities, and *micronutrients* or *trace elements,* used in very small amounts. Plant growth can be retarded if these occur in insufficient quantities, are not balanced by other nutrients or do not become available sufficiently quickly. Micronutrients are required by plants in only small amounts, but they are no less important than macronutrients.

While many nutrients are derived from the weathered mineral soil, organic matter can also be an important source, particularly in the case of nitrogen, phosphorus and sulphur, nutrients which are often in short supply. Nutrient content will therefore depend on the type of mineral and organic components which make up a soil.

For example, potassium, calcium and magnesium will be plentiful in soils rich in feldspars, mica and hornblende, while nitrogen will occur in higher quantities in organic material rich in proteins and amino acids. Sulphur can occur in minerals such as pyrite and gypsum, in addition to organic matter, and phosphorus can occur in the mineral apatite as well as in nucleic acid and other organic forms.

Consequently, nutrient-rich soils are generally associated with base-rich parent materials such as limestone, marl, basalt and some sandstones, and also with grassland or broadleaf trees; soils developed from parent materials such as granite, gneiss and quartz-rich sandstones, or those supporting heathland or needleleaf trees are usually nutrient-deficient.

However, nutrient availability can be equally important as nutrient content. For example, nutrients may be easily lost by leaching in a coarse textured soil, even if developed from a nutrient-rich parent material. Nutrient availability is also determined by the quantity held in an exchangeable form on clay and organic colloids, from which they can be removed by roots.

For example, calcium is normally held in this form in much larger quantities than potassium and magnesium, but consequently it is also more prone to loss by leaching, especially in freely draining soils. In the case of nitrogen, phosphorus and sulphur, their availability will also depend not only on the type of organic matter but on its rate of decomposition; if this is slow, due for example to low temperatures or poor drainage, these elements may not be released in sufficient quantities for certain plants.

Nutrient deficiency can be manifested in plants in four main ways. These are *chlorosis* (yellowing of leaves due to a lack of chlorophyll), *purpling* (darkening of leaves due to the accumulation of anthocyanin pigments), *local necrosis* (death of tissue) and *stunting* (reduced growth).

Chlorosis can result, for example, from a deficiency in nitrogen, phosphorus, magnesium, iron, zinc or copper, while purpling is often associated with phosphorus deficiency. Local necrosis can result from potassium or molybdenum deficiency, and stunting of leaf growth can occur due to zinc deficiency.

Nutrients also exert an important influence on soil mesofauna and micro-organisms. In terms of mesofauna, the principal source of nutrients is organic matter occurring as either living or dead plant or animal material. In the case of herbivorous fauna, the nutrient status of the plant-derived organic matter will determine population levels, with higher levels usually present in soils with high nitrogen content organic matter, such as those associated with many grassland and broadleaf tree types.

Carnivorous mesofauna will occur in greatest quantities in soils which have large populations of animals on which they prey, and these are therefore usually the same types of soil. For soil micro-organisms, carbon and nitrogen are the nutrients which are most commonly deficient, and various adaptations have been developed in order to overcome this problem; these include the fixation of atmospheric nitrogen, the ability to survive long periods without nutrition, and the release of antibiotics which inhibit the growth of competitors. Another adaptation is autotrophy, or by the oxidation of ammonia, sulphur or sulphide.

Moisture and Aeration

Soil moisture is required for the growth of most plants, and the moisture content of a soil is determined to a large extent by drainage conditions. For example, a coarse-textured soil, or one with a well-developed open structure, will usually drain more easily than a finer-textured or compacted soil because of its greater hydraulic conductivity, and will therefore be more prone to drought.

However, clay-rich soils such as Vertisols, which contain expanding lattice minerals, may also have low moisture contents because of large vertical shrinkage cracks which form during periods of dry weather, causing rapid infiltration and percolation through the upper part of the profile. Similarly, shallow soils or those on steep slopes often have low moisture contents. In arid environments, slope position is particularly important, with footslope soils usually providing the best water supply. The availability of moisture to plants is, however, determined not only by the moisture content of a soil, but also by the suction at which the water is held. For a given moisture content, the forces holding water in the soil, which must be overcome if a plant is to obtain water, are lowest for coarse-textured soils which have relatively large pores, and greatest for fine-textured soils in which the matric suction effects of the small pores are much higher.

For any given texture, the forces also decrease as moisture content increases; sand has the least water available to plants (only 7 per cent) while silt loam has the most (16 per cent). In order to survive in moisture-deficient soils, plants have developed a variety of adaptations, for example deep and closely spaced roots to maximise water acquisition, and narrow, waxy or hairy leaves to minimise water loss from transpiration.

In addition to plant uptake, moisture is required to allow penetration of roots and burrowing organisms; when soils are dry they often become hard and compact, thus restricting these activities. In contrast to plants and mesofauna, micro-organisms do not generally experience problems in dry soils, and indeed some types, especially spore-forming bacteria, can withstand drought conditions for many years, although bacterial growth rates usually increase when dry soils become moist.

Excessive moisture can cause different, but equally severe problems for biota. Soils can be prone to excessive wetness for a variety of reasons; for example, heavy rainfall, low permeability or poor drainage due to topographic location. Under such conditions, most pores are filled with water and therefore the root zone becomes anaerobic, which can have serious effects on the metabolism and growth of plants not adapted to such conditions, even if the conditions last for only a few days.

Root systems of these plants require oxygen partial pressure of at least 0.02 bar, with nutrient uptake increasing as partial pressure increases, up to a value of around 0.2 bar. When root cells become affected by a lack of oxygen, they start to convert glucose into ethanol (C_2H_5OH), causing cell membranes to become prone to leakage, and ion and water uptake to be impaired. Consequently wilting can occur, accompanied by chlorosis.

Also under waterlogged conditions, anaerobic bacteria produce ethylene (C_2H_4), which can inhibit root growth at concentrations of 1 ppm or more, and volatile fatty acids can also be produced during carbohydrate decomposition, which can accumulate to harmful concentrations.

Anaerobic micro-organisms can cause denitrification, reducing nitrate to nitrites, or in more extreme cases to nitrous oxide and nitrogen, therefore causing the loss of this important nutrient. An additional problem with this reaction is that an accumulation of nitrite nitrogen can be toxic to plants.

In poorly drained gley soils, manganese and iron become reduced, in which form they are readily soluble and can be taken up by plants in toxic concentrations. Soils which experience periodic or prolonged flooding can pose an additional problem to plants, but those adapted to such conditions are able to transfer oxygen rapidly to their roots via spongy stem tissue, thus maintaining an aerated rhizosphere.

Soil micro-organisms are also inhibited by a restricted oxygen flow, as can occur in clayey soils, although certain groups are adapted to anaerobic conditions, but their growth is usually slower than those found in aerobic soils. Soil meso- and macrofauna, however, experience problems in extremely wet soils due to a lack of oxygen and instability of burrowing passages; in such soils the fauna are therefore concentrated near the surface, with few, if any, burrowing animals.

CONCLUSION

Soils not only respond to environmental conditions, but can also influence these conditions and therefore play an important role in the operation of environmental systems. These systems can be divided into four main types—the hydrosphere, atmosphere, geosphere and biosphere. Although these environmental components are in many ways interrelated, the role of soil in each system separately in order to provide a clearer understanding of the processes involved.

We will start by examining the hydrosphere because soil moisture is one of the main keys to soil-environmental influences, and many of the aspects relating to the hydrosphere are therefore relevant to those of the other systems. Having first introduced some basic concepts, various components of the hydrological cycle will then be considered.

Closely related to the hydrosphere is the atmosphere, with soils exerting an important influence on climate at or near the ground surface, and this is considered next. The geosphere is then examined with respect to the role of soil in geomorphic processes and landform development. Finally, we consider the biosphere, in which soils are an important factor in controlling the habitation and distribution of biota.

4

Soil and its Management for Wheat Production

BAND PLACEMENT

Band placement away from the seed row can be used to avoid toxicity and to improve fertilizer use efficiency. Banded N fertilizer is usually more efficient than broadcast incorporation because concentrating fertilizer in the band reduces the contact with soil and microbes, and reduces losses due to denitrification and immobilization. The banding benefit varies between soils and years mainly due to differences in moisture and susceptibility to loss. Also, if the band is located near the seed row rather than random, fertilizer loss due to weed uptake is reduced.

Pre-plant banding (or deep banding) involves placing granular, liquid or gaseous fertilizer N in a ribbon several inches below the soil surface before seeding. Banding is often done in the fall on fields that tend to be wet in the spring. Fall banding can spread the workload without significantly lowering N efficiency, and often allows growers to buy fertilizer at lower cost than in the spring. The banding depth often is 8 to 10 cm (3 to 4").

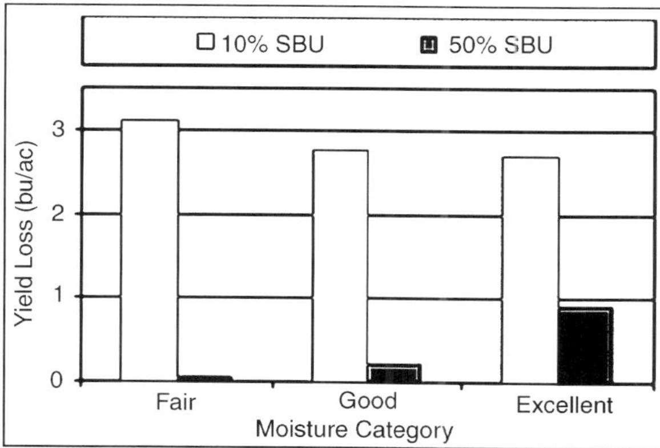

Fig. Effect of Moisture and SBU on Canola Yield [101 kg N/ha (90 lb N/ac)] in Alberta

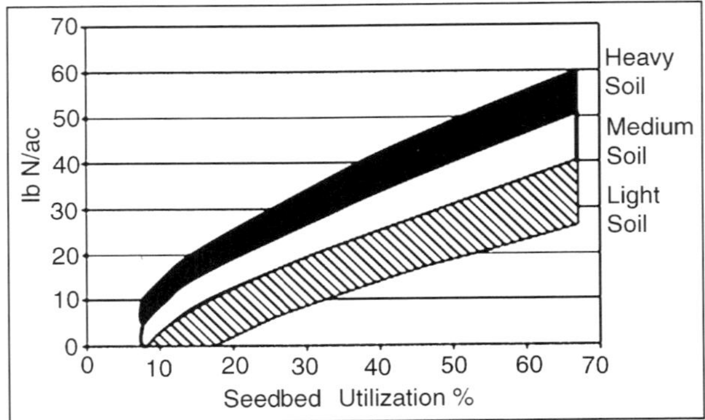

Fig. Approximate Maximum Rates for Seed Row Placement in Canola with Good to Excellent Seedbed Moisture

Under dry or sandy soil conditions, ensure anhydrous ammonia is placed deep enough to prevent visible gaseous loss, and ensure the soil flows well around the openers to permit a good seal behind the shanks. Spring banding can be shallower due to better moisture and tilth. However, in dry springs the banding operation can reduce seedbed moisture and quality.

Make spring band spacings narrower than in the fall. Spring banded anhydrous ammonia can be immediately followed by seeding, providing there are several inches of vertical separation between the injection point and seed depth. Canola emergence directly over the bands may be slightly reduced, but yields are generally not affected. Figure illustrates research conducted at the Agriculture and Agri-Food Canada (AAFC) Scott, SK Research Centre on the safety of seeding directly after banding anhydrous ammonia 10 to 15 cm (4 to 6") deep. Yields varied between seeding dates following anhydrous banding over the years with no relationship between yields and plant stands. Proper soil packing over the seed row to firm the soil disrupted by the banding operation was deemed more important to avoid stand reduction than was potential injury from the banded ammonia. No difference was found between banding ammonia parallel to and perpendicular to the seed row.

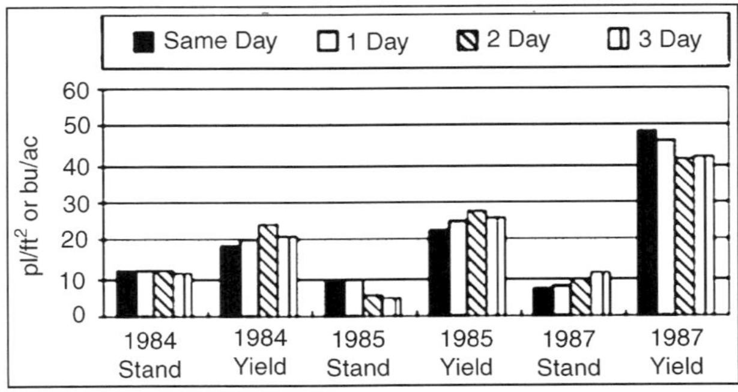

Fig. Effect of Waiting Period after Anhydrous Ammonia Banding on Canola Stand and Yield

Side banding involves placing the fertilizer band to the side and often below the seed during the seeding operation. This method has good N use efficiency and avoids seed row toxicity if the separation is maintained under field conditions. However, side banding at seeding does involve more complicated, costly openers, increased draught and wear.

Mid-row banding involves placing fertilizer between every second seed row or between a paired row during the seeding operation. This method also has good N use efficiency and avoids seed row toxicity if the separation can be maintained under field conditions. Both side and midrow banding can improve N efficiency and yield response by favouring crop versus weed access to the fertilizer. In recent years, openers have been designed that allow anhydrous ammonia to be side-row or mid-row banded during seeding. Ensure the anhydrous ammonia is horizontally separated from the seed by at least 5 cm (2").

SEED ROW PLACEMENT

Although seed row placement of N fertilizer is an efficient method for uptake, canola is sensitive to seed row N and this limits application rates. Canola seedlings are injured by excessive seed row N by the salt effect that reduces water uptake by the seed, and by ammonia toxicity. Greenhouse research at the Agriculture and Agri-Food Canada (AAFC) Beaverlodge, AB Research Centre in 1960 showed that rapeseed was sensitive to seed-placed N. Subsequent field research confirmed that rapeseed was more sensitive to seed-placed N than cereals. During this period, seed row placement was limited to P and very low rates of N. The common drills during this time were double disc and hoe press drills, which give a very narrow seed spread.

The adoption of conservation tillage seeding systems and air-seeders has greatly influenced fertilizer placement. Development of pneumatic delivery implements (air seeders and air drills•) has facilitated both dry fertilizer banding and direct seeding. Conservation seeding systems limit tillage passes in order to retain surface residues and this reduces the options for fertilizer application.

Table. Seedbed Utilization (SBU) of Various Openers*

	Spread Width of Fertilizer in Seed Row											
	2.5 cm (1") Disc or Knife			5 cm (2") Spoon or Hoe			7.6 cm (3") Sweep			10 cm (4") Sweep		
Row spacing cm	15	23	30	15	23	30	15	23	30	15	23	30
Row spacing "	6	9	12	6	9	12	6	9	12	6	9	12
SBU %	17	11	8	33	22	17	50	33	25	67	44	33

However, newer machines designed for direct seeding have resolved the fertilizer placement issue by either placing the fertilizer away from the seed or increasing the seedbed utilization. Seedbed utilization is the spread width of fertilizer and seed relative to the row spacing. For example, a 7.6 cm (3") spread with 15 cm (6") row spacing creates 50 per cent seedbed utilization. Table outlines the seedbed utilization (SBU) obtained with different openers and row spacing.

- Although some openers also vertically spread seed and fertilizer, this is not considered in the table since seed should be placed at a consistent depth for uniform germination and emergence. The actual spread width varies with air flow, soil type, moisture, residue and speed, and should be checked under prevailing field conditions.

Numerous trials have examined the safe seed row N amounts with various openers and configurations. Research conducted in Alberta from 1992-96 illustrates the effect of seed row N on canola emergence and yield. The research involved 32 site years with canola at various Alberta locations. The highest N rate and least SBU reduced canola emergence and yield 90 and 45 per cent of the time, respectively. Sites with limited moisture (due to sandy texture, low seedbed moisture or dry conditions) two weeks after seeding experienced the greatest reduction in emergence and yield with 101 kg N/ha (90 lb N/ac) as urea and low SBU.

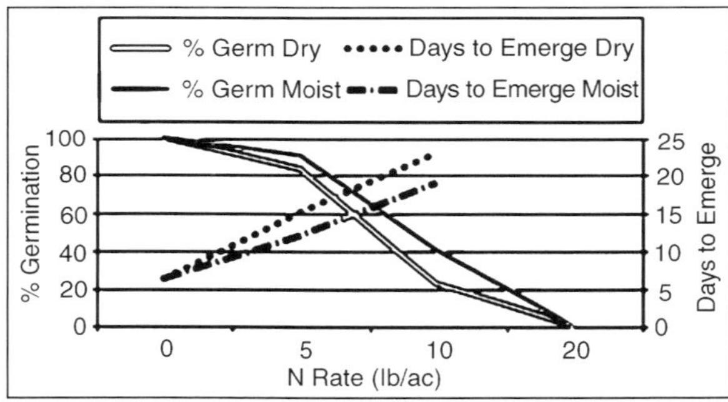

Fig. Seed Row N Fertilizer Effect on Canola Emergence

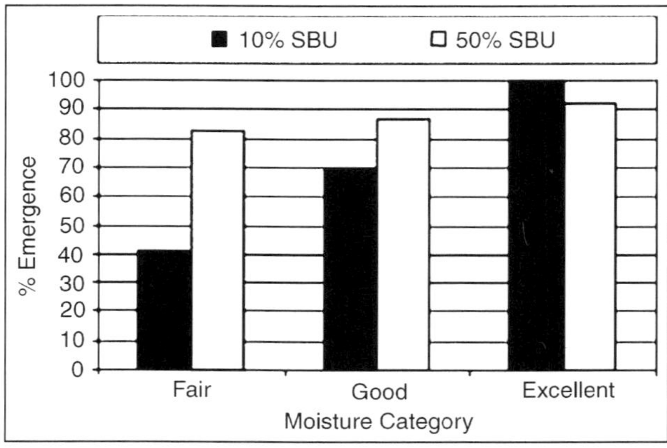

Fig. Effect of Moisture and SBU on Canola Emergence [101 kg N/ha (90 lb N/ac)] in Alberta

Figure shows the approximate safe rates of seed row granular N fertilizer based on prairie research to date. In canola, there is no significant difference in seed row safety between urea (46–0–0), ammonium sulphate (21-0-0-24) or ammonium nitrate (34–0–0). Anhydrous ammonia (82–0–0) must be placed separately from the seed. If moisture conditions are dry, reduce the safe amount of seed row N by half. The N rates are in addition to N contained in seed row phosphate fertilizer.

BROADCAST INCORPORATION

N fertilizer can be spread onto the soil surface then incorporated into the soil with a tillage implement. Although this method can be time and labour saving, fertilizer efficiency is usually sacrificed. Under dry conditions, broadcast-incorporated fertilizer can be stranded in dry surface soil, and not accessible by plant roots growing down into moisture. In wet conditions, broadcast-incorporated fertilizer is more vulnerable to losses due to denitrification, immobilization and leaching than banded fertilizer. Broadcasting without incorporation is the least efficient method of applying N fertilizer due to increased loss due to run-off, erosion and volatilization. Broadcasting after crop emergence or topdressing generally has low efficiency, but it can serve as a rescue treatment when poor fertility was not corrected prior to seeding.

FOLIAR APPLICATION

Foliar application is possible, but only a limited amount of fertilizer N can enter through leaves without significant leaf burn. The only practical instance of foliar fertilization for canola on the prairies occurs under irrigation where up to 20 per cent of applied N can be supplied in irrigation water early in the growing season.

TIME OF NITROGEN FERTILIZER APPLICATION

Nitrogen fertilizer efficiency is greatly affected by the placement method and the application date. However, differences between placement methods or timing varies widely between years or fields due to variability in weather, soil type and drainage. Generally, N fertilizer applied at or near seeding is the most efficient.

The major disadvantages of applying all the fertilizer N at seeding are:

- Increased time required to seed and fertilize during the short seeding season
- Higher fertilizer prices during the spring season compared to fall
- Increased risk of reduced seedbed quality due to extra tillage/soil disturbance at or near seeding
- Seed row N limits
- Fertilizer applicators being unavailable during the busy spring season

Considerable fertilizer placement/timing research has been conducted on the prairies. Table gives approximate relative efficiencies of the various placement and timing methods.

Table. N Fertilizer Effect on Pod, Seed Number and Yield

Time and Placement	Relative Efficiency (%)		
	Dry*	Medium	Wet
Spring broadcast-incorporated	100	100	100
Spring branded	120**	110	105
Fall broadcast-incorporated	80	75	65
Fall banded	120	110	85

- These are soil-climate categories based on typical conditions expected in the spring.
- Extra tillage associated with spring banding can dry out the seedbed, reduce emergence and yield in some cases.

An extensive survey conducted by Alberta Agriculture Food and Rural Development in 1982 on the practices of above average growers confirmed the banding benefit (top yields were associated with farmers who fall-banded fertilizer N).

DATE OF FALL BANDING

Nitrogen fertilizer banded too early in the fall increases potential for losses on wet, poorly drained soils. Delay banding until soil temperatures have cooled to less than 10°C on well drained sites, and less than 5°C on poorly drained sites.

This will significantly reduce losses by reducing the rate of change from ammonium to the more vulnerable nitrate form. Unfortunately, the opportunity to fall band fertilizer into cold soil is quite short since snow or frozen soil can occur without much warning. On average, fall soil temperatures on the prairies drop 1°C every 5 days. In the Black and Gray soil zones, fall soil temperatures usually decline to 10°C by the last week of September. In these zones, since the banding benefit is significant due to typically wet spring conditions and moderately high N rates, begin fall banding in late September. This is earlier than the normal practice.

NITRIFICATION INHIBITORS

Since only the nitrate form is susceptible to losses through denitrification and leaching, methods to delay the conversion from ammonium to nitrate (nitrification) would be beneficial. Banding achieves this delay to some extent. However, interest in developing chemical inhibitors of nitrification has stimulated research programmes for many years. A variety of chemicals have been tested under prairie conditions, but none has achieved commercial success.

For example, nitrapyrin (N-Serve) was recognized as a nitrification inhibitor in the early 1960s. Nitrification inhibitors have not been successful to date for various reasons, including cost, potential toxic effects on the soil environment and inconsistency.

UREASE INHIBITORS

The widespread adoption of direct seeding has resulted in interest in seed-placed fertilizer. While increasing seedbed utilization increases the safety of seed-placed fertilizer, the higher disturbance and draft is not always desirable. Openers with banding capability add cost and draft. Even with sidebanding, seedling damage can sometimes occur, particularly with wide row spacing, high rates of N application or insufficient separation between seed and fertilizer. Therefore, there is interest in safening urea fertilizer so that seedling injury is reduced, allowing more freedom for seedplaced N.

Agrotain (N-n-butyl-thiophosphoric triamide) safening urea by inactivating urease enzymes in the soil adjacent to the granule. This slows the breakdown of urea to ammonia, reducing the potential for seedling damage from seed-placed or sidebanded applications of urea or urea ammonium nitrate. As long as N remains in the urea form, the risk of damage is minimal. In addition, since Agrotain delays the release of ammonia, there is more time for the uncharged urea to move away from the seed-row in the soil water or with rainfall. Movement of the urea away from the seed, combined with a slower release of ammonia from the urea, will decrease the concentration of ammonia in contact with the germinating seedling, thereby reducing seedling damage.

Fig. Canola Seedlings with Sufficient N (left) versus those with an N Deficiency

In field studies, Agrotain was effective in reducing seedling damage from side-banded urea and urea ammonium nitrate, where soil and environmental conditions led to seedling damage from the untreated fertilizer. The improved stand did not always lead to a higher crop yield because canola has the ability to compensate for reduced stands. The studies also showed that canola oil and chlorophyll content were often improved by using Agrotain. While Agrotain appears effective for use with side-banded N applications, more information is needed to determine if the safening effect is great enough to allow for seed row placement of the full rates of urea or urea ammonium nitrate needed for a high-yielding canola crop.

PHOSPHORUS (P)

Phosphorus is an important plant macronutrient, but it is required in smaller amounts than nitrogen. Western Canadian soils are commonly P deficient and fertilization usually increases yield and economic returns. Good P fertilizer management is important to optimizing canola production.

ROLE OF PHOSPHORUS IN THE CANOLA PLANT

Phosphorus functions in the plant as a structural element and also in energy transfer. The structural components that rely on P include nucleic acids (the building blocks of DNA) and phospholipids (fats and oils), which are important membrane constituents.

Phosphorus plays a significant role in energy transfer in all living organisms. The P energy transfer compounds are phosphate esters about 50 different esters have been identified. ATP (adenosine triphosphate) is the principal phosphate energy compound used for starch synthesis and nutrient uptake. Energy produced during respiration and photosynthesis is captured by these phosphate compounds, which then are transported to areas that are building plant tissue.

Fig. Progressively Less N Deficient Leaves (left to right)

Fig. N-sufficient (Left) Flowering Plants and Deficient Plants

Fig. N-sufficient Raceme (left) and Deficient Raceme

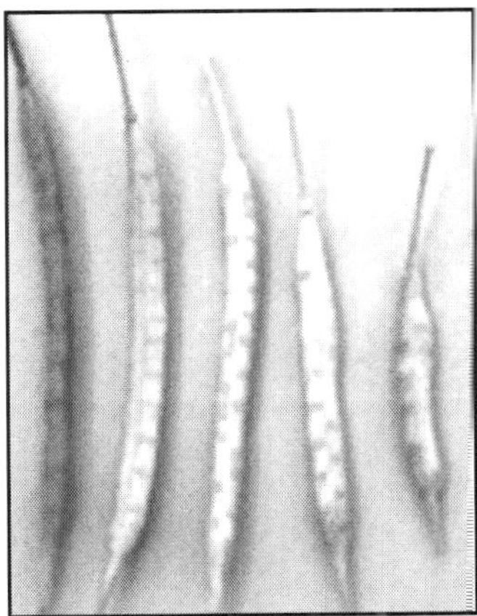

Fig. N-sufficient Pods (left) Progressing to Deficient Pods

The energy stored in the phosphate compound is released, and the molecule is recycled back to be recharged. This recycling of phosphate energy compounds is accomplished at extremely fast rates, and a small amount can satisfy the plants energy needs.

CHARACTERISTICS OF PHOSPHORUS UPTAKE BY CANOLA

The main P forms taken up by roots from the soil solution are the primary and secondary phosphate ions ($H_2PO_4^-$ and HPO_4^{-2}). These phosphate anions exist transiently in the soil solution due to rapid removal by roots and microbes, or reaction with other soil minerals. The P level is highest in young vegetative material and in the canola seed. Figure illustrates the P uptake and level over the 1998 season at the AAFC Melfort, SK Research Centre. Canola seedlings take up P rapidly during early growth, but not as rapidly as N. Studies conducted in Manitoba in the 1960s showed that canola P uptake in early growth stages was more rapid than oats, flax and soybeans. The P level remains fairly high in the leaves (0.3 to 0.4 per cent) until late flowering when significant translocation occurs into developing pods and seeds. By maturity, 75 to 80 per cent of the P in above-ground dry matter is in the seed. Canola seed contains 0.7 to 0.8 per cent P, about double that of cereal grains. Canola stems and pods at harvest contain only 0.1 to 0.2 per cent P.

Canola is an efficient scavenger of soil P even though Brassica species are non-mycorrhizal (mycorrhizae are symbiotic associations between certain soil fungi and plant roots where the fungi contribute to the P nutrition of the plant). Many cereal crops can form these beneficial relationships. In spite of canola being non-mycorrhizal, research has shown that canola takes up more P than cereals. Canola has several mechanisms to achieve this efficient P uptake. Canola has abundant fine roots with the ability to branch and proliferate in zones of higher nutrient content such as around fertilizer bands or granules. In addition to root proliferation in fertilizer zones due to branching, canola roots can increase the root hair number and length in response to low P conditions.

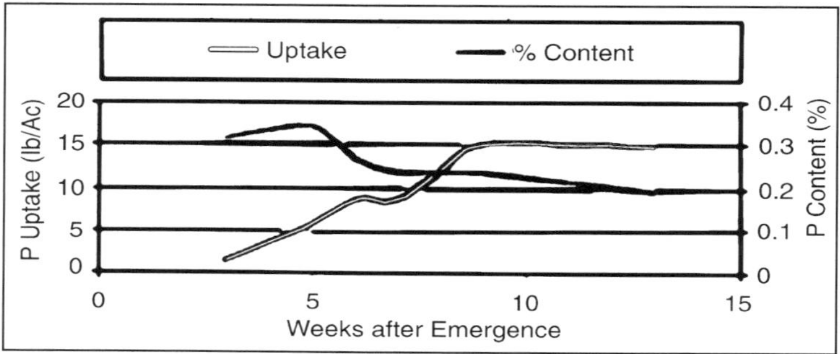

Fig. Phosphorus Content (per cent) and Uptake by Canola Over the 1998 Growing Season, Melfort, SK

The second mechanism in canola roots that enhances P uptake is solubilization of relatively insoluble mineral P forms. Canola has the ability to acidify the rhizosphere just behind the root tip near the zone of root hair formation. In a recent western Canada growth chamber experiment, the pH of the canola rhizosphere fell up to 0.8 units over five weeks compared to a drop of less than 0.4 units for wheat rhizosphere. Canola absorbed more of the relatively insoluble P forms than wheat. The acid generated by canola roots is predominantly caused by exudation of organic acids such as malic and citric acid. Canola roots also release enzymes (phosphatases) that mineralize phosphate from organic P pools. Cation-anion uptake imbalance may also contribute to rhizosphere acidification when the main form of N uptake is ammonium. However, under western Canadian field conditions, canola takes up the majority of N in the nitrate form.

After phosphate enters the root, there are three barriers to cross before reaching the xylem system that feeds aboveground growth. These barriers are the cell plasma membrane, vacuole membrane and the xylem loading site. The rate of phosphate transport across these membranes is affected by the plants P status. As the plant P content increases, the P transport rate decreases (feed back regulation). The P uptake rate is often more related to shoot than to root P level. This regulated transport system requires energy. Factors that influence root respiration will affect root P uptake. For example, cold soil or low oxygen content in a saturated soil reduces root respiration and consequently P uptake. There is competition for the phosphate transport system by arsenate. This can impact P nutrition in soils high in arsenate.

The xylem loading system is usually regulated separately from the systems at the plasma and vacuole membranes. Phosphate ions typically are rapidly transported from the roots to the shoots. Unlike N, P is absorbed and transported throughout the plant in the inorganic form (mainly H_2PO_4). Similar to N, phosphate is readily remobilized from aging tissue such as leaves to more active growing points. Phosphate stored in cell vacuoles can also be readily mobilized. Immature plants adequately supplied with P have 85 to 95 per cent of the total inorganic phosphate stored in the vacuoles. In contrast, in P deficient plants, almost all the phosphate in leaves is found in active pools (cytoplasm and chloroplasts). By maturity, most of the plant P is stored in organic form as phytate in the grain. Phytate serves as a readily accessible P source for the germinating seedling. Animal nutritionists are interested in seed phytate (including canola meal) since these compounds interfere with absorption of minerals such as zinc, iron and calcium. Considerable attention has been given to reducing phytate levels in grains, including canola, and some success is being reported.

PHOSPHORUS EFFECTS ON CANOLA GROWTH AND DEFICIENCY SYMPTOMS

Canola plants suffering from strong P deficiency can experience slow leaf expansion, smaller and fewer leaves. Deficiency symptoms appear by the second week of growth since canola seedlings are able to obtain sufficient P from seed reserves for the first week of growth. Figure (of field research results from five sites in western Canada in 1991) illustrates the significant increase in early season growth with P fertilization. Phosphorus deficient leaves may have a dark green, bluish green to purplish colour since chlorophyll and protein formation are less affected than cell and leaf expansion. Under severe P deficiency, purple colouration arises from accumulation of anthocyanin pigments. Mildly deficient plants may look normal but are small. Above-ground plant P content at flowering should be above 0.24 per cent.

Root growth is less affected by P deficiency than shoot growth, leading to a typical decrease in the shoot-root ratio. With a more severe deficiency, root development is restricted, but not as dramatically as stem and leaf growth. Although overall root branching is restricted in P deficient soils, root hair length and density usually increase.

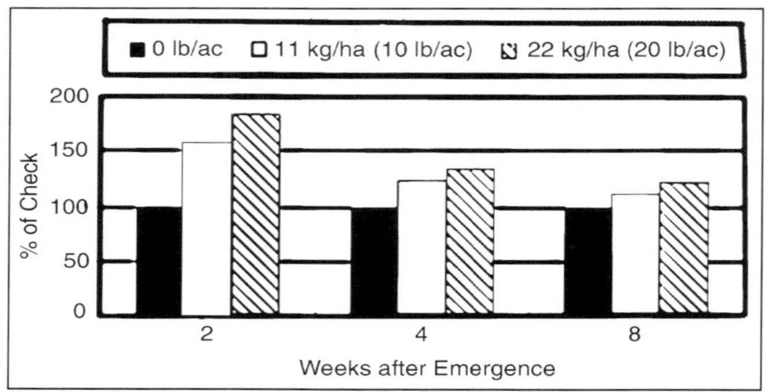

Fig. Effect of P Fertilizer (P2O5) on Canola Dry Matter after Emergence

P deficiency affects the maturity and development of reproductive tissue. Even a mild P deficiency can result in maturity delays of several days compared to plants with adequate P. In addition to a flowering delay, a P deficiency can reduce the number of flowers and seeds per pod. Also, a P deficiency can cause leaves to die and drop early, which contributes to the overall yield loss.

CANOLA RESPONSE TO PHOSPHORUS FERTILIZER

Most agricultural soils in Canada have inadequate P for producing canola crops. However, the canola yield response to P fertilizer on deficient soils usually is much less than the average response to N fertilizer on N deficient soils. Research in the 1960s showed that rapeseed often responded more to P fertilizer than wheat or flax. Subsequent research established that yield response could be predicted from soil test values. Central Alberta research in the 1980s found that 23 of 48 sites responded to P fertilizer. A recent Alberta P study from 1991-1993 found a statistically significant response to P fertilizer at 42 site-years while 81 site-years had no response. Economic analysis of the results suggested that 70 per cent of the canola sites responded to 7 kg (15 lb) P_2O_5/ac and 53 per cent responded ec"and \$0.75/kg (\$0.34/lb) P_2O_5. Data from such fertilizer experiments are compiled into databases to predict fertilizer response. Figure is an example of canola response to P fertilizer based on soil test P in Alberta.

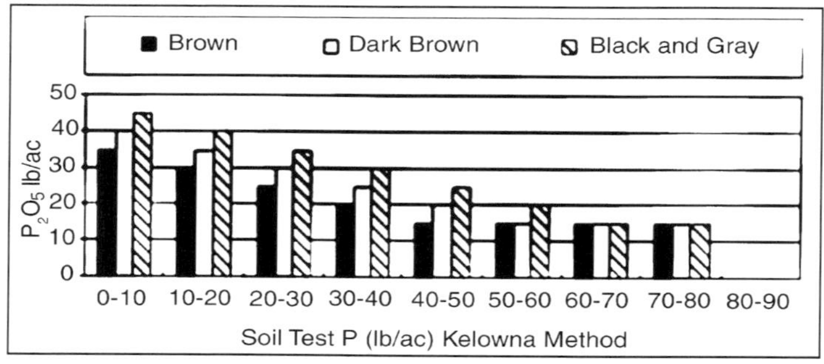

Fig. Phosphate Fertilizer Recommendations for Canola on Medium to Fine Textured, Neutral Soil under Medium Moisture

Canola response to P fertilizer depends mainly on the amount of plant available P in the soil but is also influenced by moisture and temperature. In cold soil, P availability and movement is reduced. Canola response to P fertilizer is greater under these conditions. Phosphorus fertilization often slightly advances maturity of canola crops by one or two days. This slight difference may be important in short growing seasons.

PHOSPHORUS FERTILIZER EFFECT ON CANOLA QUANTITY

P fertilization generally has negligible effects on canola quality. Experiments in western Canada have found that P fertilizer increased, decreased or did not affect oil content. Canola protein content has occasionally been slightly raised by P fertilization. A recent field experiment in Manitoba on two very deficient sites found that P fertilizer significantly increased both protein and oil content.

PHOSPHORUS CYCLE

Prairie soils contain significant amounts of total P450 to 907 kg P/ac (1,000 to 2,000 lb P/ac). However, most of this soil P is relatively insoluble with limited availability to plants. Canola roots obtain P by absorbing phosphate dissolved in the soil water. Since the amount of phosphate dissolved in the soil water is very small at any given moment, there must be constant replenishment into the soil water from the insoluble forms. This replenishment of soil solution P around roots arises from slightly soluble minerals, P desorption from surfaces, organic P mineralization and fertilizer. Figure depicts the P cycle.

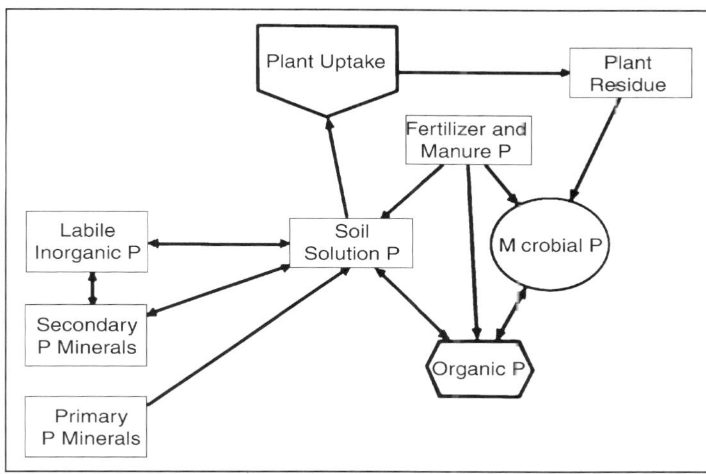

Fig. The Soil Phosphorous Cycle

Both organic and inorganic P forms occur in soil, and both are important sources of plant available phosphate in soil water (soil solution P). Primary and secondary phosphate ions ($H_2PO_4^-$ and $H_2PO_4^{-2}$) can be present in soil solution, with $H_2PO_4^-$ the major form at pH <7.2. This solution P has several possible fates: it may be absorbed by roots, adsorbed to mineral surfaces, precipitated with various cations such as Ca^{+2}, or immobilized into microbial biomass and soil organic matter. Soil phosphate supply is usually highest in the pH range of 6.5 to 7.0. At high pH levels (>7.5), calcium and magnesium cations can precipitate with phosphate to form salts with low solubility. In contrast, in acidic soils (pH<6), iron and aluminium cations react with the phosphate to form insoluble compounds. Phosphate is not a mobile nutrient in soil due to these soil constituent reactions.

The natural soil weathering process causes acidification and this encourages the eventual conversion of primary P to secondary minerals and unavailable forms (occluded P). This transformation to unavailable forms takes centuries.

As phosphate is removed from the soil solution, the lower level stimulates phosphate release from exchangeable and labile inorganic pools. As labile pools are depleted, non-labile secondary P minerals slowly dissolve to maintain the labile and solution pools.

The organic P pool also contributes to the maintenance of phosphate in the soil solution. Organic P in prairie surface soil constitutes about 25 to 55 per cent of total P and is a large pool of potential plant available P. Microbial processes drive the organic section of the P cycle. Phosphate from organic matter can be released through decomposition and then incorporated into new microbial biomass or enter into the soil solution.

Most organic P compounds released during decomposition are quickly degraded and exist briefly in soil. Some organic P compounds can be stabilized in soil through adsorption to soil constituents or by physical isolation within aggregates. Tillage decreases the soil organic P content by exposing stabilized forms to new or more vigorous microbial attack. Organic P can be degraded to phosphate by enzymes (phosphatases) released by soil microbes and by canola roots.

Significant seasonal fluctuations occur in both organic and inorganic P pools. However, reports conflict on the direction and magnitude of the fluctuations. For example, several experiments reported decreases in organic P during the summer growing season and gains over the winter, while another experiment measured major declines over the winter. Inorganic P (soil test extractable P) can also vary widely and inconsistently between fall and spring.

PHOSPHORUS FERTILIZER MANAGEMENT FOR CANOLA

The majority of P fertilizer is not absorbed by canola in the application year. Instead, most of the fertilizer P reacts with soil constituents to form relatively insoluble salts or stabilized P compounds. Timing and placement are two management strategies used to maximize P fertilizer uptake and yield response.

TIMING OF PHOSPHORUS FERTILIZATION

Since phosphate reacts with soil constituents to form insoluble compounds over time, P fertilizer efficiency can be increased by limiting the time from application to crop uptake. However, the P fertilizer application date affects canola yield much less than placement method. Most growers apply P fertilizer at seeding, minimizing P availability losses by reducing reaction time. Research in western Canada has shown the effectiveness of fall and spring banding of P fertilizer is similar.

PHOSPHORUS FERTILIZER PLACEMENT

Phosphorus supply during the first two to six weeks of canola growth is critical to achieve optimal yield. Therefore, place P fertilizer to maximize early season access.

Seed row placement is an effective placement method when soil P levels are low to moderate and spring soil conditions are cold. Cold soil decreases phosphate solubility and diffusion in the soil solution, slowing P movement to roots and root uptake rates. This condition increases the likelihood of response to readily accessible seed row P (the popup or starter effect). Unfortunately, canola seedlings are sensitive to seed row fertilizer and this limits seed row P rates. The maximum safe rate of seed row P fertilizer for canola depends on seedbed utilization and soil moisture conditions. Pot experiment results conducted at AAFCs Beaverlodge, AB Research Centre in the 1960s illustrate canolas sensitivity to seed-placed P fertilizer. The seedbed utilization in this experiment was very restricted (3 per cent), even less than a double-disc press drill. The dry and moist soil corresponded to 30 and 50 per cent of available water in a sandy loam.

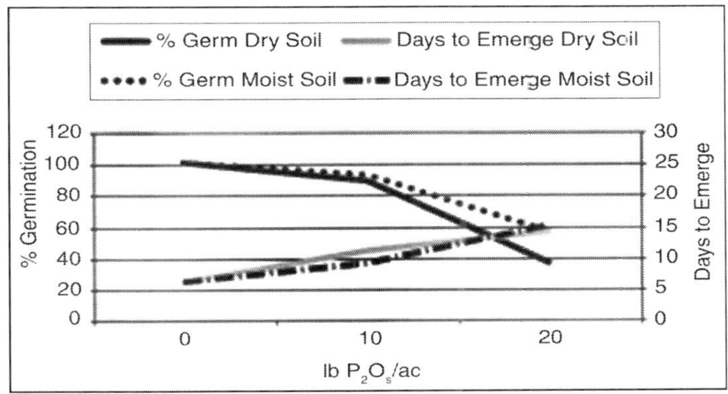

Fig. Effect of Seed-placed P Fertilizer on Canola Emergence

Subsequent field research has confirmed that excessive P fertilizer placed in the seed row can reduce plant populations and yield. High seed-placed P fertilizer rates have lowered plant populations in some cases but did not affect yield. However, lower plant populations due to excessive seed row fertilizer will increase yield variability and usually lowers high yield potential under optimal growing conditions.

Under dry soil moisture conditions with low seedbed utilization (such as disc opener), the maximum safe P_2O_5 seed-placed rate is approximately 22 kg/ ha (20 lb/ac). The rate can be safely increased to 28 kg/ha (25 lb/ac) under good moisture conditions with low seedbed utilization. As seedbed utilization increases, proportionally increase seedplaced P fertilizer rates. Some research suggests that the larger seed of *B. napus* will tolerate slightly more seedplaced P than *B. rapa*. Due to the significant emergence and yield reductions caused by moderate to high rates of seedplaced P fertilizer, place these rates separately from the seed. This fertilizer/seed separation can be achieved by increasing the spread width in the seed row or by placing the fertilizer in bands away from the seed row.

Pre-plant band placement is an effective method since it reduces fertilizer contact with sensitive canola seed and with soil constituents that will fix P over time. Also, the deeper fertilizer placement tends to be more accessible to roots as they normally grow down to moist soil. Banding fertilizer prior to seeding reduces the fertilizer handled during seeding and can provide time and labour benefits. Pre-plant band placement is currently a common method for placing all fertilizer. Phosphorus fertilizer can be banded in late fall or spring prior to seeding.

Side-banding places fertilizer near the seed row during seeding. The fertilizer normally is banded 2.5 to 5 cm (1 to 2") below and beside the seed row. Several direct seeding machines use a mid-row or paired-row method of banding fertilizer. Air seeders with shovels or knives are used with shank spacings ranging from 20 to 36 cm (8 to 14"). The fertilizer is usually banded to a depth of 5 to 13 cm (2 to 5"). No consistent agronomic benefits accrue to banding deeper than 8 cm (3") and fuel costs increase significantly with deeper depths. In high P-fixing soils, place fall P bands deeper than subsequent tillage depths to avoid mixing the band with soil.

Split application methods refer to combinations of band and seed row placement. Split application takes advantage of the consistent benefit of seed-placed P fertilizer up to 22 kg/ha (20 lb P_2O_5/ac), and avoids seedling injury by placing the remainder of the P fertilizer in a band (usually with N and S).

Broadcast-incorporated placement involves spreading P fertilizer on the surface followed by cultivation to work it into the soil. This method is significantly less effective than seed-placed or banded P fertilizer due to increased contact between the P and reactive soil constituents. Application rates with broadcast-incorporated P fertilizer usually have to be two to four times seed-placed or banded rates to get an equal response. Therefore, broadcast-incorporated methods are less economical. Research comparisons of P fertilizer placement methods at typical rates show that highest yields are frequently obtained with seed-placed and split applications, followed closely by pre-plant band methods. Broadcast-incorporated methods produce significantly lower yield responses.

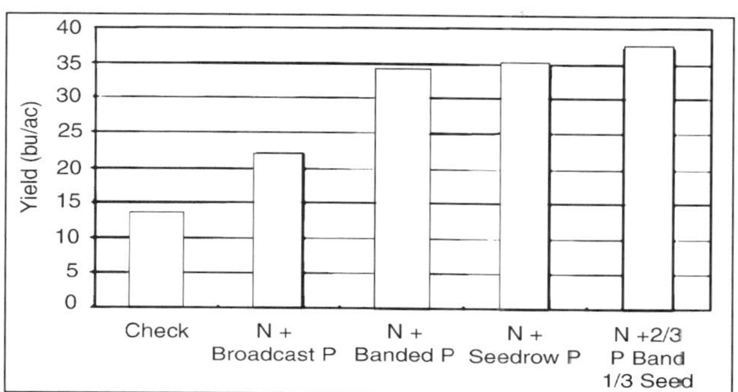

Fig. Canola Yield Response to Different P Fertilizer Placements

Research by Westco Ag research illustrates the relative usefulness of various P fertilizer placements. Differences between placement methods are largest under conditions of low soil test P (such as after forage breaking), as well as cold spring soil. On typical prairie farms that have received P fertilizer for many years, canola yield response differences between seed-placed and banded methods tend to be minimal.

Phosphorus fertilizer placement issues have been largely resolved by ground opener development with increased seed row spread or side-band capability. Split P application between a band and seed row appears to be the most consistent method due to reduced seed row toxicity, less P uptake interference from high N bands, and a dual location that hedges against poor access in either cold or dry surface soil conditions.

The application of both N and P fertilizer in a single band is called dual banding. At low to moderate rates of N, the uptake efficiency of P is sometimes increased from a dual band. At higher N rate above 90 kg/ha (80 lb/ac) the concentrated N in the band can reduce early season P uptake due to ammonia and nitrite toxicity that hinders root entry into the band. This P uptake interference appears to be strongest in recent band applications, and could be a problem with dual spring banded N + P fertilizer immediately before or during seeding.

NON-TRADITIONAL SOURCES OF PHOSPHORUS NUTRIENTS

Phosphorus deficiencies on the prairies are normally corrected with annual applications of commercially refined P fertilizereither dry blends of mono-ammonium phosphate (12-51-0) or liquid blends of ammonium polyphosphate (10-34-0). Manure also serves as a traditional source of P and other nutrients.

ROCK PHOSPHATE

Canola has an ability to absorb native soil P through acidification of the rhizosphere. Pot experiments have demonstrated that canola can utilize more rock phosphate than other crops, apparently due to the rhizosphere acidification. This has prompted promotion of rock phosphate as a viable alternative P fertilizer for canola. Rock phosphate is the relatively insoluble, gray-black powdery material that is refined by fertilizer manufacturing plants into soluble phosphate fertilizer.

Idaho is a common source of rock phosphate marketed in western Canada. Research on the prairies indicates that rock phosphates do not perform satisfactorily compared to fertilizer phosphate. The poor performance is due to poor solubility, lower P_2O_5 content, and the predominance of neutral, calcareous soils on the prairies. While high rates of rock phosphate do slightly improve canola yields on some soils, this is not cost effective compared to fertilizer phosphate. Typically, rock phosphate application rates need to be six to eight times that of fertilizer phosphate for equivalent yield response. Research by Alberta Agriculture Food and Rural Development at Ellerslie, AB illustrates the poor performance of rock phosphate compared to fertilizer P. All the P sources were seed placed and 112 kg N/ha (100 lb N/ac) was pre-plant banded.

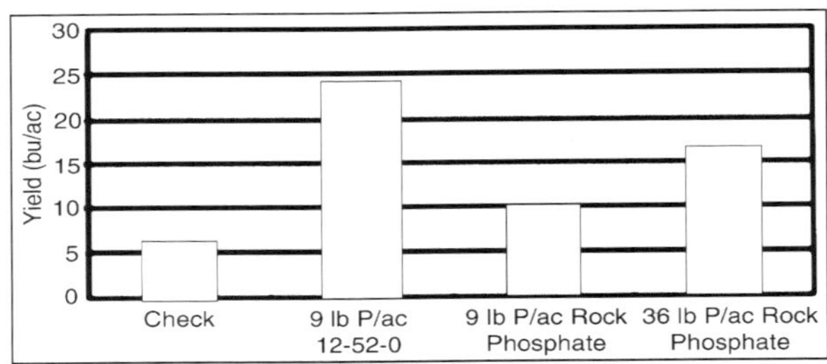

Fig. *B. rapa* Yield Response to Rock Phosphate and P Fertilizer

Another non-traditional means of P nutrition recently developed for canola is biologically based. Although canola does not form symbiotic associations with mycorrhizae that improve P nutrition in other crops, other rhizosphere microbes exist that increase P solubilization and subsequent plant P uptake. AAFC, Lethbridge, AB researchers identified an organism (*Penicillium bilaii*) that solubilized P minerals and improved the P uptake of cereals and canola. This organism was then commercialized as a seed inoculant (Provide).

Field experiments conducted on the prairies have shown that inoculating canola with Provide increases early season P uptake and vegetative growth, and results in higher yield with and without P fertilizer. The canola yield response to inoculation with Provide at 15 P-responsive sites in western Canada is summarized in Figure.

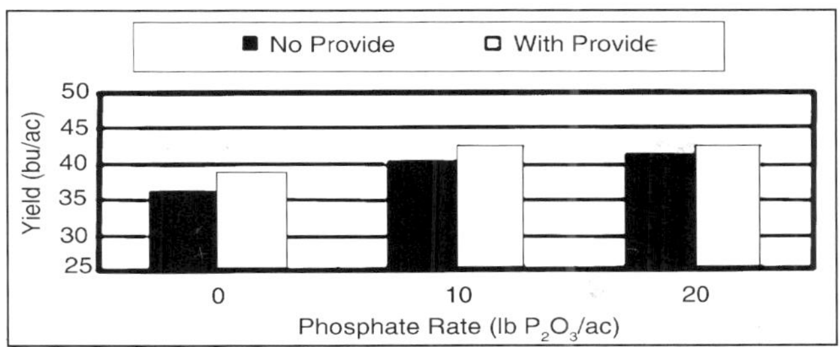

Fig. Canola Yield Response to Inoculation with Provide at P-Responsive Sites

On average, growers can expect to apply 11 kg less P_2O_5/ha (10 lb less P_2O_5/ac) when canola is inoculated with Provide. The adoption of Provide inoculation by canola growers has been limited, perhaps due to the inconvenience of inoculation, the short viable period after inoculation, inconsistency and cost relative to simply using more P fertilizer. In the future, biological fertility enhancing microbes will likely become more common.

Fig. P-sufficient Canola (left) Compared to P-deficient Canola

Fig. P-sufficient Canola at Flowering (left) Compared to P-deficient Canola

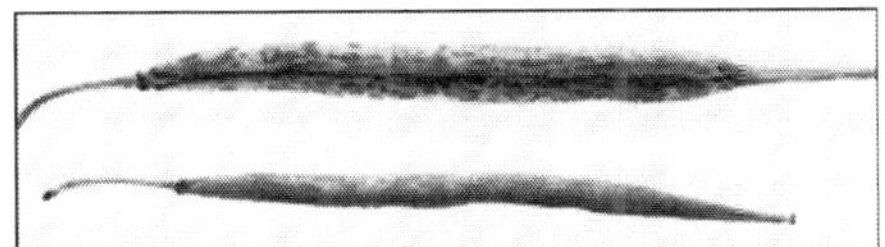

Fig. A P-sufficient Canola Pod (top) Compared to a P-deficient Canola Pod

POTASSIUM (K)

The macronutrient potassium (K) is required in large amounts by canola similar to nitrogen. In spite of the large requirement, canola yield responses to

K fertilizer (potash) are infrequent, due to ample soil K reserves on the prairies, and canola's strong ability to absorb K.

ROLE OF POTASSIUM IN THE CANOLA PLANT

Potassium is different from most other essential nutrients since it does not become part of structural components in the plant. Instead, most of the K in plants remains dissolved in the cell sap and performs several major functions.

One major function for K is that of enzyme activation. Enzymes are protein complexes that catalyze chemical reactions. More than 60 enzymes need to be activated by K. This activation occurs when potassium cations (K^+) bind to the enzyme surface, changing the enzyme shape, and allowing the enzymes active site to attach to its substrate more rapidly or accurately. For example, K stimulates the activity of an enzyme (starch synthase) that catalyzes starch formation from glucose. While other cations can also stimulate this enzyme, K^+ is the most effective. In K deficient plants, the lack of stimulation of the starch synthase results in an accumulation of soluble sugars and N compounds, and a decrease in starch.

Another major function of K is in water relations. Potassium helps to maintain a favourable water status in plants in several different ways. Potassium cations dissolved in cell sap perform major osmotic functions. Osmosis is the tendency for water levels to equalize between different areas separated by a porous membrane. Dissolved ions such as K^+ attract water and thus are osmotically active substances. Potassium is the major dissolved ion in cell sap and provides most of the osmotic pul that draws water into roots.

Potassium cations also maintain the water relations in plants through their crucial role in regulating water loss (called transpiration) from pores (stomata) in the leaves. Although the stomata must open to allow movement of carbon dioxide and oxygen in and out of the leaves, water loss also occurs. This transpiration creates a gradient that pulls water and nutrients up through the xylem to the leaves. However, plants cannot afford excessive water loss and need to regulate the stomata opening. For example, photosynthesis stops during darkness, and the need for nutrients and water decreases greatly during night. Plants have developed a system that closes stomata during the dark or during drought. Potassium cations, in combination with chlorine, calcium and certain hormones, are responsible for governing the opening and closing of the stomata. Upon receiving a signal induced by darkness, K^+ and Cl^- are pumped from the two guard cells surrounding the stomata, which causes a loss of turgidity of the guard cells and thus allows the pore to close. Potassium deficient plants often have higher transpiration rates and display wilting.

Potassiums osmotic activity also provides the physical force that expands cells during growth. New cells accumulate K+ and associated anions like Cl^- in the large central vacuole that occupy 80 to 90 per cent of the cell volume.

The K^+ ions attract water and inflate the cell, stretching it to a new larger size. Potassium-deficient plants can exhibit low growth rates and small cells.

Energy relations in the plant are influenced by K. Potassium affects photosynthesis at several levels. K^+ is the main ion that counterbalances the H^+ flux during photosynthesis in the chloroplasts. Potassium also maintains a favourable pH gradient in the chloroplasts for making phosphate energy compounds. Potassium helps the translocation of photosynthate sugars by maintaining a high pH in phloem tubes needed for loading, and by maintaining osmotic gradients needed for sap flow.

Potassium is needed for N uptake and protein synthesis. K^- cations are the major counter ions that balance nitrate during transport and storage in vacuoles. Many steps of protein synthesis require high K^+ levels.

The K level is highest in seedling canola, then declines steadily up to maturity as shown in Figure, *B. napus* canola grown at the AAFC Research Centre in Melfort, SK in 1998. Canola K uptake is rapid during the early growth stages and tapers off by the end of flowering. Under high K fertility and good growth, canola can absorb more K than apparently needed, a situation termed â□œluxury consumption.â□• As canola matures, the K level in leaves declines while the stem level increases. By harvest, the stem and straw material contain about 1 to 2 per cent K. In contrast to N and P, the K content of the seed (0.8 to 1 per cent K) is low relative to the stem. Unlike K^+ in the vegetative parts, seed K is probably complexed with phytate as a salt.

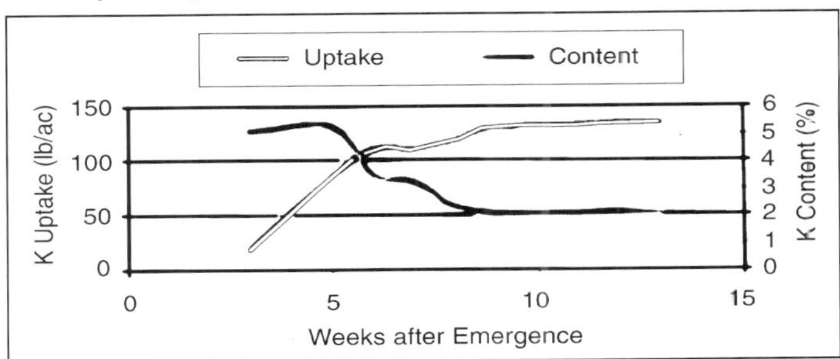

Fig. Potassium Content (per cent) and Uptake by Canola over the Growing Season

POTASSIUM EFFECTS ON CANOLA GROWTH AND DEFICIENCY SYMPTOMS

Potassium deficiency reduces overall canola growth but to a lesser degree than N or P deficiency. Since K is mobile within the plant, deficiencies are first visible in older leaves. The edges and areas between veins of older leaves tend to turn pale green or yellow, followed by withering. The yellowing can occur first in middle leaves before older ones if observed at bolting to flowering stages. In severe cases, leaves die but remain attached to the stem. Small white spots can develop on leaves. Plants are prone to wilting during midday. Potassium deficiency symptoms in canola are rather nondistinct and can be easily confused with other problems. Fortunately K deficiency in canola is rare on the prairies.

CANOLA RESPONSE TO POTASSIUM FERTILIZER

Although canola absorbs large amounts of K, responses to fertilizer K are rare on the prairies. In fact, canola or rapeseed responses to fertilizer K are infrequent around the world testament to the crops strong ability to absorb soil K.

Numerous fertilizer research studies on the prairies and in Ontario have established that canola rarely responds to applied K, even under conditions where cereals normally respond. Although the K soil test is adequate for cereals, the usefulness declines for canola. Critical levels are often stated to be around 280 kg K/ha (250 lb K/ac) or 112 ppm in the top 15 cm (6"), but research indicates that canola will not consistently or economically respond to fertilizer K unless the soil test is very low 78 to 112 kg K/ha (70 to 100 lb K/ac) or 35 to 50 ppm. Very sandy or peaty soils are the most likely soil types to have very low K soil test values.

Other factors that increase the likelihood of K deficiency are:

- Free lime in the rooting zone
- Acid soil
- Poor drainage
- Cool temperatures
- Soil compaction
- Shallow root zone

Unlike cereal responses, potash applications have not been shown to help with canola disease resistance, lodging or seed quality (oil content or meal protein content).

POTASSIUM SUPPLY FROM THE SOIL

Western Canadian soils generally contain ample plant available K due to an abundance of K minerals (such as mica and feldspar) in the parent material (3 to 4 per cent K). There is often 17,000 to 56,000 kg K/ha (15,000 to 50,000 lb K/ac) in the top 15 cm of prairie mineral (non-peat) soils. The weathering of these minerals slowly releases K^+ held in crystal structures typically only about 1 per cent of total soil K is available for plant uptake.

This available K is mostly (90 per cent) exchangeable K^+ adsorbed to clay surfaces and organic matter, while the other 10 per cent is found dissolved in the soil solution. Approximately 10 to 20 per cent of the total soil K is slowly available from smaller mica particles and certain clays. Figure outlines these K pools. Losses due to leaching or erosion are ignored in this figure, as they are usually small.

Figure shows that the various pools are in dynamic equilibrium. As K^+ is removed by plant uptake and through leaching on sandy soils, additional K is released from the mineral soils to become available. Available K moves to plant roots by diffusion through the soil only up to 6 mm (1/4"). Therefore, the equilibrium process that repeatedly moves K from the slowly available to readily available pool is very important for K nutrition.

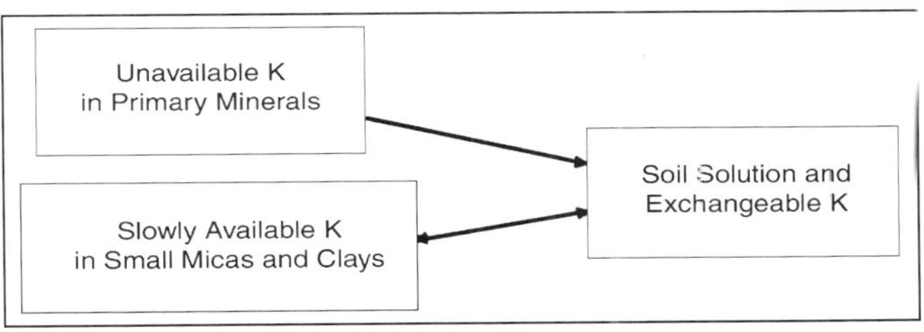

Fig. Potassium Soil Cycle

The rate of movement from the slowly available to readily available pool varies among soils due to differences in minerals and clays. This variation in K dynamics creates problems for soil testing. An extractant that measures plant available K in soil solution and exchangeable K over a short time period does not assess the replenishment power. Unfortunately, tests that measure the replenishment power are time-consuming and cost-prohibitive.

POTASSIUM FERTILIZER MANAGEMENT FOR CANOLA

Potassium is relatively immobile in the soil since the K^+ cations are readily adsorbed to the negative surface charges on clay particles and organic matter. Potassium can also be fixed into the clay lattice structure of certain clay types. Potassium is more mobile in sandy soil and thus can be leached in these soil textures. In most soils, K^+ is much less mobile than nitrate but somewhat more mobile than phosphate. This relative immobility means that fertilizer placement will greatly affect uptake efficiency. Ensure application methods minimize contact with soil and increase root contact. Banding and seed-placed methods can achieve good uptake efficiency. Since canola responses to K fertilizer are rare on the prairies, there has been limited K placement research in this crop. Seed-placed K fertilizer is an efficient application method but the high salt index of potash fertilizer limits the amount that can be safely applied near the seed. Canola has a much lower tolerance to seed-placed potash than cereals, and stands will be reduced if seed-placed K rates exceed 17 kg K20/ha (15 lb K20/ac) with drills that have low seedbed utilization (such as double disc drills). Higher rates of potash fertilizer can be safely seed placed as the seedbed utilization is increased.

If other nutrients such as N or P are also seed placed, this reduces the safe rate of seed-placed K. Good seedbed moisture, higher clay and organic matter contents help reduce the severity of seedling damage from seedplaced K fertilizer. However, most K deficient soils are sandy, and are sensitive to seed-placed K.

Due to canola's sensitivity to seed-placed K fertilizer, a band placement away from the seed row is more advisable. Sideband placement is an efficient method and the separation between fertilizer and seed reduces the risk of germination damage. Openers with side-band capability are becoming available, especially for direct seeding implements. Deep banding prior to seeding should also be an efficient and safe method of K fertilization. Potash fertilizer can be banded together with other nutrients. Banding efficiency should not differ greatly between fall and spring unless the soil is very sandy and subject to leaching loss under conditions of high snowmelt or spring rainfall.

The broadcast-incorporation application method is less efficient, and probably requires rates double that of banding to achieve a similar crop response. However, in situations where banding equipment is not readily available and seed placement is too risky, broadcast incorporation may be useful and not overly expensive due to the relatively low cost of potash fertilizer.

The higher fertilizer rates necessary for broadcast K may also benefit subsequent crops with a higher K response than canola.

Fig. Leaf Edge Scorch Due to K Deficiency

Fig. Scorched Leaf Due to Severe K Deficiency

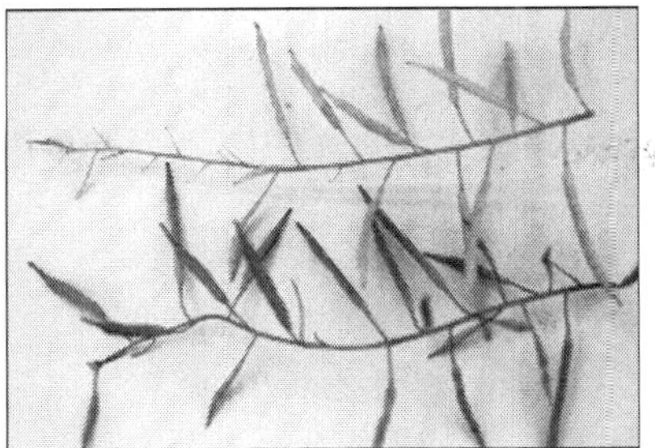

Fig. Severely K Deficient Pods (top) and Sufficient (bottom)

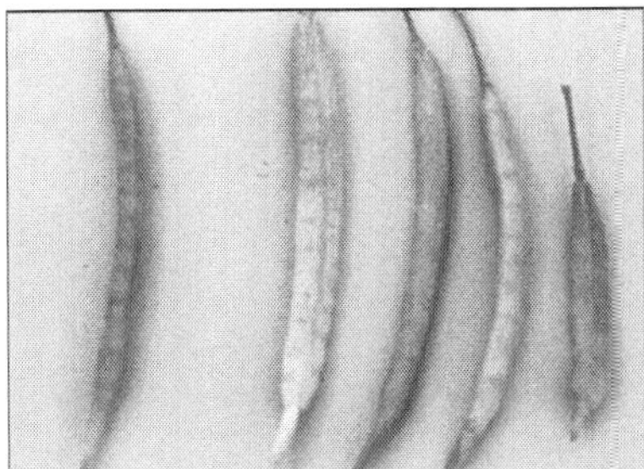

Fig. Sufficient Pod (left) and Severe K Deficient Pods

SULPHUR (S)

Sulphur (S) is the fourth macronutrient, but ranks as the third most limiting nutrient on the prairies. Sulphur deficiency in western Canada was first identified in 1927 on Gray Wooded soils in Alberta. Canola is more sensitive than cereals to S deficiency and frequently responds to fertilizer S addition. Therefore, pay equal attention to N, P and S.

ROLE OF SULPHUR IN THE CANOLA PLANT

Canola contains large amounts of S. Sulphur is part of structural and enzymatic components. Sulphur is a key component of two essential amino acids (cysteine and methionine) and is needed for protein synthesis. Chlorophyll synthesis also requires S. Both of these amino acids are also precursors for coenzymes and secondary plant substances.

Glutathione, an important antioxidant in plants and animals, is synthesized from cysteine. Glutathione contents are higher in leaves than roots. It's found primarily in the chloroplasts where its anti-oxidant ability is needed to detoxify free radicals generated during photosynthesis.

Glutathione also functions as transient S storage, and a precursor of phytochelatins (compounds which detoxify heavy metals in plants).

Thioredoxins, another important group of S compounds related to glutathione, help activate several enzymes in carbon metabolism. Sulphur also is part of several enzymes and coenzymes such as ferrodoxin, biotin (vitamin

H), coenzyme A, urease, and thiamine (vitamin B1). An important group of secondary plant S compounds in canola are glucosinolates. Plants contain over 100 different glucosinolate compounds. These secondary compounds, although not well understood, probably have a number of functions. Glucosinolates are stored in cell vacuoles, and can be broken down by an enzyme (myrosinase) to yield glucose, sulphate and volatile compounds such as isothiocyanate. Glucosinolates contribute to defence or attractant systems for certain insects and diseases. When plant cells are destroyed by insect feeding, glucosinolates are broken down, releasing various deterrents/attractants.

Glucosinolate levels are highest in growing points, roots, and youngest leaves, all of which are most vulnerable to insects and diseases.

The role of glucosinolates as S reserves to maintain plant S during periods of high demand (such as bolting, flowering, podding and seed fill) is controversial.

However, recent research in Europe showed that glucosinolates comprised a small S pool in leaves, and under induced S deficiency, sulphate (SO_4^{-2}) mobilization from storage in cell vacuoles was about 10 times greater than contributions from glucosinolates. Sulphur is also a constituent of sulpholipids, which are membrane components.

CHARACTERISTICS OF SULPHUR UPTAKE BY CANOLA

The main S form absorbed by canola roots is sulphate. In industrial areas, atmospheric S compounds dissolved in rain can be absorbed by leaves. However, this amount is quite small and is decreasing with better air pollution control. sulphate absorption is accomplished with active transport systems across membranes. The uptake rate increases as the sulphate level increases in the soil water. Low plant S contents also increase the root uptake rate. Negative feedback signals for S uptake may be sulphate or glucosinolate levels in vacuoles, or the levels of organic S compounds such as cysteine, methionine or glutathione. Sulphate uptake faces competition from molybdenum and selenium. Therefore, soils high in these minerals can experience antagonism with S uptake. The S level in canola plants is highest in the early seedling stage when young leaves comprise most of the dry matter (Figure). As plants develop, the overall S level declines but not as dramatically as with N. By maturity, canola straw contains approximately 0.3 to 0.4 per cent S while pod chaff contains slightly more S (0.5 to 0.6 per cent). Canola seed contains about 0.4 to 0.6 per cent S. At harvest, canola straw and pod chaff contain roughly twice as much S per acre as that in the seed.

There has been limited research on the complex S partitioning into the various compounds of different plant parts over the growing season. Most of the plant S ends up in protein and stored sulphate. As leaves senesce, protein-S is readily remobilized, while stored sulphate remobilization is slow and more limited.

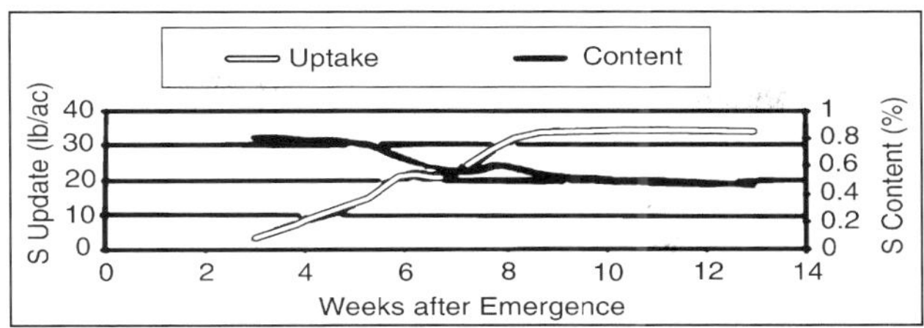

Fig. Sulphur Content (per cent) and Uptake by B. napus over the Growing Season

Therefore, overall, S has medium mobility. In canola quality varieties, glucosinolate biosynthesis has been blocked in pod walls, and sulphate has been found to accumulate in contrast, glucosinolates accumulate in rapeseed pod walls. Glucosinolates and glutathione account for a small fraction of plant S, usually around 10 per cent in young vegetative tissue. Glucosinolate contents will vary over 10 fold in the same tissue at different ages. Glucosinolate contents also vary several-fold between different plant organs. Also, growing conditions, genetics, and S supply all affect glucosinolate contents.

SULPHUR EFFECTS ON CANOLA GROWTH AND DEFICIENCY SYMPTOMS

Sulphur has several effects on canola growth. Since chlorophyll synthesis requires S, deficiency will affect visible leaf colour and photosynthesis. Protein synthesis requires S containing amino acids, and, therefore, S deficiency affects rapidly growing parts, especially reproductive structures. Mild S deficiency often does not result in noticeable symptoms, but still can reduce yield. Medium deficiencies do not show symptoms until bolting, flowering and podding. Under severe deficiency, symptoms show up about two weeks after germination. Growth chamber research conducted at the University of B.C. shows that S deficiency affects shoot growth more than roots (Figure). By the bolting stage, S deficiency begins to affect yield parameters such as branches per plant, fertile flowers per plant, seeds per pod and individual seed weight. Under mild to moderate S deficiency, thousand kernel weight is normally not significantly affected since plants compensate by reducing the seed number per pod. Nitrogen deficiency affects pods per plant more than S deficiency, while the opposite is true for seeds per pod.

Sulphur deficiency symptoms vary depending upon the severity and timing of the deficiency relative to crop growth stage. In the vegetative stage, foliar symptoms show up under severe S deficiency (Figures). Since S has a low mobility within the plant, symptoms are observed more readily on the youngest leaves, which are greenish-yellow compared to the normal bluish-green in B. napus. The yellowing (chlorosis) starts from the leaf edges and the tissue around leaf veins remains green. Subsequently, the leaf edges and bottoms may turn purple. Besides the leaf colour, S deficiency in young plants causes smaller leaves as well as upward cupped leaves.

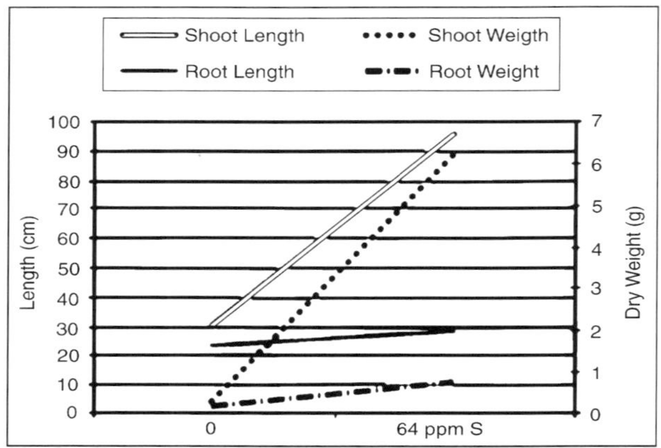

Fig. Effect of S on Rapeseed Shoot and Root Growth

By the bolting stage, new leaves of S-deficient plants show chlorosis, purpling and spoon-like leaf cupping. The purple colour is caused by enhanced pigment (anthocyanin) synthesis due to sugar accumulation resulting from S limited amino acid and protein synthesis. The degree of leaf cupping is highly dependent on the timing of the S deficiency. There is significant cupping when S deficiency occurs before half of the leaf weight is attained.

By flowering, S-deficiency symptoms can show up in the petals. If severe S deficiency occurred in the vegetative stage, symptoms can be found both in foliage and flowers petals can be smaller and lighter yellow. However, if S deficiency occurs around flowering, leaf symptoms may not be obvious, but flower petals may become paler. Yellow and white petals may even exist side by side on a single flower. The life span of S-deficient petals is shortened to one day from two or three, and pollen production is greatly reduced. In addition, S-deficient petals are egg shaped compared to more round petals on sufficient S plants. Flowering is often delayed and prolonged. Reports suggest that S-deficient plants do not attract honey bees, perhaps due to lack of pollen.

SOIL FERTILITY

Profitable canola production relies heavily on adequate plant nutrition, which in turn is affected by management of soil fertility. In addition, the nutritional level of the plant will affect the crop response to stress factors such as disease and adverse weather. Balanced, effective fertilizer management not only contributes to profitable canola yield but also helps to maintain the productivity of the soil resource.

BASIC PLANT NUTRITION

The living plant depends on a number of basic factors for normal growth:

- Light
- Air

- Water
- Nutrients
- Physical support

Soil plays an important role in all these factors except for light. If any of these basic factors are limiting, plant growth will be reduced or the life cycle may not be completed" this is called the principle of limiting factors. In other words, plant growth potential is limited by the factor in shortest supply. Yield may be reduced when one nutrient reaches excessive levels that cause toxicity, so the proper balance of nutrients is important. Also, other factors such as improper management or pests can lower yield. Therefore, a systems approach is necessary to integrate all the factors in the best combination to achieve the most economic yield.

ESSENTIAL PLANT NUTRIENTS

The plants mineral composition does not simply reflect the elements needed for growth. Plants can selectively absorb required elements for their growth. But they also can take up elements not needed for growth. The terms essential plant nutrient or essential mineral element were formed to describe the minerals needed by plants to grow and complete life cycles. Essential plant nutrients must be directly involved in some aspect of the plant metabolism such as structural material, enzymes or hormones, and they must not be totally replaceable by another mineral element. For higher plants such as canola, there are 14 essential nutrients (besides CO_2, oxygen and water).

Table. Essential Plant Nutrients

Macronutrients	N, P, K, S, Mg, Ca
Micronutrients	Fe, Mn, Zn, Cu, B, Mo, Cl and Ni

Above table also indicates that plant nutrients are often classified by the relative amounts needed.

Table. Approximate Amounts of Nutrients in the Above- Ground Portion of a 1,960 kg/ha (35 bu/ac) Canola Crop

Element	kg/ha	lb/ac
Nitrogen (N)	112-134	100-120
Phosphorus (P)	1-28	15-25*
Potassium (K)	67-134	60-120*
Sulphur (S)	22-28	20-25
Calcium (Ca)	45-67	40-60
Magnesium (Mg)	13-20	12-18
Iron (Fe)	~1	~1
Chlorine (Cl)	~0.8	~0.7
Manganese (Mn)	~0.2	~0.2
Zinc (Zn)	~0.2	~0.2
Boron (B)	~0.2	~0.2
Copper (Cu)	~0.7	~0.06
Nickel (Ni)	~0.004	~0.004
Molybdenum (Mo)	~0.004	~0.004

- $P \times 2.3 = P_2O_5$; $K \times 1.2 = K_2O$
- Crop uptake of nutrients is greatly affected by conditions in the soil or weather (dry, wet, cold, compaction, nutrient imbalances, salinity, etc.).

Macronutrients are needed in large amounts relative to micronutrients. Table shows the relative amounts of nutrients contained in a typical canola crop. Some nutrients can be accumulated in plants much higher than necessary for growth.

GENERAL NUTRIENT UPTAKE

The following discussion outlines nutrient movement into and through the canola plant. The level of most nutrients in the plant sap is much higher than in the water surrounding the roots. For example, a typical N content in a canola plant at the rosette stage would be 5 to 6 per cent N, whereas a fertile soil in the spring would contain about 0.0002 per cent N in a plant available form on a dry weight basis. The N level in the soil solution would be in the range of 0.00002 per cent. Therefore, plant nutrient uptake must be highly selective.

Nutrient uptake begins when plant available forms move from the soil water through pores in the root skin (exodermis) into the free space of the roots. This free space comprises about 5 to 10 per cent of the roots internal volume. This movement is a passive process (does not require energy from the plant) driven either by diffusion (movement due to differences in concentration) or mass flow (simply carried by water flowing into the roots). The movement is selective since pores into the free space act as a size filter. Many nutrient ion diameters are much smaller than the pores. For example, potassium and calcium are only 10 to 20 per cent of the pore size, and have easy access to the free space. Large diameter substances such as metal chelates, viruses and fungi are restricted from entry by the small pore size.

As plant roots grow, the soil volume and surface area explored increases, which increases the capacity for nutrient absorption. In addition, roots possess a cation exchange capacity (CEC) due to negative charges in cell walls. This root CEC attracts positive ions (cations) like ammonium (NH_4^+) but repels negative ions (anions) such as nitrate (NO_3^-).

After entry into the free space, nutrients move into the cell interior by crossing a plasma membrane found on the inside of cell walls. Another similar membrane is found surrounding a large central storage compartment (vacuole) that usually fills more than 80 per cent of the total cell volume. The plasma and vacuole membranes are effective barriers and are the main sites for nutrient uptake selectivity. The membranes contain carrier systems or ion pumps that transport certain nutrients.

Such systems are called active since they require energy from the plant to work. This energy demand for ion uptake by roots is considerable, taking up to

1/3 of the energy during rapid growth. The energy for root activity arises from respiration, which requires carbohydrates and oxygen. This explains why nutrient uptake often stops in flooded soils" there is a lack of oxygen.

Some active uptake systems are constant while others have a rate that can be regulated. As the plant level of nutrients and related compounds increases, the root uptake rate can decrease (negative feedback). In contrast, as plants build tissue, the level of nutrient "building blocks" • decreases, and the roots are signalled to increase the uptake rate (positive feedback) for nutrients. Passive ion channels through the membranes allow for selective nutrient movement.

The selectivity of the various transport systems across the membranes is not absolute. There is often competition between ions of similar size and charge. For example, chloride (Cl^-) competes with nitrate (NO_3^-). This competition between Cl^- and NO_3^- is important in certain saline soils with Cl^- as a major component of the salt. Most prairie soils contain salt with sulphate as the main anion.

Since cation and anion uptake are regulated differently, plants must be able to compensate for differences in electrical charges that arise from disproportionate uptake of cations and anions. Plant cells maintain a pH in the range 7.3 to 7.6 by either releasing or consuming hydrogen cations (H^+), which is achieved by formation or removal of organic acids.

The nutrient journey continues in a path from cell to cell through tiny connecting tubes (plasmodesmata), although some nutrients can continue to move between cells through the free space. The next barrier occurs at the waxy layer (Casparian band) that surrounds the central vascular tissue (phloem and xylem). The phloem and xylem are special tissues that act like highways for nutrient transport from roots to leaves (xylem) and from leaves to growing points and roots (phloem). In young root tips, the Casparian strip is not well formed and thus is an incomplete barrier. The mechanism how ions pass through the Casparian strip and into the xylem (xylem loading) is not well understood. There is probably a combination of active (ion pumps) and passive channels for ion movement into the xylem. Xylem loading is regulated separately from root uptake, thus creating a control system for nutrient movement. Nutrients are carried by water up through the xylem. Water flows up through xylem tissues due to a suction-like force created when water evaporates from the leaves, and from slight pressure produced by roots. Once inside the xylem sap, nutrients can be unloaded and reloaded before reaching the end growing points.

Once nutrients reach their targets, there often is considerable recycling, especially for the mobile nutrients such as N. For example, a normal feature of plants appears to be simultaneous import and export of nutrients from leaves. This dynamic nutrient cycling is termed remobilization or retranslocation. In young vegetative plants, nutrient recycling occurs from the mature leaves to roots and young leaves through the phloem. The remobilization ability of different nutrients affects where deficiency symptoms occur. Deficiency symptoms of mobile nutrients such as N will first appear in old tissues. In contrast, deficiency symptoms of nutrients with limited mobility such as sulphur and copper will occur in young tissue, and can hinder flower/seed development.

Nutrient remobilization is particularly important when seeds are forming. At this stage, mobile nutrients are being exported from ageing leaves while nutrient imports are decreasing. Also, root activity and nutrient uptake generally decrease by this stage due to drying soils, nutrient depletion in the soil and a relative shift in the energy supply from roots to developing pods and seeds. Plant parts with a strong energy or nutrient demand are called â□œsinks. As a result, old leaves are sacrificed to supply pod and seed growth. Mobile nutrients in the seed have mostly been transferred from other plant tissue. In canola the sources are pods, stems and leaves.

Roots are not the only sites where nutrient uptake can occur. Some nutrients can be absorbed by leaves and other above ground plant parts. Nutrients in the gas form NH_3 (ammonia), NO_2 (nitrogen dioxide) and SO_2 (sulphur dioxide)can enter leaves through leaf pores (stomata) and then be changed into organic forms. These gases are major air pollution components and in some areas contribute considerably to plant nutrition. In areas with intensive livestock operations, NH_3 uptake can contribute 10 to 20 per cent of the nitrogen for adjacent crops. SO_2 is readily absorbed by leaves. In a European field experiment, almost half of the total sulphur (S) taken up by vegetative rapeseed came from atmospheric S compounds, probably SO_2. This may partly explain why S deficiencies have increased in western Canada after environmental regulations enforced cleanup of S emissions from gas plants.

SOIL PROPERTIES THAT AFFECT PLANT NUTRITION

Soil is a complex mixture of non-living substances (minerals, organic matter, gases and liquids) and living organisms (bacteria, fungi, insects, worms, etc.). These factors influence soil fertility either directly or indirectly.

Soil solids consist of mineral particles, organic matter in varying stages of decomposition and living organisms. Solids make up about half the soil volume, while water and gases make up the other half in the pore space.

Soil mineral particles vary widely in size and are classified by size:

- Rocks are larger than 2 mm (0.08") in diameter
- Sand particles range from 0.05 to 2 mm (0.002 to 0.08") in diameter
- Silt particles range from 0.002 to 0.05 mm (0.00008 to 0.002") in diameter
- Clay particles are smaller than 0.002 mm (0.00008") in diameter

These particles are made from various mineral types with different elemental composition, which affects weathering processes and thus the release of certain nutrients. Two soils with identical texture could be drastically different in fertility due to differences in mineral composition. Potassium is an example of a plant nutrient whose supply arises from mineral weathering in soil.

The soil colloidal fraction refers to microscopic particles of clay and organic matter. The surface of the colloidal fraction is where most soil chemical reactions occur and it is very important in nutrient supply.

The proportion of sand, silt and clay determines the texture of a soil. Soil texture is grouped into five or more classes. The texture influences fertility by affecting moisture holding capacity, air exchange and the CEC. Adequate moisture is key to fertilizer response and potential yield for canola in western Canada.

The CEC is an important property that influences the soil storage of many plant nutrients. Most nutrients are present in the soil water as positively charged cations. A few are negatively charged anions. The CEC indicates a soil's ability to hold or store cations. Prairie soil particles typically have a negative charge. The process of electrical attraction that holds cations to negative surfaces of soil colloids is called adsorption (not absorption). The cations are not permanently stuck to the colloidal surface and can be exchanged with other cations. With time certain cations may become 'fixed'• into forms that are not easily removed from the exchange complexes. Adsorbed cations are not removed by water moving through the soil and can be accessed by plant roots. Cations with a higher positive charge (for example Ca^{+2}) are held more tightly than those with a lower charge (for example K^{+}).

Soil negative charges arise due to substitutions in the mineral crystals by elements with smaller positive charge, and due to reactions at the edges. Organic particles also contain a significant number of negative charges. The total particle surface area in a soil increases as particle sizes get smaller. Therefore, a soil high in clay has a much greater surface area than a sandy soil.

A high clay soil also has a bearing on the surface area and negative charge. Clay minerals are microscopic layers of aluminium and silicon crystals formed by weathering of other minerals. Thus clays are called secondary minerals. The type of clay depends on the original minerals and the weathering extent. Fairly young clays common in western Canada (such as montmorillonite) have a 2:1 arrangement of silica:alumina crystal sheets, while older clays have a 1:1 arrangement. Generally, 2:1 clays have 10 to 100 times more surface area, negative charges and, consequently, a higher CEC than 1:1 clays. Organic colloids have 10 to 100 times more negative charges and higher CECs than the 1:1 clays. Therefore, soil organic matter levels greatly influence the CEC.

The CEC strongly influences soil fertility. A higher CEC means that more cations, including plant nutrients, can be loosely stored in a plant available form, giving the plant a greater pool of nutrients to draw from. Since most cations are not highly soluble, only small quantities can be dissolved in the soil solution at one time. The CEC soil property allows a reservoir of nutrients to be stored then released to plant roots. This continuous replenishment of nutrients in soil water is very important for several nutrients, including potassium. A high CEC also means that fewer cations will be lost through leaching out of the root zone.

Since soils are predominantly negatively charged, anions [such as nitrate (NO_{3}_) and sulphate (SO_4^{-2})] are repelled by soil colloids and tend to stay in the soil water. They will flow with water and are potentially subject to leaching loss.

Soil organic matter (OM) plays an important role in soil fertility as a plant nutrient storehouse. Not only does OM adsorb many cations due to a high CEC, it also stores nutrients as part of its structure. As the OM is decomposed by soil microbes, nutrients are released from the organic structure into plant available forms this process is called mineralization. Mineralization from OM is the primary natural source of plant available N and S in prairie soils, and also influences P availability. Mineralization of individual nutrients will be described in later sections. Soil OM also plays a secondary role in soil fertility by improving physical properties such as water holding capacity, infiltration, aggregation (tilth) and buffering pH.

SOIL TESTING FOR NUTRIENT CONTENT

Plant nutrient content in soil varies over years, between fields and even within fields that appear uniform. Soil sampling and analysis methods (soil testing) were developed to assess the fertility level and to predict crop response to applied fertilizer or manure. Soil testing is not an exact science due to nutrient variability inherent in most fields and the inability to predict growing season weather. Although soil testing is not exact, it can help estimate soil fertility and give reasonable guides for profitable fertilizer application.

Spatial nutrient variability in fields creates problems for soil testing and fertilizer application. The variability makes it difficult to obtain representative soil samples. Using single fertilizer rates across variable fields results in over-fertilized and under-fertilized areas within the field. Although variable rate fertilization is being researched and developed, most fields still are fertilized with a single rate. In addition, fertilizer response calibrations developed from research sites with low variability will under-predict the optimum fertilizer rate for larger farm fields with more variability.

For meaningful soil test results proper soil sampling is necessary. A sampling error in the field is usually much greater than the analytical error in the lab. Ensure soil samples accurately reflect the overall field. However, intensive soil sampling is not convenient, cost effective or practical. Research shows that the accuracy of a composite soil sample increases with the number of sub-samples taken. Accuracy refers to how similar the soil sample value is to the true field average (is the sample representative). Precision describes how often the same value can be obtained when repeating the procedure (reproducibility). The common recommendation to sample a field 20 times and mix all the samples into one composite, produces an accuracy of about ±17 per cent for NO_3^-, assuming 80 per cent precision. This is an acceptable level of accuracy and precision for fertilizer recommendation purposes in most cases.

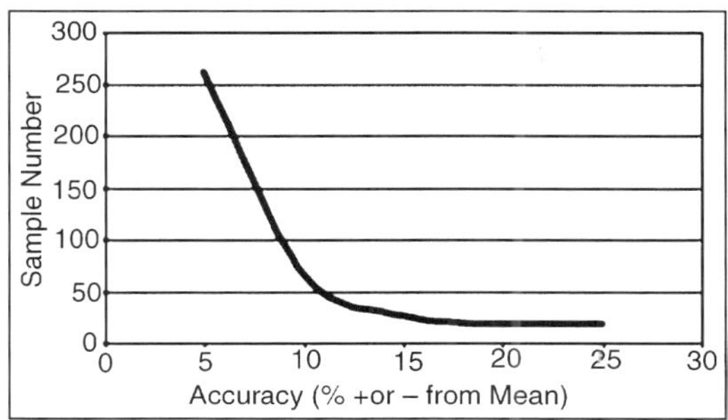

Fig. Effect of Sub-sample Number on per cent Accuracy of Composite Sample for Nitrate-N with 80 per cent Precision

In addition to adequate sample numbers, proper soil sampling techniques include:

- Where and how to sample each field (sampling plan)
- Proper equipment
- Proper sampling time
- Correct depth
- Proper sample handling

SOIL SAMPLING PLAN

Sample individual fields separately. The first step is to assess the field variability and identify representative areas. The type and level of variability can influence the choice of sampling plan.

Consider the four basic sampling plans:

1. Random soil sampling uses a random pattern across a field, generally avoiding unusual or problem areas such as hilltops and potholes. Bulk together 20 soil cores into one composite sample, air-dry then send to a soil test lab. This common method is adequate for smaller, relatively uniform fields.
2. Topographic sampling involves separating sets of samples based on topography. Identify the dominant topographic features such as hilltop, midslope and bottom slope, and take 20 core samples for each type. Bulk sub-samples from each type into one composite sample for each landscape type. Send several samples to the testing lab. This method can provide more meaningful results for variable fields, but at additional expense and labour. You must be willing and able to apply different fertilizer rates in the separate landscape areas.

3. Benchmark sampling expands on the topographic sampling concept by considering unique areas based on topography, soil texture and type and typical crop growth. Once these unique areas are identified, each year the samples are taken from the same spot within each benchmark area. There may be one or more benchmark sites within a field, depending on the variability. Each benchmark area then becomes a reference area on which fertilizer recommendations are based. The benchmark sampling area is much smaller than the whole field, and together with sampling in the same spot from year to year, this is assumed to reduce variability of the test results. If the benchmarks are carefully selected to represent the majority of the field, then good soil test results can be obtained.
4. Grid or systematic sampling follows an organized grid pattern, perhaps every 0.2 to 2 ha (0.5 to 5 acres). This method can reveal field nutrient variability and allow for variable rate fertilization and precision farming techniques. However, the sampling and analysis costs are not economic with current field crop prices on the prairies.

PROPER SOIL SAMPLING EQUIPMENT

Soil sampling to a 60 cm (2') depth can be done with a probe or auger. Do not use flight or screw sampling augers if samples need to be separated by depth since mixing will occur. If testing for micronutrients, ensure the sampling tool is chrome plated or stainless steel and rust-free. Do not bulk samples in metal pails. Clean plastic pails labelled for location and depth work well.

Information sheets, sample bags and shipping boxes are available from soil testing labs and most fertilizer dealers.

PROPER SAMPLING DATE

The most accurate sampling time is just prior to seeding. However, this is practical because time is needed to purchase and perhaps place the fertilizer before seeding. Sampling is commonly done in the spring or fall. Spring sampling is done after the soil has thawed and is no longer saturated from snow melt. Fall sampling can begin once the soil has cooled to 5 to 7Â°C. This helps reduce nutrient content changes due to microbial activity.

The old assumption that N availability does not change over the winter has been proven wrong. Research in Alberta found that available N increased by 56 kg/ha (50 lb/ac) in stubble fields from fall to late winter while the soil was frozen. Summerfallow fields increased by 73 kg/ha (65 lb/ac). This overwinter gain in available N is apparently due to death of soil microbes and subsequent release of available N forms from their ruptured cells. However, the available N gained over the winter was temporary as it was mostly lost in the spring, probably through denitrification. The stubble fields lost 45 kg/ha (40 lb N/ac) in the spring while summerfallow fields lost 73 kg/ha (65 lb/ac). Therefore, the net change from late fall to late spring was minimal.

Late fall sampling tends to more accurately reflect spring NO_3^- (nitrate) contents than early fall sampling, especially for Black soils. Alberta research in the 1980s compared soil samples taken in the fall (early October and early November) to spring samples for 26 stubble fields. Early fall samples averaged 34 kg/ha (30 lb/ac) less nitrate N than spring samples, while late fall samples averaged only about 17 kg/ha (15 lb/ac) less. The early fall samples were also more variable in relation to spring samples. Overall, late fall samples more accurately predicted spring nitrate contents and grain yields than early fall. Spring samples were slightly better than late fall samples for predicting grain yield and N uptake. In contrast, recent research in Manitoba measured very little change in soil nitrate levels in cereal stubble from early September to freeze-up. North and South Dakota extension soil scientists recommend early fall sampling in view of time constraints, but reduce the N recommendation by 0.2 kg (0.5 lb) for each sampling day prior to September 15.

Another experiment in central Alberta compared the effects of sample timing on phosphorus (P) soil test values. On average over 27 sites, the extractable P increased from 28 kg/ha (25 lb/ac) in early October to 49 kg/ha (44 lb/ac) by early November, and to 50 kg (45 lb) by spring (late April to early May). The relationship between early fall and spring P values was not close and, therefore, it was not possible to simply correct the early fall values. In contrast, research on a Brown soil in Saskatchewan over 24 years found both overwinter increases and decreases in soil P tests, but relatively few were significant. An experiment on irrigated alfalfa on a Dark Brown soil near Lethbridge, AB found significant overwinter increases in organic P. The conflicting results may be related to the differences in soil organic matter content and biological activity, and, therefore, potential for microbial changes to the plant available P pool.

In conclusion, early fall sampling can create higher than necessary fertilizer N and P recommendations due to an underestimation of spring nitrate N and available P, especially in the Black and Gray soil zones. Fertilizer response curves have been calibrated only against spring nutrient contents.

One disadvantage to late fall sampling after soil has cooled to 5 to 7°C is that fall fertilizer banding opportunities become more limited. By the time the samples are taken, dried, sent to the lab, analysed and results returned, the soil may have become frozen or covered with snow. On average over the prairies, soil cools by 1°C every five days in the fall.

CORRECT SAMPLING DEPTH

The appropriate sampling depth depends on the nutrients to be tested. For mobile nutrients such as N and sulphur (S), sampling to the 60 cm (2') depth is usually the most accurate according to research conducted in the 1960s. The fertilizer response database for N was developed with 0 to 60 cm (0 to 24") samples. However, recent research in Saskatchewan and North Dakota indicates that the 0 to 30 cm (0 to 12") depth may be more accurate for N than either 0 to 15 cm (0 to 6") or 60 cm samples. This recent research and the fact that sampling 60 cm is considerably more difficult, supports the 0 to 30 cm depth as a reasonable recommendation for these areas. In contrast, research in Manitoba has documented that the 0 to 15 cm depth is inferior. Separate samples for 0 to 15 cm, 15 to 30 cm and 30 to 60 cm was often recommended in the past, but was rarely done due to additional expense and time.

For immobile nutrients such as P and potassium (K) and most micronutrients, the ideal depth is 0 to 15 cm because the fertilizer response calibrations are based on that depth. If only the 0 to 30 cm depth is sampled, the soil testing labs must use a correction factor to estimate the value for the 0 to 15 cm depth. Since these nutrients are relatively immobile, they tend to remain at the fertilizer application depth. Therefore, the 0 to 30 cm depth may underestimate these nutrients and lead to high fertilizer recommendations.

PROPER SAMPLE HANDLING

Handle samples carefully to prevent accidental mixing and contamination. Mix each sample and spread on separate pieces of clean paper to dry at room temperature. Use a fan if required and avoid additional heat sources. Once dry, fill soil sample cartons or bags with about 0.5 kg (1 lb) of soil, and label with field number and depth.

CONSISTENCY OF SOIL TEST LAB RECOMMENDATIONS

Soil sampling, analysis and interpretation is not an exact science. However, reasonable precision and accuracy is needed in order to make costly fertilizer decisions. Over the years, various growers, agencies or companies have sent duplicate samples to different labs to compare their analysis and recommendations. Unfortunately, widely differing results and recommendations have occurred, causing growers to question the credibility of the labs or the soil testing practice.

Two fundamental reasons contribute to differences in results. First, labs may be using different analytical procedures to measure soil nutrient content. For example, there are several methods to extract soil phosphorus. Or the technique may differ slightly when using a particular method. Over time, labs are harmonizing their test methods and in time this problem may disappear.

Variations in soil test recommendations arise mainly due to the differences in each labs interpretation. The recommended fertilizer amounts at the same soil test level can vary significantly from lab to lab.

This may be due to using:

- Different critical (deficient) soil levels being used
- Regional fertilizer response (calibration) data but modifying recommendations to fit a particular philosophy of fertilizer use or economic payback
- Recommendations from other regions or countries
- A unique system of fertilizer recommendation not based on regional calibration data or economics

Although consistency among labs has been improving since the first comparison studies in the 1980s, further improvements are needed. Calibration data must continue to be collected to account for changes in fertilizer application techniques and changes in other agronomic practices.

Overall, soil testing is a useful agronomic practice. Use labs that base fertilizer recommendations on economics using regional calibration data. Be prepared to question unusual recommendations based on experience and the local knowledge of qualified agronomists. Keep in mind that the accuracy of fertilizer recommendations will always be limited by sampling challenges and the inability to predict the weather of the upcoming growing season.

HOW TO DIAGNOSE NUTRIENT DEFICIENCY SYMPTOMS

In moderate to severe nutrient deficiencies, visible symptoms can indicate the specific nutrient that is lacking. Nutrients that are only slightly limiting often do not show visible symptoms, a situation that has been termed hidden hunger.

A systematic diagnosis of visible symptoms is needed to correctly identify limiting nutrients. Symptoms usually appear on either old or young leaves depending on the mobility of the nutrient in question. Chlorosis (loss of green colour, yellowing) and necrosis (death of plant tissue, often leading to white or brown colour) are important visible symptoms. Diagnosis under field situations can be complicated by high field variability, multiple deficiencies, and other causes such as weather, pests and herbicide injury. For example, a sulphur deficiency can easily be confused with Group 2 herbicide injury due to similar symptoms.

PLANT AND TISSUE TESTING

Crop nutritional status also can be assessed by plant and tissue analyses. These methods can supplement, but not replace, soil testing. Plant and tissue analyses measure the nutrient content of above ground plant parts during growth. The values are compared to established ranges for inadequate, adequate and excess levels.

Plant tissue testing is suitable for diagnosing crop problems that may be nutritionally related and to identify any nutrients that may be limiting yields.

Plant analysis can determine if the fertilizer rate and method of application were adequate to meet crop needs. The disadvantage of tissue sampling is timing after tissue samples are taken and analysed, it may be too late to correct deficiencies in the current crop. No reliable interpretative criteria exist for nutrient ranges in seedling canola. Also, nutrient contents usually differ greatly between different plant parts and ages. Therefore, the proper part must be sampled at the proper growth stage.

An adequate sample will contain 50 to 80 plants, depending on the nutrients to be tested and plant part/age. Avoid unusual, dead or stressed plants, as well as those covered with soil or recent sprays. Cut samples with a clean, rust-free knife or scissors.

Dry the plant samples on clean paper or plastic at room temperature (do not oven dry). After drying, keep the samples in a paper bag. The following table shows sufficiency levels for most plant nutrients in flowering canola.

Table. Plant Tissue Analysis Interpretative Criteria for Canola (whole above ground plant at flowering)

Nitrogen (N) %	> 2.4
Phosphorus (P) %	> 0.24
Potassium (K) %	> 1.4
Sulphur (S) %	> 0.24
Calcium (Ca) %	> 0.49
Magnesium (Mg) %	> 0.19
Zinc (Zn) ppm	> 14
Copper (Cu) ppm	> 2.6
Iron (Fe) ppm	> 19
Manganese (Mn) ppm	> 14
Boron (B) ppm	> 29
Molybdenum (Mo) ppm	> 0.02

Comparing the plant analysis results from two areas of a field that differ visibly in growth can be difficult to interpret because nutrient content differences can be confounded by growth differences. If the two areas differ mainly in deficiency symptoms, then comparative sampling can be useful. In this case, collect the samples soon after the symptoms appear and before major differences in growth and maturity occur. Plant and tissue analyses need to be interpreted by experienced individuals.

NITROGEN (N)

Nitrogen is the most common limiting nutrient (other than water) for canola production. Therefore, a good understanding of this nutrient is needed to efficiently manage fertilizer N and maximize economic returns.

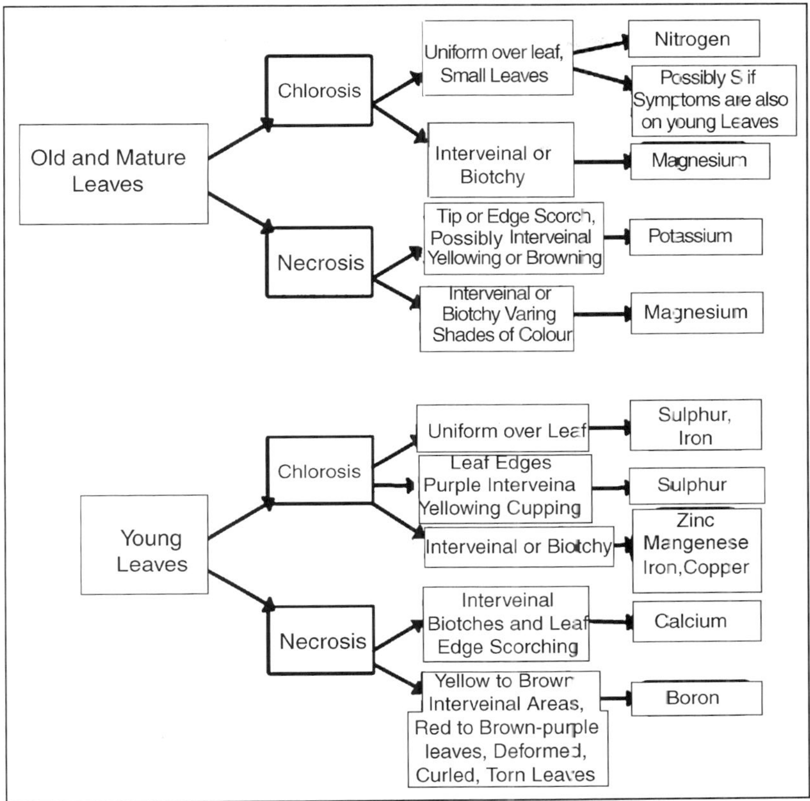

Fig. Diagnosing Nutrient Deficiency Symptoms

ROLE OF NITROGEN IN THE CANOLA PLANT

As shown in the previous section (Table), canola, like most crops, contains large amounts of N. Nitrogen is a part of many critical plant components: amino acids and proteins (which form enzymes); genetic material (nucleotides and nucleic acids); and other components found in membranes (such as amines), co-enzymes and others.

The majority of the N in green plant tissue is present as enzyme protein in chloroplasts where chlorophyll is located. By harvest, the majority of the N in a canola plant is found as seed protein. The relative N proportions in the plant changes over time and growth stage. The N proportioning closely resembles the dry matter partitioning as shown in Figure. The N level in canola plants is highest in the early seedling stage when young leaves are the majority of the plants dry matter. As the plant grows into flowering stages, the overall N level declines due to stem material and leaf loss. By maturity, canola straw contains just 0.5 to 1.5 per cent N while the seed contains 3.4 to over 4 per cent N.

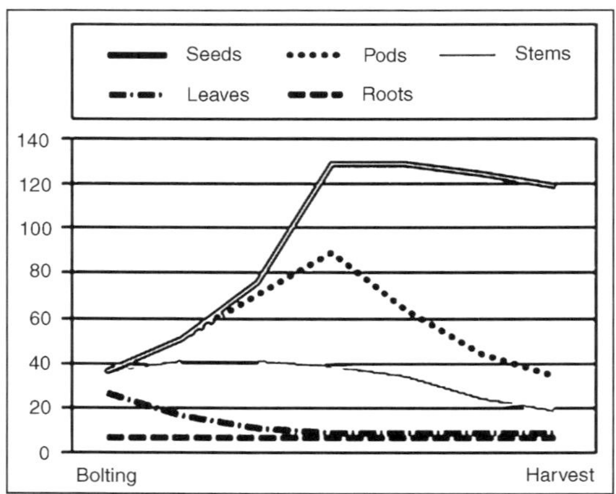

Fig. Nitrogen Partitioning in Canola

NITROGEN EFFECTS ON CANOLA GROWTH

The most obvious N effect is an overall increase in plant growth (height and dry matter). This stimulation occurs early in the vegetative stage and continues into the reproductive stage. Research in England illustrates the canola leaf stimulation, leaf area and pods/seed production by fertilizer N (Figures). Other research generally confirms that N fertilizer mainly increases canola leaf area index, leaf duration, plant weight, growth rates, number of flowering branches, plant height, number of flowers, number and weight of pods and seed yield. Therefore, good N fertility is necessary to produce a large, photosynthetically efficient leaf area that will support high numbers of flowers, pods and seed yield.

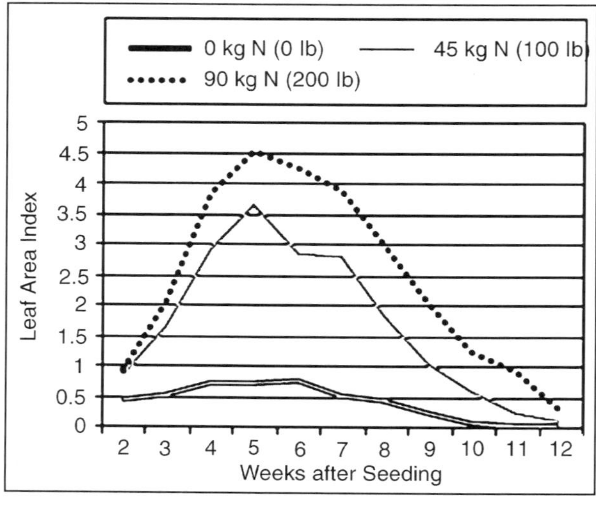

Fig. Canola Leaf Area Response

CANOLA NITROGEN DEFICIENCY SYMPTOMS

Healthy canola plants with adequate N have dark green leaves. Nitrogen is mobile within the plant and can be moved from older to younger leaves and pods.

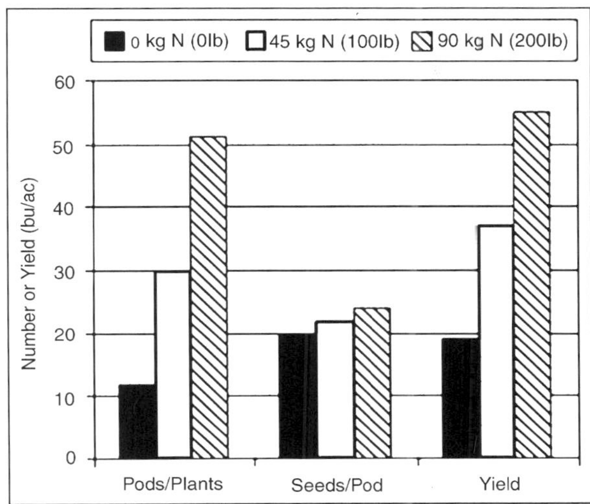

Fig. N Fertilizer Effect on Pod, Seed Number and Yield

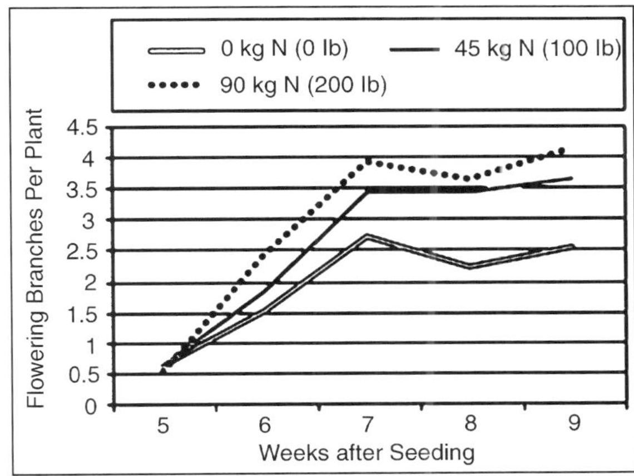

Fig. Effect of N on Flowering Branches

Therefore, N deficiency symptoms first show up in older leaves as pale green to yellow colouring, and sometimes purpling. These older leaves tend to die early, turn brown and drop off prematurely.

Overall plant growth is slow, with short thin stems, small leaves, and few branches. The amount and time of flowering is restricted, and pod numbers are low. Nitrogen-deficient canola pictures are shown in Figure. In healthy canola, plant tissue tests of above ground material at flowering will show more than 2.5 per cent N.

CANOLA RESPONSE TO FERTILIZER NITROGEN

Canola responds well to applied fertilizer N on deficient soils. Most stubble fields have insufficient N for high canola production and thus require fertilizer or manure application. Research data has been collected that describes crop response to fertilizer in relation to initial soil reserves in western Canada and around the world. This data is used by soil testing labs to predict fertilizer requirements. Research on the prairies has found that profitable dryland canola yield response to fertilizer N is unlikely when the soil contains more than 34 to 45 kg nitrate N/ ha (75 to 100 lb nitrate N/ac) in the top 60 cm (2'). Higher yielding winter canola types typically respond to more N fertilizer than winter wheat. In contrast, spring canola types have a similar N fertilizer response to high yielding CPS wheat in western Canada. An example of typical canola response to fertilizer N on a deficient stubble black soil in Alberta is shown in Figure. The vertical line represents the economical level of N fertilizer under medium moisture with $265/ tonne ($6/bu) canola, $0.88 kg ($0.40 lb) N and a 2:1 risk ratio.

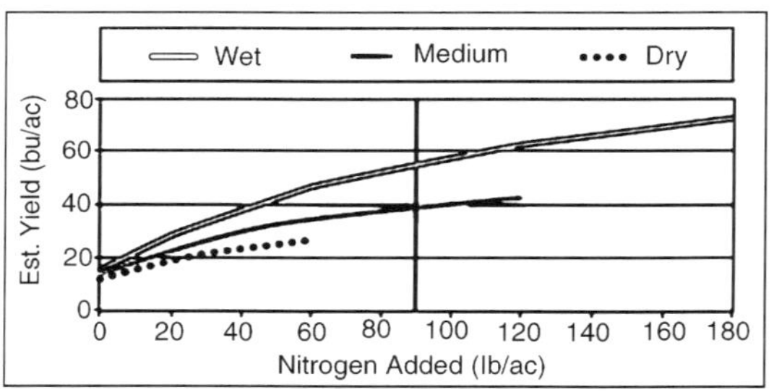

Fig. Canola Response to N Fertilizer

The previous crop influences the crop response to N fertilizer. Different preceding crops vary in the amount of available N removed from the soil and the amount released from crop residue. In addition, the disease break with different rotational crops can affect the yield potential and, therefore, the economic response to fertilizer. Moisture availability greatly affects the yield response to N fertilizer. Conversely, adequate N is needed for crops to respond to moisture. Canola yield is reduced under extremely dry or wet conditions. Research at Outlook, SK illustrates the synergistic response of canola to moisture and N over eight years (Figure). Under dry soil conditions, root growth and activity are reduced, resulting in less N uptake. In addition, soil microbial growth is slower, which reduces N release from soil organic matter, but also reduces temporary tie-up by soil microbes. Normally, more plant available N is left in the soil after a dry growing season than after wet seasons.

Drought and high temperature stress near flowering and podding dramatically reduce both water and N fertilizer efficiency. This is true for both stubble and fallow crops, although moisture is more often limiting for stubble crops. Unfortunately, crops use the most water during the flowering and podding period when soil moisture reserves and rainfall are usually low.

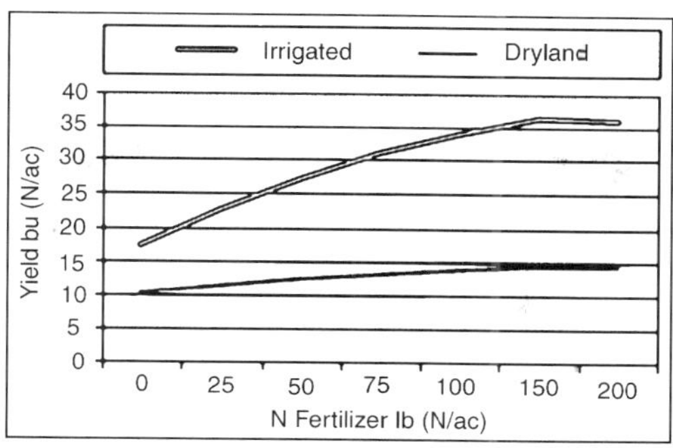

Fig. Canola Response to N and Moisture at Outlook, SK

Therefore, the amount and timing of rainfall are important. Studies in western Canada generally find that growing season precipitation increases grain yield two to three times more than equivalent amounts of stored soil moisture. Heavy N fertilization can reduce canola yields when excellent spring moisture conditions are followed by drought. Under this condition, the N stimulates larger leaves, increases transpiration and moisture use. As a result, soil moisture can be depleted, leaving little for flowering, podding and seed fill. Excessive N fertilization can also reduce yields by promoting lodging, delaying maturity that may increase fall frost damage, and increase foliar disease due to the dense canopy and lodging.

EFFECT OF NITROGEN FERTILIZER ON CANOLA QUANTITY

N fertilization generally increases the protein content of canola seed and meal. However, when N fertilizer is added under conditions of S deficiency, there may not be a protein increase but a rise in free amino acids due to hampered protein synthesis.

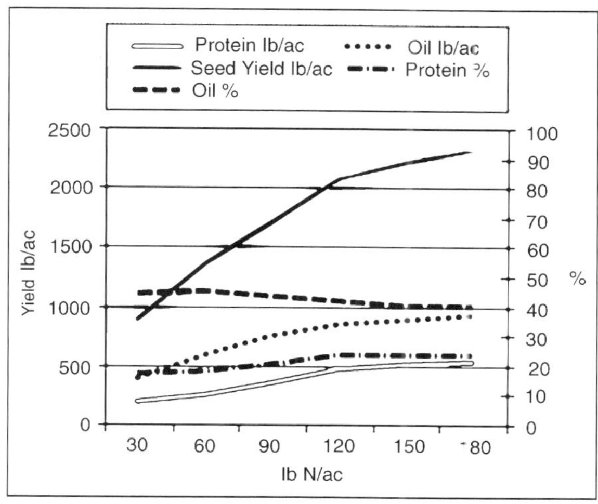

Fig. Canola Response to N and Moisture at Outlook, SK

In contrast, N fertilization may slightly decrease the canola seed oil content, especially at higher rates. Plant breeders report that oil and protein contents are often inversely related any attempt to change one causes an opposite change in the other component. Although the seed oil content may decrease at high N rates, the total oil yield per acre still increases because yield increases more than oil level decreases. The overall effect of N fertilization on yield and canola quality based on research in Manitoba is illustrated in Figure.

SUPPLY OF NITROGEN BY SOIL AND FERTILIZER TO CANOLA

Canola plants obtain N from several sources:

- Soil supply of nitrate and ammonium arising from soil organic matter decomposition and fertilizer residues of previous years
- Nitrate and ammonium released through mineralization during the current growing season
- Fertilizer or manure N additions to the current crop
- Other minor sources such as available forms deposited during rainfall and lightning, nonsymbiotic N fixation in the root zone, and ammonia uptake by foliage

Since canola plants obtain most of their N from the soil environment, understanding the N cycle is very important to managing N fertilizer effectively.

NITROGEN CYCLE AND TRANSFORMATIONS

Nitrogen is subject to many different processes in soil due to microbial activity and physical forces (Figure).

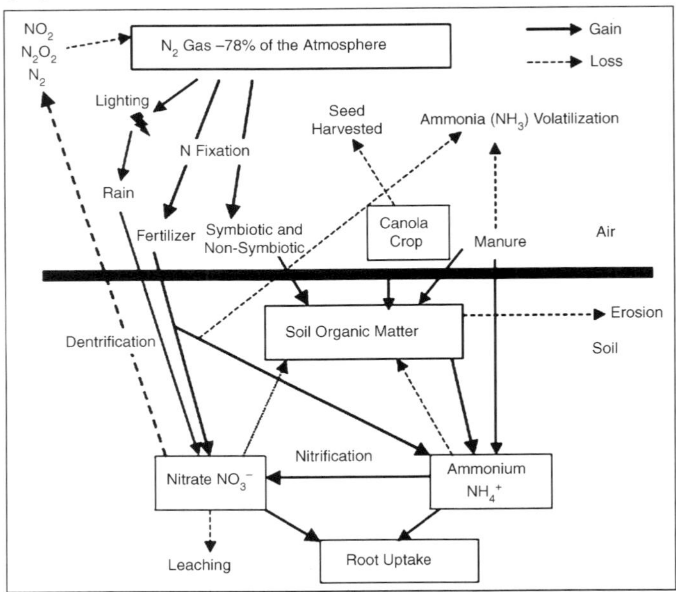

Fig. Nitrogen Cycle and Transformations

TRANSFORMATIONS THAT SUPPLY PLANT AVAILABLE NITROGEN

Soil organic matter is a major N reservoir containing several thousand pounds of organic N per acre. This large organic N storehouse needs to be decomposed by soil microbes before becoming available for root uptake. The decomposition process is called mineralization.

The decomposition rate is fairly slow and variable, ranging from about 0.4 to 5 per cent per year. As shown in Figure, organic N is first slowly changed to ammonium by bacteria. Then different bacteria rapidly change the ammonium to nitrate, in a two-step process called nitrification. The ammonium is first changed to nitrite (NO_2^-), then to nitrate. Under normal soil conditions, the bacterial oxidation of ammonium to nitrite is much slower than nitrite to nitrate. Therefore, very little nitrite is normally found in soil. This is fortunate since nitrite is toxic to plants. The result of these rate differences is that most of the plant available N in the soil tends to be nitrate. This is also why most soil testing labs analyse soil for nitrate and dont include ammonium.

Since N transformations result from soil microbial activity, soil conditions such as temperature, moisture and acidity will strongly affect the rates at which these processes occur. If soil is cold, saturated or very acidic, then mineralization and other microbial activities will proceed slowly. Cultivation stimulates organic matter decomposition and mineralization of N by improving aeration and physical mixing that gives soil microbes access to new organic matter supplies. There are several sources of plant available N to the soil system, in addition to soil organic matter decomposition and fertilizer additions (Figure). The atmosphere contains 78 per cent N_2 and thus is a huge source of N. But it first must be changed to plant available forms through biological or industrial processes called N_2 fixation. Biological fixation of atmospheric N_2 is performed by certain bacteria or blue-green algae species.

Biological N fixation in soil falls into three general types:

1. Symbiosis with legumes
2. Associative
3. Free-living N fixing bacteria

Symbiotic N fixation is much larger relative to the other types. Canola is not a legume and cannot form the symbiosis with rhizobia to fix atmospheric N. However, canola can benefit from residual N fixed by previous legume crops.

Associative N fixation occurs when bacteria just inside the root or on the root surface use root exudates for energy to fix atmospheric N. The plant benefits indirectly when the bacteria dies and its N is released through mineralization. The results of many experiments in Russia and throughout the world on cereal inoculation with associative N fixing bacteria have shown varying and unpredictable responses. This suggests that the interactions between plants and the bacteria are complex, unstable and can vary greatly depending on genotypes. Much less research has been done on canola inoculation with associative N fixing bacteria. The limited research does show that some strains will successfully colonize the roots of *Brassica* species but often do not significantly influence harvest dry weight or N accumulation.

TRANSFORMATIONS THAT REDUCE PLANT AVAILABLE NITROGEN

Several different mechanisms contribute to N loss from soil, and, therefore, lost opportunity for increasing yield. Understanding the conditions that promote such losses can be valuable in avoiding such conditions in the field and improving the N fertilizer efficiency.

DENITRIFICATION

One major N loss from soil occurs through a microbial process called denitrification. As shown in below Figure, nitrate can be changed by certain bacteria to gases such as N_2O and N_2, which escape back to the atmosphere. These soil bacteria have the ability to switch their respiration from using oxygen to nitrate. Since respiration is more efficient using oxygen, these denitrifying bacteria will only switch to nitrate if oxygen is absent, such as in waterlogged soil. Therefore, denitrification becomes significant when soils become saturated. Other secondary factors that encourage denitrification include carbon availability (crop residues), warm soil, and neutral to alkaline pH.

Denitrification accounts for 10 to 50 per cent of the available N losses in prairie soil. Research on the prairies has shown that considerable N denitrification losses occur during spring thaw. For example, research near Edmonton, AB found that 16 to 60 per cent of annual denitrification loss occurred immediately following snowmelt. During spring thaw on the prairies, frozen subsoil is often a barrier for water drainage and the overlying thawed soil becomes saturated. The second major period of denitrification loss occurs in late spring and early summer during rainfall events that cause soil saturation. Remember that denitrification occurs regardless of the nitrate source from fertilizer, manure or from decomposition. Effective N fertilizer management strives to avoid having large amounts of nitrate present during spring melts. Summerfallow is especially prone to large denitrification losses since large amounts of nitrate and moisture are stored during the fallow year, which increases the denitrification potential in the next spring thaw. Summerfallow also is a major contributor to leaching losses of nitrate.

IMMOBILIZATION

Immobilization is the second major N transformation that reduces the plant available N supply. Figure below indicates that soil bacteria may use either nitrate or ammonium for their own growth, temporarily tying up the N in the soil organic N storehouse. Immobilization essentially is the reverse of mineralization, and occurs when residues with low N content (like cereal straw) are being decomposed. Since these residues don't contain enough N for the microbes to make their own protein, they need to use the nitrate and ammonium. Soil microbes thus compete with plant roots for the available N, and plant growth suffers when N supplies are inadequate for both microbial and plant growth needs. The poor crop growth in heavy chaff rows is due in part to immobilization of N and other nutrients by the decomposing microbes. One effective fertilization strategy is to place the fertilizer away from residues and thus avoid immobilization losses. The remaining transformations that reduce plant available N are usually relatively minor.

LEACHING

Leaching of nitrate can occur since this form is not adsorbed to the soil and moves readily with soil water. Leaching losses can be significant in sandy soils in high rainfall areas or under summerfallow, but overall leaching probably contributes to less than 10 per cent of the available N losses on the prairies. To reduce leaching losses time fertilizer applications to avoid prolonged exposure to wet conditions, and consider band placement to delay the conversion to the vulnerable nitrate form.

VOLATILIZATION

Volatilization occurs when ammonia escapes from the soil to the atmosphere. Such losses happen in a variety of ways. One obvious loss occurs when anhydrous ammonia fertilizer is improperly applied (too shallow or into a too dry or wet soil). Broadcasting urea fertilizer on the surface without incorporation can also lead to significant volatilization losses if significant rainfall (more than 6 mm (1/4")) does not occur soon after application. All ammonium based fertilizers are subject to volatilization if broadcast on the surface of soils with high pH, surface lime salts, low soil organic matter, warm temperatures and dry conditions. To reduce volatilization loss, ensure proper fertilizer placement into the soil.

WEEDS

Weeds can contribute to poor N fertilizer efficiency by competing with crops for uptake. The competitive ability of the crop for fertilizer uptake can be improved by placing the fertilizer near crop roots rather than broadcasting or random banding.

EROSION

Erosion of topsoil carries significant N and other nutrients away from the field. Use soil conservation techniques to minimize such losses. A final minor loss mechanism occurs when ammonium is fixed into the crystal structure of certain clays. Some soils contain expanding type clays that allow ammonium to enter within the plates of the crystal structure and become fixed. Such ammonium trapped within the crystal lattice is held tightly and unavailable for root uptake.

SOIL PREPARATION

ROTATIONS AND CROP SEQUENCE

Crop rotation is an ancient agronomic practice compared to technology such as herbicides. Since rotation significantly increases yield, it is a cornerstone of farm management. The benefits to rotation are numerous, complex and still not fully understood. Studying rotation effects is difficult because long-term experiments are needed to exclude shortterm influences of weather, soil productivity and management changes.

Good crop rotation has a beneficial impact on weeds, diseases and insects. When a crop is grown continuously, pest populations adapted to that crop will increase.

Rotating crops tends to reduce pest build-up, especially immobile types. For example, rotations with high canola frequency have more cruciferous weeds such as stinkweed, but less grassy weeds such as green foxtail or wild millet (Table).

Table. Effect of Crop Rotation on Stinkweed at Scott, SK (Based on 1981, 1985 and 1990 Plant Counts)

Rotation	Stinkweed Plants/m^2 in Canola Phase
Fallow-canola	190
Fallow-canola-wheat	129
Fallow-canola-barley	27
Fallow-canola-barley-hay	23
Fallow-canola-wheat-barley-hay-hay	50

The impact of crop rotation on weeds is partly due to different herbicides available in canola compared to cereals. Alternate crops in the rotation allow different herbicides to be used and thus can provide better overall weed control in the long term.

Rotating crops and herbicides also helps to rotate herbicide groups and thus reduce the build-up of herbicideresistant weeds. The recent introduction of herbicide-tolerant canola systems (Roundup Ready, Liberty Link, Clearfield and Navigator) has improved weed control in canola and provided more options for rotating herbicide group use.

Similarly, crop rotation can lower the risk of diseases. Short rotation canola tends to have more problems with blackleg and seedling root rot.

Table. Effect of Rotation on Blackleg

Rotation	Blackleg (% of Plants with Basal Lesions)
Fallow-canola	43.5
Fallow-canola-wheat	9.5
Fallow-canola-barley	9.0
Fallow-canola-barley-hay	8.0
Fallow-canola-wheat-barley-hay-hay	4.0

Proper rotation with cereals will reduce canola disease, provided that volunteer canola and cruciferous weeds are controlled in the cereal break crops. Also, rotation with canola can reduce many cereal diseases such as common root rot, take-all and tan spot. Australian research has documented a significant reduction in take-all disease of wheat following canola, indicating that canola has a "biofumigation" effect.

Rotation also affects water use. Oilseed crops like canola and flax may use less water than cereals, and thus improve subsequent crop yields in dry areas. Water use is more efficient when crops with different rooting systems are rotated.

In a diverse rotation of cereals, oilseeds, pulses and forages, the different kinds of residue returned to the soil can improve soil fertility over the long term.

This fertility increase is due to differences in nutrient cycling, soil aggregation and complex interactions between plants, soil and microbes. Chemicals naturally present in plants and released during growth or decomposition can negatively or positively affect subsequent crops (this phenomenon is termed allelopathy).

Carefully plan crop rotation to avoid a serious build-up of disease, insects or hard-to-control weeds. Another factor to consider is volunteer growth from previous crops that may cause serious competition and seeding problems.

Herbicide carryover from previous crops may also adversely affect canola. No single crop rotation will suit all circumstances. The choice of which crops to grow, and in what sequence, depends to a large extent on the soil and climatic conditions of a particular farm and also on the grower's management skills. In areas with adequate moisture, rotations that include cereals and broadleaf crops maintain pests at lower levels and produce consistently higher yield. An added benefit is that yields are often more stable because crops differ in their response to stress at various stages.

Therefore, stress may damage one crop seriously, but others in the rotation may be less affected. Use the information in Table as a guide in selecting a field for canola.

Table. Guide to Field Selection for Crops

Crop Before Canola	Wait Period	Remarks
Barley,	None	No diseases in common with canola. Can be grown
canary seed, spring rye, fall rye, oats, wheat, triticale, winter wheat		the year before or after canola. Control stinkweed, cleavers, mustard, volunteer canola and other problem weeds in these crops. Consider potential herbicide residue carryover problems. A low probability of a volunteer problem for one year for canola on spring rye and triticale.
Buckwheat	One year	A high probability of a volunteer buckwheat problem exists for one year. A low probability of a damping-off and root-rot problem exists for one year. Where wild mustard was a problem in the buckwheat, a medium probability of a contamination problem for one year.
Canola	Three years	A high probability of a disease problem for sclerotinia for four years, blackleg for three years, white rust in B. rapa varieties for three years, alternaria for two years; and a low probability for root rot for two years and damping off for one year. If wild mustard was present, a medium probability of a contamination problem for several years. Consider potential herbicide carryover problems. A high probability of a root maggot problem for one year following a B. rapa crop.
Corn	Two years	Consider potential herbicide carryover problems. Soil test to a depth of 60 cm (24") to monitor nutrient levels and avoid over fertilization as corn is usually heavily fertilized.
Flax	One year	A high probability of a volunteer problem for one year. A low probability of a disease problem for root rot for two years and damping off and sclerotinia for one year.
Forage legume (alfalfa, clover)	One year	Consider potential herbicide carryover problems. A low probability of a disease problem for three years for sclerotinia, two years for root rot and one year for damping off. Available nitrogen may approximate summerfallow nitrogen levels if legume breaking is done before June 30. Perennial legumes have a relatively high sulphur requirement and may deplete soil supplies. Soil test to determine sulphur status.
Mustard	Two to	A high probability of a volunteer problem for one
	to three years	year, followed by a medium to low probability for two years. Mustard is a contaminant of canola, therefore, control volunteer mustard in the wait period. If wild mustard was present, a medium probability of a contamination problem for one year. A high

		probability of a disease problem for sclerotinia for four years, white rust for three years, and a low probability for root rot for two years and damping off for one year.
Potato	One year	Consider potential herbicide carryover problems. A low probability of a disease problem for three years for sclerotinia, two years for root rot and one year for damping off. If wild mustard was a problem in the potatoes, a medium probability of a problem for one year. Soil test to a depth of 60 cm (24") to monitor nutrient levels and avoid over fertilization as potatoes are usually heavily fertilized.
Pulse (pea, bean, lentil)	One year	Consider potential herbicide carryover problems. A low probability of a disease problem for three years for sclerotinia and one year for root rot and damping off.
Sugar beet	Three years	Cyst nematodes infest sugar beets causing root deformity, reducing production and sugar content. Nematodes may lie dormant in soil for several years as cysts (also attacks canola/rapeseed and mustard). Three- to four-year rotations among sugar beets, canola/rapeseed and mustard reduce nematode populations in the soil. A low probability of a disease problem for one year for root rot and damping off.
Sunflower	Three years	A high probability of a volunteer problem for one year. Consider potential herbicide carryover problems. A high probability of a disease problem for sclerotinia for four years, and a low probability for root rot and damping off for one year.

EFFECTS OF PRECEDING CROP ON CANOLA

Growing canola on canola stubble usually results in reduced yield compared to canola on cereal, pulse or flax stubble. Studies by Agriculture and Agri-Food Canada at Melfort and Aylsham, SK in the Moist Black soil zone have shown a large yield reduction in growing canola on canola (Table).

Table. Relative Yield of Crops Grown on Various Stubble Types at Melfort and Aylsham

Stubble	Peas	Flax	Canola	Wheat	Barley
% Yield of Check (Crop on own Stubble = 100%)					
Cereal	125	111	152	98	109
Other oilseed	114	109	177	131	138
Pea	100	142	196	147	152

Alberta Management Insights summarized the effect of stubble type on canola yields using crop insurance records. Three million acres of data are included in this study.

Table. Effect of Stubble Type on Canola Yield

Soil Zone	Stubble Type	1992-98 Avg. Yield % Canola on Canola	Range of Annual Avg.
Black (south-central)	Wheat	107	96-130
	Barley	115	99-127
	Canola	100	-
	Summerfallow	118	99-139
Black - Dark Grey (north-central)	Wheat	106	87-133
	Barley	125	109-158
	Canola	100	-
	Summerfallow	121	97-155
Dark Grey - Grey (north-central)	Wheat	138	94-172
	Barley	129	113-150
	Canola	100	-
	Summerfallow	134	112-157
Thin Black	Wheat	109	96-120
	Barley	115	103-131
	Canola	100	-
	Summerfallow	125	107-136
Dark Brown	Wheat	116	96-152
	Barley	119	97-164
	Canola	100	-
	Summerfallow	152	123-219
Dark Grey - Black (Peace region)	Wheat	103	94-109
	Barley	101	91-122
	Canola	100	-
	Summerfallow	110	105-119

In addition, similar data from Manitoba is summarized in Table.

Table. Relative Yield of Major Crops Sown on Selected Stubble Types in Rotation

	Stubble Type					
	Wheat	Barley	Oats	Brassica Napus Canola	Flax	Peas
Relative % Yield (Crop on Own Stubble=100%)						
Wheat	100	109	110	118	114	120
Barley	115	100	110	119	122	122
Oats	114	103	100	124	123	115
Brassica napus canola	114	115	117	100	118	128
Flax	148	148	146	133	100	**

In General

- Canola yield is generally better on cereal stubble than canola. The exception appears to be in the Peace River region, perhaps due to the longevity of established canola diseases there (blackleg and brown girdling root rot) and the prevalence of canola.

- In Alberta, canola on barley appears to be slightly better than on wheat except in the Peace and Dark-Grey zones. Although the increased canola yield on cereal stubble is less than found in scientific experiments, this may be due to limiting factors of weeds, diseases and fertility in some commercial fields.
- In Manitoba, canola on flax and pea stubble had the highest yields.
- Yearly yield variation in the Alberta study is very large, indicating a strong weather influence on rotation response.

ROTATIONAL IMPACTS

Growing canola on canola stubble promotes disease and insect build-up. Light disease or insect levels will usually do little damage in the first crop of canola. However, their build-up in that crop can result in significant damage and yield losses in following canola crops. Yield reductions may occur when canola is grown on other crops susceptible to the same diseases and insects, such as flax, mustard, sweet clover, soybeans, field beans, lentils and sunflowers. As shown in the above tables, canola does well on cereal stubble if volunteer plants are controlled. Also, if cruciferous weeds or volunteer canola were a problem in the preceding crop, disease and insect problems could persist into the canola crop. If the preceding crop is a hay or pasture crop, use an early partial fallow period to facilitate seedbed preparations in the next spring. This will allow some weed control and soil moisture build-up. Late summer timing of sod-breaking may limit subsequent canola production because of depleted soil stored moisture, poor fertility and weed control, and the difficulty in preparing a good seedbed with sod clumps.

Another rotational impact of other crops on canola is the potential for toxins that inhibit the growth and yield of subsequent crops (allelopathy). Research around the world has documented that leachates from wheat, alfalfa and some forage grasses can be toxic to canola seedlings. Recent Australian research has found that wheat varieties can differ in the toxicity of their residues to canola. Wheat residue phytotoxicity may partly explain occasional poor emergence and vigour of canola direct seeded into heavy wheat straw. To reduce residue toxicity, spread wheat straw and chaff evenly behind the combine, and in cases of extreme straw production, bale wheat straw. Growing semidwarf cereal varieties may also reduce straw load.

EFFECTS OF CANOLA ON SUBSEQUENT CROPS

Positive Impact

Canola is an excellent break crop for cereals. However, when pre-emergent herbicides are used in the canola crop, sow oats and small seeded grasses next. Some studies have found higher soil moisture and nitrogen reserves on canola stubble than cereal stubble. Cereal root and foliar diseases are usually reduced after canola. The reduction in cereal diseases arises because canola is not a host of cereal diseases and thus their field populations tend to decline.

Also, recent Australian research indicates that canola actually kills certain cereal pathogens such as take-all root rot due to release of inhibitory compounds such as isothiocyanates during decomposition. This direct disease reduction has been termed biofumigation.

Commercial field data from provincial crop insurance departments confirm cereal yield advantages after canola. The benefit of canola stubble on subsequent wheat yield has also been confirmed in Manitoba from crop insurance records.

Cereal yield response to canola stubble from these crop insurance records is lower than found in research experiments. This may be due to limiting factors of weeds, disease and fertility in some commercial fields. Rotations with canola have been shown to reduce year-to-year yield variability. In areas with adequate moisture for continuous cropping, a good rotation is cereal-broadleaf-cerealbroadleaf- for example, wheat-canola-barley-peas.

Negative Impact

The phytotoxic effect of canola residues has been documented in Canada by several researchers and is attributed to toxins leached from residues as well as toxins produced during microbial decomposition.

Phenolic compounds are considered to be the main class of phytotoxic compounds that reduce the growth rate of roots and shoots. Isothiocyanate, a breakdown product of glucosinolate, has been shown to reduce weed seed germination, especially with small weed seeds such as spiny sow thistle and flixweed. The phytotoxicity is greatest with fresh residues and young tissue (herbage).

Mature canola residue does not normally cause subsequent crop germination and growth reductions except for nitrogen deficiency due to immobilization by the decomposing residue and colder soil temperature. This is likely due to leaching of the phytotoxins out of the residues by rain and snow before the next crop is seeded.

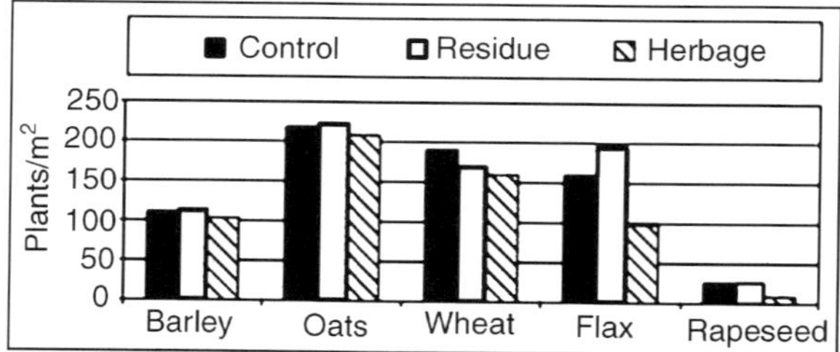

Fig. Effect of Incorporated Rapeseed Mature Residue or Fresh Herbage on Various Crop Plant Densities

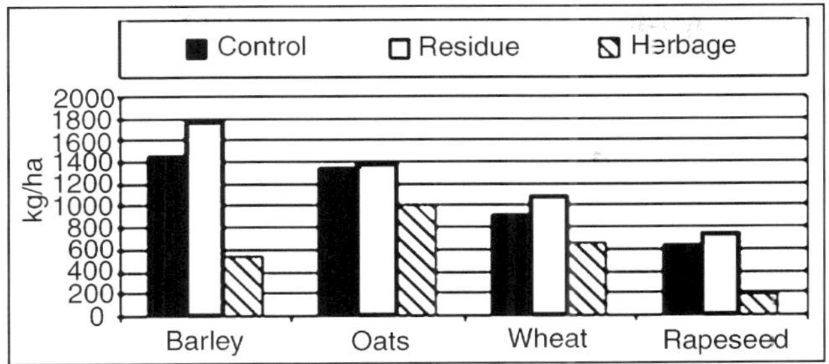

Fig. Effect of Incorporated Rapeseed Mature Residue or Fresh Herbage on Various Crop Yields

Table. Effect of Stubble Type on Barley and Wheat Yield in Various Soil Zones

Soil Zone	Stubble Type	1992-98 Avg. Yield % of Barley on Barley	1992-98 Avg. Yield % of Wheat on Wheat
Black (south-central)	Wheat	105	100
	Barley	100	120
	Canola	111	119
Black - Dark Grey (north-central)	Wheat	98	100
	Barley	100	127
	Canola	104	120
Dark Grey - Grey (north-central)	Wheat	115	100
	Barley	100	**
	Canola	111	**
Thin Black	Wheat	105	100
	Barley	100	106
	Canola	112	105
Dark Brown	Wheat	94	100
	Barley	100	107
	Canola	104	105
	Summer fallow	118	117
Dark Grey - Black (Peace region)	Wheat	93	100
	Barley	100	107
	Canola	122	128

Therefore, problems may occur in situations where thick volunteer canola is worked under in the spring before seeding flax or a cereal crop. Also, poor straw and chaff spreading with the combine may result in an N deficiency and toxicity problems in the chaff row for the next crop, especially if heavy volunteer canola growth occurs in the chaff row. Stress conditions such as an N deficiency and high temperatures can increase the allelopathic effect. The allelopathic nature of canola and other Brassica species have potential to be used in green manure systems for weed control. Canola may reduce beneficial organism populations such as rhizobia and vesicular-arbuscular mycorrhizal fungi (VAM) in following crops. These microorganisms do not colonize canola and thus their populations will decline after a canola crop. However, the breakdown products of glucosinolates may also reduce the populations with a result similar to the effect on take-all disease. A recent Australian study found that VAM colonization in wheat and flax was generally lower followingBrassica crops, but this did not negatively affect crop uptake of P or Zn, or affect crop yield. Further research is needed. An average canola crop provides less trash cover than an average cereal crop because it produces slightly less residue and breaks down more quickly. Recent research at the Agriculture and Agri-Food Canada (AAFC) Lethbridge, AB Research Centre found that crop residue losses during fallow were lentil>canola>rye>barley>wheat>flax. Lentil and canola residues often break down twice as fast as wheat. Thus, wind or water erosion risks increase with canola especially when it is followed by summerfallow.

When land is frequently cropped to canola, there is a slight risk that soil structure may deteriorate because of the short residue life. However, crop rotations with canola are much less damaging to soil quality than rotations with summerfallow. Such cropping practices lead to poor soil tilth and an increased tendency to crust after heavy rains, especially on soils with high clay and low organic matter contents (such as Grey-Wooded soils).

Studies at AAFC Melfort, SK and Beaverlodge, AB Research Centres have shown an advantage for a grain-forage rotation on degraded and Grey-Wooded soils. Several years of other crops like cereals and forages will build up or maintain soil organic matter due to crop residues with slower decomposition rates. This will promote better soil structure and tilth, and reduce soil erosion when a trash cover is maintained. Researchers at Beaverlodge have also reported that canola is a suitable companion crop for grass establishment.

Frequent canola production can also increase hard-tocontrol volunteers and weeds like stinkweed, cleavers, wild mustard and Canada thistle. Ensure fields considered for conventional (non-herbicide-tolerant) canola production are as free as possible of wild mustard, stinkweed, cleavers and cow cockle. B. napus varieties have little or no seed dormancy and usually pose little problem for long-term volunteering. However, seeds of B. rapa varieties, buried by deep tillage and exposed to certain temperature and moisture conditions, may develop dormancy similar to wild oats. The seed can stay dormant for many years then germinate, causing volunteer and contamination problems. One way to reduce this problem is to keep the seed on or near the soil surface to promote germination and mortality and prevent induced dormancy.

SEEDBED REQUIREMENTS AND PREPARATION

Canola emergence is greatly influenced by seedbed conditions. Good seedbed conditions are more important for canola than cereals due to the shallow planting depth necessary for this small-seeded crop.

A good seedbed will:

- Supply enough moisture for germination and seedling establishment
- Provide adequate warmth and aeration
- Have minimal physical resistance for the seedling to emerge
- Be relatively free of weeds and disease
- Offer some resistance to erosion

The wide range of soil characteristics, residue levels and weather across the prairies has caused a corresponding evolution of a wide range of tillage implements, seeding systems and ground openers.

CONVENTIONAL TILLAGE SEEDBED PREPARATION

Prior to the widespread adoption of conservation tillage or direct seeding, canola seedbed preparation involved numerous tillage operations for the following reasons:

- To bury previous crop residues that interfere with herbicide/fertilizer application or seed placement
- To control weeds which have germinated
- To place soil-applied herbicides or fertilizers
- To create a fine soil structure in the zone of seed placement that balances water infiltration and storage, and for adequate air movement

Although tillage can help prepare a good seedbed and control weeds, there are disadvantages.

Excessive or untimely tillage can:

- Deplete seedbed moisture
- Degrade soil structure that encourages crusting
- Contribute to soil compaction
- Create large lumps
- Increase soil organic matter losses and erosion

Unnecessary tillage adds unnecessary fuel, machinery and labour cost. Use just enough tillage with conventional tillage seeding systems to achieve a reasonably level, uniform, wellpacked, granular surface structure with a mix of granules in the 1 to 5 mm size. The small granules will provide good seed to soil contact for water absorption, while the larger granules will provide some wind erosion protection. Achieving this mix of soil granules without increasing the soil erodability is a challenge since granules smaller than 0.84 mm are susceptible to wind erosion. Leave enough residue on the soil surface to reduce wind and water erosion without interfering with seeding operations.

Several factors must be considered to achieve a good seedbed in conventional tillage systems:

- The tillage implement is suitable for the soil and residue conditions
- The soil moisture content is suitable ("timeliness")
- The implement is properly adjusted for depth and speed

Considerable experience is needed to select the most suitable implement, to properly adjust it for the correct uniform depth and to begin tilling at the right soil moisture content in order to achieve a granular seedbed rather than a powdery or lumpy one. Sandy soils are easily worked into a fine seedbed with minimal tillage. However, the workability of sandy soils leaves little margin for error-overworked sandy soils quickly become very fine structured and susceptible to water and wind erosion.

In contrast, clay-textured ("heavy") soils cannot be worked under wet conditions because large lumps or clods develop which prevent good seed to soil contact. Subsequent tillage to break up the lumps can pulverize the remaining soil and thus make the soil prone to crusting. Crusting of low organic matter clay soils (Grey-Wooded) is a major challenge for canola germination and establishment. Work clay soils at moisture contents slightly drier than field capacity-at this stage, the moist clay can be squeezed by hand into a pliable ball but no free water appears on the soil or hand.

Medium textured ("loam") soils are more forgiving than clay or sandy soils and are best worked when moist. Loam soils worked wet can still create clods, while excessive tillage can reduce them to a very fine structure that can crust or be vulnerable to erosion.

Ensure the spring tillage depth is shallow-2.5 to 5 cm (1 to 2")-since soils tend to dry out quickly to the depth of tillage. Canola is different than cereals. You can't seed to moisture if the top 5 to 7.6 cm (2 to 3") of soil have dried out. A common mistake is to cultivate deeply or an excessive number of times in the spring, then try to firm up the seedbed by several harrow/packing operations. This dries out the seedbed and often pulverizes the surface structure, creating a significant crusting and erosion risk. Grower surveys have reported that minimum shallow tillage (one to two operations) before seeding results in the highest yields on stubble.

A firm, well-packed seedbed will:

- Provide good seed to soil contact for moisture absorption during germination
- Retain moisture in the seed zone
- Provide adequate aeration
- Facilitate uniform shallow seeding depth

Excessive tillage can create loose, dry seedbeds that are difficult to firm up and susceptible to erosion. Ensure the seedbed is firm enough that footprints are not deeper than the thickness of the sole of a boot. Packing operations mainly reduce the granule and pore sizes in the surface soil, which will reduce moisture loss. However, do not pack the soil too much because the granules can be pulverized, restricting water infiltration, aeration and predisposing the soil to crusting and erosion. On-row packing during seeding is beneficial, especially for canola. The decision about how much extra packing should occur when seeding canola is a difficult one. Too little packing could result in poor emergence if dry conditions prevail after seeding while too much could result in erosion or crusting if wind or heavy rain follows seeding. This variable effect of extra packing on canola germination, establishment and yield has been shown in research trials. The Canola Council of Canada Crop Production Centres have conducted many experiments comparing post- and pre-packing prior to 1997 and found varying results. Generally, pre-packing is more desirable than post-seeding packing in conventional tillage systems because it will enable more uniform, shallow seeding. Post-seeding packing by rolling is occasionally used to firm up the seedbed and push down rocks. However, the extremely smooth surface left after rolling makes these fields very prone to wind erosion.

Fall tillage has declined dramatically over the past decade as producers strive to increase water capture from snow, retain surface residues for erosion control, and to lower fuel, labour and machinery costs. One fall tillage operation to place fertilizer, control weeds or work in heavy residue can still be beneficial if it avoids extra spring tillage that can dry out the seedbed. Fall fertilization also can take advantage of lower fertilizer prices and reduce the workload in the busy spring seeding period. If canola is planted on summerfallow, enough residue must be left on the soil surface to reduce erosion and crusting potential.

The reasons to summerfallow include:

- Weed control
- Soil moisture conservation
- Increased short-term nutrient availability
- Reduced residue-borne plant diseases
- Reduced risk of crop failure due to drought

Conservation summerfallow maintains sufficient plant residues on the soil surface to prevent soil erosion while controlling weeds and increasing stored soil moisture. Tillage operations are reduced in number or intensity, and are replaced with herbicides. By using residue-conserving practices, adequate cover can be maintained through the fallow period until the next crop is sufficiently established to protect the soil from erosion. Use a minimum residue cover of 1,513 kg/ha (1,350 lb/ac) to protect most soils from serious wind or water erosion.

This is roughly equivalent to the residue left after harvesting a wheat crop yielding 785 kg/ha (14 bu/ac) of grain. In practice, most cereal fields will have crop residues that exceed this level. Residue levels are reduced through natural decomposition (sunlight, oxidation, and microbial activity). Canola residue breaks down about twice as fast as wheat, and this is why summerfallow after canola is not a wise practice. Residue cover declines after each tillage operation.

Table. Residue Left on the Soil Surface after Various Tillage Operations

Tillage Implement	% Residue Left after One Pass	% Residue Left after Four Passes
Wide-blade cultivator	90	60-65
Chisel plow with low-crown shovel	85	40-45
Chisel plow with normal shovels	80	35-40
Chisel plow with normal shovels plus mounted harrows	60	10-15
Heavy tandem or offset disc	35-65	5-15
Moldboard plow	0-10	0

Tillage operations can be managed to maintain surface residues while preventing serious erosion. Residue left standing will help trap snow and increase spring soil moisture. Usually 45 per cent of soil moisture conserved in an 18- month fallow period is received over the first fall and winter. By trapping snow more effectively, more soil moisture can be conserved. This increased moisture conservation can reduce the need for summerfallow and allow more stubble cropping.

CONSERVATION TILLAGE SEEDING SYSTEMS

Conservation tillage seeding systems (direct seeding, no-till or zero-till, minimum or reduced tillage) aim to improve or maintain soil quality and conserve soil moisture. The development of these tillage-seeding systems has been one of the major changes in agriculture during the past two decades.

The major advantages of these conservation tillage systems are:

- Less soil erosion by wind and water due to retention of surface residue
- Maintained or increased soil organic matter contents
- Increased soil microbial and faunal populations
- Increased soil moisture storage and infiltration rate
- Improved soil tilth
- Reduced N and S leaching losses
- Reduced root diseases
- Reduced salinization
- Reduced overall machinery investment
- Reduced labour needs
- Reduced energy requirements
- Comparable to better yields and net returns

In contrast, the following disadvantages of conservation tillage have been raised, although many have been resolved through new technology or management practices:

- Inadequate and expensive seeding equipment for direct drilling into heavy residue conditions
- Poor weed control
- Lower spring soil temperatures that reduce and delay seedling emergence
- Increased foliar disease from residue borne inoculum
- Poor fertilizer efficiency due to placement difficulties and increased N denitrification with higher soil moisture
- Delayed seeding in spring due to high soil moisture
- Increased surface soil compaction
- Greater management skills needed since fewer alternatives are available for weed control and fertilizer application
- Increased herbicide usage
- Poorer yields and net returns, particularly in wet years

Direct seeding is more flexible than no-till since some tillage can solve immediate weed problems and deal with high moisture and heavy clay soil conditions. In direct seeding, soil is not tilled in the spring before seeding to conserve seedbed moisture. Any fall tillage performed must leave the soil surface compact and level to preserve soil moisture. Most of the crop residue is retained on the surface with at least half the stubble remaining upright and anchored to trap as much snow as possible. Typical operations are fall fertilizer banding with knives, and redistributing crop residue and incorporating herbicides with heavy or rotary harrows.

The amount of soil disturbance during direct seeding varies with the type of opener. With low soil disturbance direct seeders, less than 40 per cent of the soil surface is physically worked by the openers to form the seedbed furrow. Some soil from the opener's action may be deposited between furrows, giving the appearance of more soil disturbance. Low soil disturbance can be expected from 75 cm (3") wide openers spaced at 22.5 cm to 30.0 cm (9 to 12"). Soil firmness, moisture conditions and planter speed affect the amount of soil disturbance. Low disturbance direct seeding systems are very much like no-till systems except that some tillage options remain available in direct seeding.

High soil disturbance direct seeders disturb more than 40 per cent of the soil surface. If fall tillage was done, then spring seeding occurs in loosened soil and most of the surface is disturbed. Wide ground openers that overlap will disturb the entire soil surface to some degree. Sweep openers produce high disturbance. They give varying degrees of weed control, so a pre-seeding herbicide application may not be needed. However, they may stimulate weed growth since weed seeds and volunteer seeds from the previous crop will be incorporated into moist soil. High disturbance openers may require additional seedbed finishing to cover the seed and to improve weed control. In no-till or zero-till systems, seeding is the only operation that disturbs the soil. Only 25 to 35 per cent of the soil surface is disturbed, just enough to place the seed and fertilizer into a seedbed. No-till is similar to low disturbance direct seeding except that direct seeding systems allow some tillage to deal with unusual conditions.

No-till aims to minimize soil disturbance and maintain as much crop residue cover as possible because:

- Low disturbance reduces soil moisture loss
- Weed seeds are less likely to survive and grow on the undisturbed soil surface
- Crop residue cover protects soil from wind and water erosion
- Standing stubble traps snow

Residue management is a crucial aspect for successful conservation tillage seeding.

Successful residue management needs to consider various factors, including:

- Crop residue amounts and condition, particularly green, lodged or damp straw
- Capability of the seeding machine and other implements to clear through the crop residue without plugging or "hairpinning"
- Combine or swather cut width compared to the spread width of straw and chaff behind the combine
- Alternative uses for straw when residue is plentiful
- Weed control methods

In direct seeding systems, begin residue management at harvest with a wide and even spread of straw and chaff behind the combine. Extra operations after harvest to manage heavy residue are time consuming and costly. Most new combines are equipped with good straw and chaff spreaders or are easily adapted with after-market units. Converting older combines to a better spreading system is usually more difficult. The cost, horsepower needs, type of drive and spread width varies between the different after-market units.

Residue clearance is the ability of seeding equipment to allow crop residue to pass through without bunching. Ground openers and shanks are shaped to prevent dragging and subsequent bunching of straw residue. The opener must prevent chaff residue from falling into the seed furrow and causing poor seed cover and poor furrow closing. Critical dimension is the distance between two points in a machine where plugging with straw is likely to occur.

The most common locations for plugging in the seeder are:

- Between the underside of the shank (may be the spring trip supporting mechanism) and the soil surface
- Between one ground opener and the next
- Between a ground opener and a wheel or some other adjacent structural member

The critical dimension may vary with the amount of crop residue on the surface, the moisture content of the residue or even the air humidity. Damp straw plugs the seeder quite easily. Fluffed up straw plugs more easily than straw lying on the soil surface. Dry straw on a warm, breezy day will pass through a seeder while damp straw may not. Changing speed and direction of travel may help a seeder to clear crop residue.

The straw handling performance of many seeders may be improved by modifying the location where plugging most often occurs. The space between the opener and "plug point" may need to be increased, sometimes greatly, to prevent plugging. Four-rank cultivator units on air-drills will plug less than three-rank units.

DIRECT SEEDING EQUIPMENT

Air seeders and air drills have become common machines on both conventional and conservation tillage farms. An air seeder uses a medium or heavy-duty cultivator, a central pneumatic seed and fertilizer delivery system and a ground opener for seed and/or fertilizer placement.

This system offers many options and adaptations to meet a variety of conditions. The seeder's mainframe is carried and controlled by wheels inside the frame. Levelling (fore-aft) is controlled by caster wheels in front of the frame (floating hitch type). This method of depth control is superior on land with sharp hills or gullies. Install seed row finishing equipment on the rear of the air seeder. Separate soil finishing passes may be needed over the seeded field to ensure good seed placement depth.

An air drill is an adaptation of the air seeder. The main difference is that air drills do not have wheels inside the frame carrying the ground opener hardware. Machine support and depth control comes from dedicated packer wheels on the rear of the drill. The front is carried and controlled by forward caster wheels as with any floating hitch cultivator.

An air drill has all the advantages of an air seeder including:

- Good seed depth control
- Wing-up convenience for transport
- A central seed and fertilizer metering system
- Excellent field efficiency and capacity

It also has two main advantages over an air seeder—the relatively constant packing force delivered by each on-row packer, and the increased residue clearance made possible by the absence of inside-the-frame wheels. One key aspect of direct seeding units is ground opener selection. The ground opener is the part that penetrates the soil to place seed and fertilizer. It has a soil-breaking wear point, a soil dividing body, and delivery tubes to guide seed and fertilizer to the furrow bottom. Also, deflecting surfaces may be present to guide soil back around the fertilizer bands and seed rows.

A ground opener must:

- Leave at least 15 mm (0.6") of soil between the fertilizer bands and seed rows
- Not allow seeds to fall in a concentrated fertilizer zone
- Create a good soil structure (fine aggregates) in the seed zone
- Have low draft requirements and resist wear
- Scour well in moist and high clay content soils-many openers tend to build up with soil, causing the furrow opening to be too large resulting in the seed not being covered with sufficient soil
- Leave the soil surface smooth enough for subsequent operations like crop spraying and harvesting
- Adequately "blacken" the soil surface over the seed row if there is a concern about soil temperature for seed germination

Opener performance is influenced by:

- Soil moisture content
- Soil texture
- Soil density
- Seeding depth
- Forward speed

Choosing ground openers to suit the conditions on a farm will be difficult but manageable. Seed must be placed into moist soil and surrounded by soil particles small enough to reduce open spaces in the seedbed. Coarse soil lumps in the seedbed increase soil moisture loss and reduce seed germination. For good seed-to-soil contact, ensure the ground opener either causes very little soil movement so that the soil profile is fractured very little, or causes enough agitation to create soil particles small enough to fall in around the seed. The latter case may require harrowing to spread the soil over the seed row. Soil cover depth is measured after the last implement passes over the seeded field.

Every ground opener design creates its own particular flow of soil around it. Therefore, each opener design results in a different furrow opening, seed placement and soil cover. The furrow from a specific opener is affected primarily by the soil's clay and moisture content. Additional passes with equipment using on-row packing wheels, harrows, or other soil levelling and packing equipment, will change the soil cover depth above the seed. Soil must be repacked around and above the seed to prolong seed contact with moist soil. Packing reduces moisture loss from the seedbed by creating a denser soil layer at the surface with fewer large air spaces.

A fine balance often exists between packing enough to reduce moisture loss and packing too much, which in some soils may promote crusting that hinders seedling emergence. Adequate packing is achieved when all the soil lumps are crushed both around and above the seed.

Most ground openers require a packer to close and pack the soil in the furrow to create a good seedbed. The packer's shape and width must conform to the furrow and the location of the seed underneath. A direct, minimum disturbance seeder requires an on-row packer. A wide sweep opener that cuts the full width of the seedbed requires a harrow, rod-weeder, packer or a combination of several systems to finish the seedbed. Fertilizer must be placed near enough to the seed to supply nutrients for good early growth but not too close for crop safety. Too much fertilizer placed too close to the seed can cause injury to the germinating seeds, resulting in reduced crop emergence.

In a double-shoot system (Figure), the soil buffer is a zone between the fertilizer band and seed row where there is neither seed nor fertilizer. The opener, since it is placing both the seed and the fertilizer, must leave a soil buffer of at least 1.3 to 2.0 cm (0.5 to 0.75"). To achieve this buffer width, the spacing between the centres of the fertilizer and seed outlets has to be at least 5.0 cm (2"). The soil buffer zone can be lost if the scatter of seed and fertilizer increases when planting in clay or wet soil, from travelling too fast and from too much fan speed on an air cart. Seed that lands in the fertilizer band may not germinate or the seedlings may emerge later.

The greater the number of tasks the opener must perform, the more complicated the opener and the seeding operation become. Complicated ground openers are usually more sensitive to varying soil conditions.

SINGLE AND DOUBLE-SHOOT SYSTEMS

The terms "single shoot" and "double shoot" refer to how material (seed and fertilizer) is delivered by the planter to the ground opener.

The differences are:

- A single-shoot system has only one delivery line going to the ground opener. The line carries seed and possibly some granular fertilizer.
- A double-shoot system has two lines going to the ground opener. These may be two airflow lines or an airflow line and a liquid fertilizer or anhydrous ammonia (NH_3) line.

Fig. Single Shoot Opener

Fig. Double Shoot, Paired-row Opener

Most often, a double-shoot system has a ground opener, which opens two separate furrows so the seed is placed in one furrow and the fertilizer in the other. This type of fertilizer placement is called double-shoot side banding. The fertilizer band is usually placed deeper than the seed, usually about 2.5 cm (1"). Seed is placed either in one row, above and to the side of the fertilizer band, called single side banding, or in two rows above and on both sides of the fertilizer band, called paired-row double shooting (Figure).

Fig. Double Shoot, Single Side-band Opener

A semi-dependent opener (Figure) is a variation on the double-shoot, single side-band opener. It forms distinct furrows and leaves a seed row slightly narrower than the opener. The seed opener follows behind and slightly to the side of the fertilizer point, ensuring that soil covers the fertilizer band before the seed is placed. There is little chance of fertilizer and seed mixing. The advantage of the semi-dependent opener is that the seeding depth more closely follows the land contours.

Fig. Semi-dependent Opener

SELECTING AND USING GROUND OPENERS

All commercially available openers work under some conditions, but few, if any, work well under all conditions. How does a grower know if a particular opener design will work well under the conditions on the farm? Begin by discussing ground opener performance with neighbours to learn more about options suited to the conditions in the area. Next, determine which openers will likely meet the requirements for seed and fertilizer placement, handle the amount of residue cover that usually exists on the farm, and meet other requirements specific to the operation (such as providing some weed control). Seedbed utilization (SBU) is an important consideration. Install one or more of these openers on the seeder, try them in several typical conditions on the farm and assess the results.

Ground opener design is often influenced by soil and crop residue conditions in the area where the opener was developed. If these conditions are similar to the grower's farm, chances are better that the opener will do a good job. Therefore, ask the manufacturer about the conditions in the area where the opener was developed.

Opener comparison tests have been conducted by government engineering research agencies. For example, the Alberta Farm Machinery Research Centre at Lethbridge, AB (now called the Ag-Tech Centre) has conducted performance tests on a wide variety of ground openers. "Testing of Double Shoot Openers" (AFMRC Report 721) provides comparative data on seed band depth, seed band width, fertilizer band width, spacing between the fertilizer and seed bands, opener wear, draft, power requirements and other characteristics for 15 ground openers. Check with provincial departments of agriculture for information applicable to your area.

Although it takes experience and perseverance to make ground openers work in direct seeding, many growers have developed successful direct seeding systems.

Here are some tips for ground opener use:

- Check the planter's adjustments in each new field and in areas of the field where conditions are very different. Compromising here can be costly in terms of stand establishment and often in yield.
- No particular opener works well on all soil textures. If the farm has a variety of soil textures, openers may have to be changed to suit certain fields.
- Double-shoot openers may require frequent adjustment of depth and forward speed to ensure good placement of seed and fertilizer.
- When soil conditions are very moist, particularly in fine clay soils, ensure the seed row furrow is sealing adequately to preserve seedbed moisture.
- When assembling components from several sources, pay special

attention to ensuring good mechanical arrangements of the planter. Poor assembly causes costly downtime and repairs.

PRE-SEEDING WEED CONTROL

The reduction in tillage with direct seeding results in a heavier reliance on herbicide weed control. Winter annual and early spring germinating weeds will compete strongly with canola and, therefore, pre-seeding herbicide application (burn-off) is a common practice. If pre-seeding herbicides are not applied, the initial weeds become large and often pass the growth stages needed for good control by in-crop herbicide applications. Each direct seeded field must be scouted for weed emergence-scouting must be done on foot, and sometimes on hands and knees, to find and identify weed seedlings. Although there are no published threshold numbers of emerged weeds for preseeding herbicide applications, experience has shown that growers tend to underestimate the tiny weed seedlings present before seeding in early spring.

The best time to control annual weeds with a non-residual herbicide is usually just before seeding. Spraying too early before seeding can allow new weed seeds to germinate before the crop emerges. In conditions of heavy weed infestations and low soil moisture, it may be necessary to spray early to stop soil moisture depletion. Spraying after seeding can be effective, but there is a risk that bad weather could prevent or delay the burn-off herbicide application, and lower crop yields. Also, soil particles on weed leaves from the seeding operation may reduce herbicide performance.

Several effective pre-seeding weed control herbicides are available. Glyphosate is the active ingredient in most preseeding burn-offs, and is sold under several different brands and formulations (for example, Roundup Original, Victor, Renegade, Roundup Transorb, Touchdown, Touchdown iQ and Glyfos).

The reasons for glyphosate's popularity in burnoff sprays are:

- No herbicide residues to harm seeded canola emergence and growth
- A wide range of annual weeds is controlled and most perennial weeds are suppressed
- Herbicide rotation options are increased with minimal resistance concerns
- Cost is very affordable

Phenoxy herbicides such as 2,4–D and MCPA are also used occasionally in pre-seeding herbicide mixes. Research has shown that significant canola stand thinning and yield loss can result if dry conditions occur between burn-off and crop emergence.

CANOLA RESPONSE TO CONSERVATION TILLAGE SEEDING

Several research reports have been published on the response of canola to different tillage seeding systems. However, firm conclusions are difficult to draw from the research for several reasons. First, some of the research was conducted on plots that were previously in conventional tillage systems and, therefore, results may not indicate the outcome on long-term direct-seeded fields. Secondly, direct seeding equipment has evolved significantly and early research will not reflect these advances. Finally, herbicidetolerant canola systems have been recently developed and widely adopted which improve weed control and thus may affect the performance under direct-seeded systems.

WEED CONTROL

In addition to equipment concerns, weed control is often stated as a major obstacle that hinders adoption of directseeded systems. Initially, a change to direct seeding was expected to bring more weed problems, especially perennials and wind disseminated species. For example, at the Agriculture and Agri-Food Canada Scott, SK Research Centre a 12-year study comparing zero tillage (ZT) with conventional tillage (CT) in two rotations found that ZT only increased yield by increasing soil moisture where weed control was adequate. Yield decreases with ZT were associated with poor control. However, during this period, suitable post-emergent herbicides to control annual broadleaf weeds such as stinkweed, wild buckwheat, lamb's quarters and red root pigweed were not available.

At the Agriculture and Agri-Food Canada Beaverlodge, AB Research Centre a study compared CT, ZT and reduced tillage (RT) on clay Grey-Wooded soils from 1989-1991. As this short study progressed, there was a trend of relatively greater weed density under ZT and a shift in species composition. The weeds that increased in density under ZT canola were stinkweed, wild oat, smartweed, field horsetail and dandelion while wild buckwheat decreased. However, canola yields were not significantly reduced under ZT compared to CT.

In contrast, other recent research on the prairies has found that clear changes in weed communities do not generally occur with adoption of conservation tillage seeding. Weed communities are more influenced by locations, rotations and years than by the tillage system. For example, an Agriculture and Agri-Food Canada study at Rycroft, AB from 1989-1993 compared the impact of three tillage systems (CT, RT and ZT) on the weed population during early crop growth. The study found that the relative contributions to the size and diversity of weed flora are likely to be greater by common species under CT and by rare species under RT and ZT. No consistent increase in the weed population occurred with time under all three systems. Similarly, a study by Agriculture and Agri-Food Canada at three Saskatchewan locations from 1986-1990 did not find any increase in perennial and annual grass weeds with ZT. Weed community changes were influenced more by location and year than by tillage system.

Potentially, certain weeds can proliferate under direct seeding if management is not careful-examples are dandelion, narrow-leaved hawk's beard and foxtail barley. Given the different environment found in direct-seeded fields, it should not be surprising that different weed species with adaptation to residue-covered habitats will become dominant over species commonly found in conventional, cultivated fields. But research indicates that weed shifts are manageable by careful attention to rotations, herbicide selection and timing (especially pre-harvest glyphosate for perennials). Herbicide-tolerant canola systems have recently added effective options for improved weed control in directseeded systems.

SOIL MOISTURE AND MOISTURE USE

Many studies in western Canada and world-wide have reported higher soil moisture under conservation-tillage seeding compared to conventional tillage. The improved soil moisture under conservation tillage is due to reduced evaporation from residue-covered soil as well as increased infiltration rates. Higher soil moisture will often improve germination, emergence, early crop growth, moisture use efficiency and yield. For example, in the 12-year study by Agriculture and Agri-Food Canada at Scott, comparing ZT with CT in two rotations with fallow, canola and wheat, yield increased with ZT where spring soil moisture was increased. In the 36 comparisons over three rotation phases, ZT increased spring soil moisture in nine cases and there were no decreases. Yield was increased in nine cases but decreased in three, and moisture use efficiency increased in six cases with two decreases. The Beaverlodge, AB tillage study found that ZT and RT increased soil moisture in the top 10 cm (4") in dry periods. In excessively wet years, canola growth, moisture use efficiency and yield can suffer under direct seeding.

SOIL TEMPERATURE

Spring soil temperatures are often cooler under conservation-tillage seeding systems than conventional tillage. Most of the western Canadian research studies have found that residue covered soil is 0 to 2°C cooler (daily average temperature) than cultivated soil. In some cases (sunny days), temperatures during midday can vary by 5°C. The colder soil is due to increased soil moisture (water is slower to warm up than air) and more heat reflectance by the residue. Although emergence is sometimes several days longer with direct seeding compared to conventional tillage, the delay usually disappears after canopy closure. Increased soil moisture usually improves yield in spite of the initial cooler temperatures. In most cases, the cooler soils will not hamper final crop stands or yield. Direct seeders often can seed shallower (which is warmer) due to better moisture, and this largely compensates for temperature differences. However, experience has shown that excessively wet springs can negatively affect canola growth and yield, partly due to temperature effects. Also, frost damage to canola seedlings has been more severe on residue-covered fields when frost followed a warm sunny day. The greater injury was likely due to lower heat radiation in the critical early morning period under residue-covered soil compared to bare soil. Pay very close attention to achieving uniform residue spreading, preferably with the combine, to reduce cold temperature problems and enable good ground opener performance and seed placement. Research in the Peace River, AB region by Agriculture and Agri-Food Canada found that a narrow strip of bare soil over the seedbed can overcome most of the cold temperature and excessive moisture disadvantage of direct seeding in unfavourable situations.

CANOLA YIELD UNDER CONSERVATION TILLAGE SEEDING SYSTEMS

Research conducted on the prairies has reported variable success with conservation tillage seeded canola-ZT or direct-seeded canola has yielded less, the same or more than conventionally seeded canola. The many changes in direct-seeding technology and variable weather effects makes it difficult to apply some of the past research findings to the farm level.

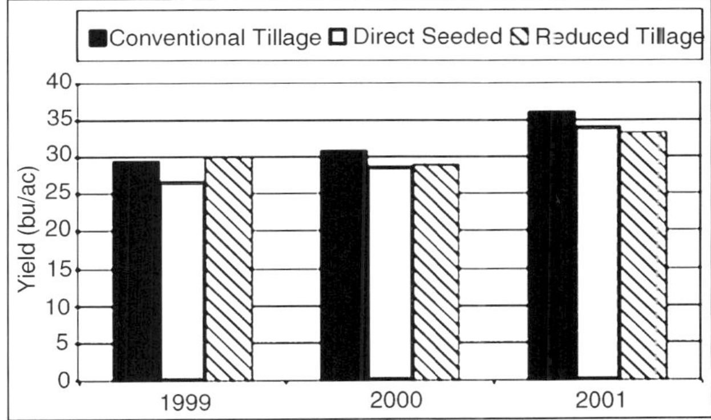

Fig. Canola Yield Comparisons Between Tillage and Direct-Seeded Systems in the Black Soil Zone of Alberta

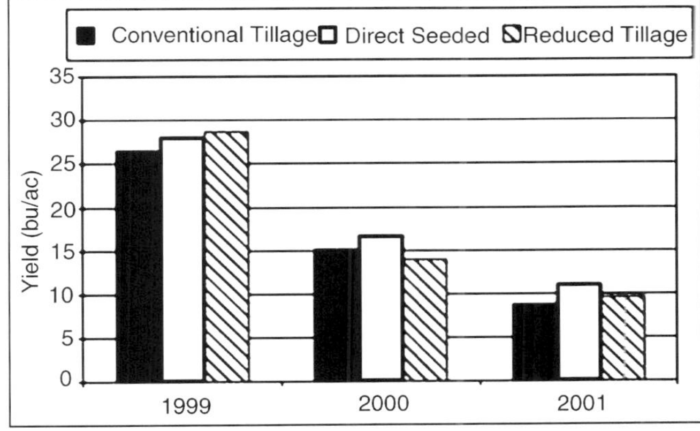

Fig. Canola Yield Comparisons Between Tillage Systems in the Dark Brown Soil Zone of Alberta

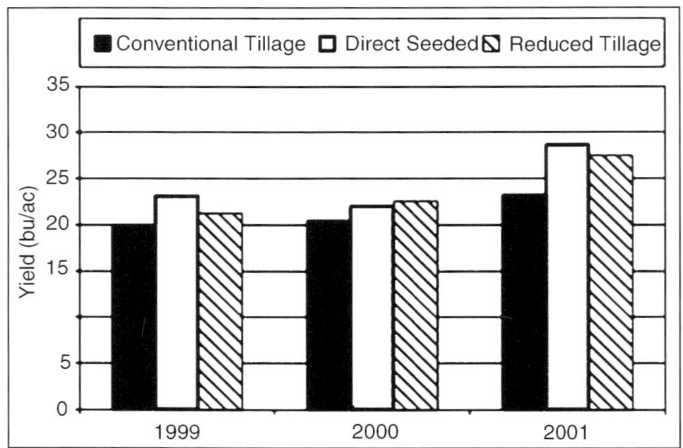

Fig. Canola Yield Comparisons Between Tillage Systems in the Peace River Region - Dark Grey and Grey Soil Zones of Alberta

Perhaps the best indication of canola performance under direct seeding compared to conventional and reduced tillage is from hail and crop agency records. Three years of recent records from the Agriculture Financial Services Corporation in Alberta shows that direct seeding and reduced tillage produced 109 per cent and 108 per cent yield of conventionally seeded canola on average in the major canola growing areas.

Based on 4.6 million insured canola acres seeded on stubble over these three years, 29 per cent, 43 per cent and 28 per cent were conventional tillage, reduced tillage and direct-seeded, respectively. This shows that conservation-tillage seeded systems are more popular now than conventional-tillage systems. Figures illustrate the yield comparisons between tillage systems reported in the major canola growing soil zones in Alberta.

NET RETURNS OF DIFFERENT TILLAGE SYSTEMS

Compared to yield, there have been fewer studies that investigated canola net returns seeded with the various tillage systems. Such studies are difficult since there are numerous combinations of machinery and, therefore, production costs can vary dramatically.

Many other confounding variables, other than the tillage seeding system, can also significantly affect the net returns. While direct seeding systems can reduce labour, fuel and some equipment costs, herbicide and other equipment costs may increase. With the widespread adoption of air-drills, the equipment cost has become less of an issue.

Alberta Agriculture Food and Rural Development conducted a survey of 185 growers in 1994 and 1995 to assess the short-term economics of conservation-tillage practices. Based on growers' costs and returns, partial budgets were compared between the systems. Machinery fixed costs were estimated through mathematical formulae.

The main report findings were:

- Zero-and reduced-tillage systems, on average, may have slight economic advantages over conventional-tillage systems.
- Zero-and reduced-tillage systems had marginally better contribution margins, and returns to land, labour and management on average, compared to conventional-tillage systems.
- Contrary to expectations, herbicide costs did not vary consistently between the tillage systems.
- Fuel, repair and depreciation costs increased as the tillage intensity increased. While this is an economic advantage for zero tillers, the more expensive machinery needed for direct seeding tends to offset this.

Overall, conservation tillage systems are suitable for canola production. The higher seedbed moisture can encourage better canola emergence. However, heavy residue fields can create significant problems for good canola seed placement, and increases the risk of frost mortality. Canola seedlings are sensitive to seed placed fertilizer, therefore, ground openers need to be chosen carefully. Further information on conservation tillage practices can be obtained from provincial and federal soil conservation agencies.

SOIL NUTRIENTS INTRODUCTION

The strong canola yield response to S fertilizer on deficient soils has been well established in western Canada. Under extreme S deficiency, canola response to S fertilizer is dramatic.

However, canola response to S fertilizer varies greatly, depending on:

- Soil sulphate levels (amount, spatial and temporal distribution)
- Availability of other nutrients (especially N, P and possibly boron)
- Soil moisture
- Amount, type, and method of S fertilizer applied

Since canola absorbs S from soil as sulphate, the soil sulphate content affects the yield response to applied S fertilizer. Researchers in the 1960-70s established a relationship between soil sulphate level and canola yield. Water soluble soil sulphate was found to be a good measure of available S for canola growth.

Canola generally responded to S fertilizer when the sulphate content to 60 cm (2') was less than 22 to 34 kg/ha (20 to 30 lb/ac). However, soil testing to determine sulphate content and the likelihood of yield response to applied S fertilizer has not been consistently successful.

For example, two field experiments in Manitoba in 1964 failed to show a rapeseed yield response to sulphate fertilizer on deficient soil while two Manitoba experiments in 1969 did find a response. The inconsistent S fertilizer response on deficient sites may be due to limiting amounts of moisture and other nutrients, or due to sulphate rich layers below the sample zone but within rooting depth, or due to S mineralization from organic matter. For example, many studies have found a significant interaction between N and S. Good yield response to either nutrient requires adequate supply of the other. An example of the N-S interaction is shown in Figure from research conducted by the AAFC Melfort, SK Research Centre in 1999. The need for balanced N and S nutrition is obvious without S, additions of N fertilizer lowered yield.

Field and controlled environment studies in southern Alberta found that rapeseed could utilize sulphate from a depth of 54 to 72 cm (21 to 28"). Subsoil sulphate salt layers are common in Brown, Dark Brown and Black soils on the prairies, and could affect the yield response to S fertilizer on fields testing low for S in the surface soil.

In contrast, S responses have been reported on fields with adequate soil test S levels. Recent research has measured high variability of sulphate contents across farm fields, which creates difficulty in obtaining representative samples. For example, Figure shows extreme sulphate variability on a 24 ha (60 ac) solonetzic field near Stettler, AB in 1994. Composite samples were taken from each 0.5 ha (acre) and tested separately. The average worked out to be 1,076 kg sulphate/ha (960 lb sulphate/ac) in the (60 cm (2') depth, which is excessive from the soil test standpoint. However, the average value is heavily skewed by the few samples with extremely high sulphate values (maximum 21,280 kg sulphate/ha (19,000 lb sulphate/ac). A better indicator of the most typical value for the field would be the value class that occurs most frequently (the mode). In this example, the mode was just 9 kg sulphate (20 lb sulphate), which is deficient.

Therefore, the majority of this field would likely respond to S fertilizer. This example illustrates that single composite soil samples from fields with high S variability can be difficult to interpret. A soil testing deficient for S is likely deficient (unless underlain by a subsoil sulphate salt layer), while soils testing medium or high for S may have deficient areas that are skewed by areas with excessive S.

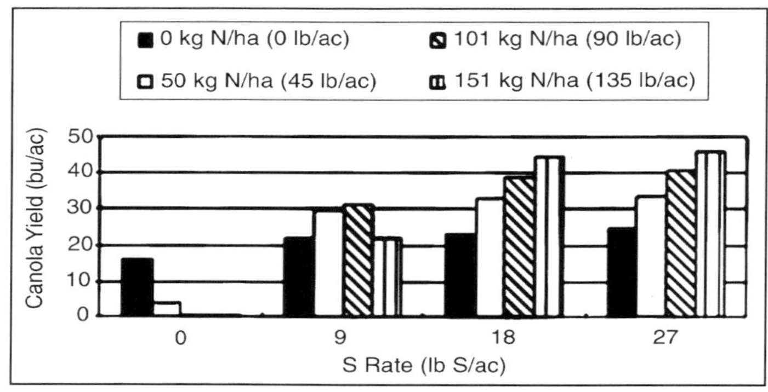

Fig. Interaction of N and S on Canola Yield at Porcupine Plain, SK

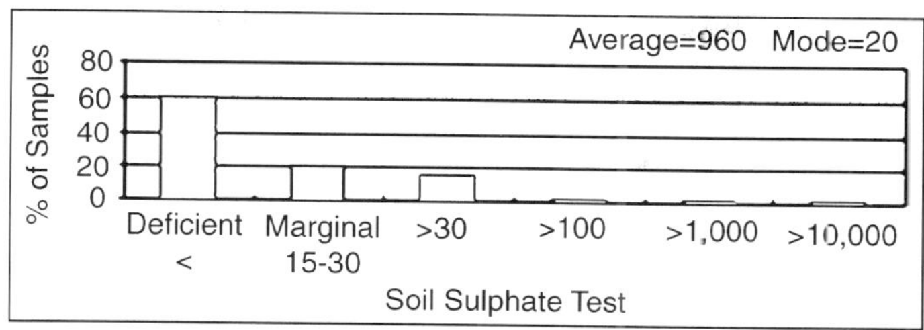

Fig. S Variability

SULPHUR FERTILIZER EFFECT ON CANOLA QUANTITY

Sulphur fertilization can affect canola quality (oil, protein and glucosinolate content). The balance between N and S has a strong influence on quality as well as yield. Significant effects on quality are usually found when S fertilizer is applied to extremely deficient soil, or when excessive S fertilizer is applied.

Sulphur fertilization can increase protein content of the meal (Table), which is desirable, but can also increase glucosinolate contents (undesirable). Glucosinolates increase with excessively high S fertilizer rates, but are usually well below the standard canola quality limit (currently 30 micromole per gram).

Oil content responses to S fertilization are inconsistent studies have reported decreases, increases or no effect. No satisfactory explanation has been given for the conflicting responses of S fertilizer on oil quality.

SULPHUR SUPPLY FROM THE SOIL

The organic portion of the sulphur cycle in soil is closely tied to N due to their association in protein. Like N, the main S reserve in soil is in organic matter. Although there is considerable variability in the relative proportions of carbon, N and S (C:N:S) in soil organic matter, the ratios are quite similar for each soil group. In a study of Saskatchewan farm soils, the C:N:S ratio ranged from 58:6:1 in Brown soils, to 63:7:1 in Dark Brown, 83:8:1 in Black, 100:8:1 in Gray Black, and 129:11:1 in Gray soils.

Many sulphur transformations in soil are analogous to Nâ□" each undergoes mineralization from organic matter, immobilization, oxidation and reduction of inorganic compounds. The soil S cycle is illustrated below in Figure.

A key component of the soil S cycle for plant growth is the mineralization path. Soil organic matter and plant residues are decomposed by soil microbes, releasing sulphate. Like N, the S mineralization rate is quite slow, and cannot match the uptake rate of growing plants. Also like N, the sulphate amounts released from residues will depend on the S content. When plant residues contain more than about 0.15 per cent S (C:S ratio about 300:1), there will be a net release of sulphate through mineralization. Below 0.15 per cent S, decomposition is slower and there will be immobilization of soil sulphate by soil microbes. The ability of soil to mineralize sulphate from organic matter has been found to be independent of the total amount of C, N or S, and of the C:N or N:S ratios in soils. However, research has found that the initial amounts of sulphate mineralized from soil is closely correlated with the initial amounts of N mineralized in short-term incubation.

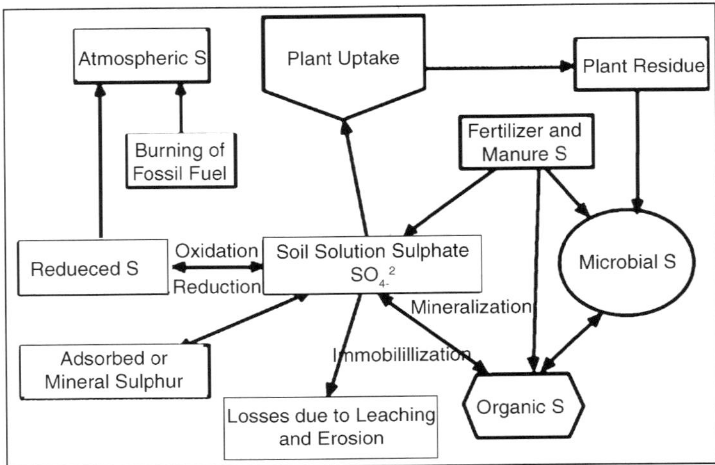

Fig. Soil Sulphur Cycle

Another important aspect of the soil S cycle is the oxidation path. In soils, sulphides, elemental sulphur and thiosulphate can be oxidized to sulphate by various soil microbes, but the main actors are bacteria from the genus Thiobacillus. The oxidation of these inorganic S compounds produces considerable sulphuric acid. Sulphur oxidizing bacteria are most active under warm, moist, well aerated conditions. It is the oxidizing ability of these bacteria that permits the agricultural use of elemental S for crop growth.

Although S reduction is shown in the soil S cycle diagram, it generally is not significant in aerated agricultural soils. In flooded soils, sulphate can be reduced by soil microbes to sulphides in a process analogous to denitrification. However, soil microbes will utilize nitrate, iron and manganese compounds before reducing sulphate.

In many western Canadian soils, there is a subsoil salt (gypsum) and/or lime (calcium carbonate) layer. This subsoil layer contains considerable sulphate, often as coprecipitates with lime. Although this subsoil sulphate solubility is reduced, it still can contribute to plant needs if it exists within the rooting zone. However, the length of time that canola grows in S-deficient topsoil before rooting to the subsoil S will affect the yield response to fertilizer S. Also, the depth to subsoil S tends to vary greatly across the field. Total S amounts (organic and sulphate) generally increase from upper to lower slope positions. In most prairie soils, sulphate is not held by organic matter and clay particles since they are both negatively charged. Therefore, sulphate is vulnerable to leaching losses.

Table. Effect of N and S Fertilizer on Canola Oil, Meal Protein and Glucosinolates (Trials in NE SK 1980-83)

N Rate		S Rate		Oil	Meal	Glucosinolates
kg/ha	lb/ac	kg/ha	lb/ac	Content %	Protein %	Micromole/g
0	0	25	22	46.1	48.4	9.5
40	36	10	9	46.7	47.2	7.9
40	36	40	36	46.5	48.1	10.1
97	89	0	0	43.9	48.9	7.3
97	89	25	22	45.0	49.9	9.2
97	89	49	44	45.5	49.7	9.1
160	143	10	9	43.7	50.7	6.7
160	143	40	36	44.3	51.9	10.1

SULPHUR FERTILIZER MANAGEMENT

The optimal method and timing of S fertilizer depends on the fertilizer form sulphate versus elemental forms.

Sulphate S

Sulphate fertilizers are highly soluble and will move easily with water in the soil. Ammonium sulphate (21-0-0-24) is a common sulphate-based fertilizer. Liquid ammonium thiosulphate (12-0-0-26) is used less frequently, and requires a short time period for oxidation to sulphate. Highest fertilizer use efficiency generally results when sulphate fertilizer is placed near roots for easy access, and just before the period of plant uptake. Under dry spring conditions, broadcast sulphate fertilizer can be stranded and result in poor uptake. However, under such dry conditions, canola germination and establishment will also be severely affected. Under average to good moisture conditions, sulphate fertilizer can be broadcast-incorporated in the spring with good results. On sandy soils, sulphate leaching can occur during wet periods. Therefore, apply fertilizer just prior to crop needs.

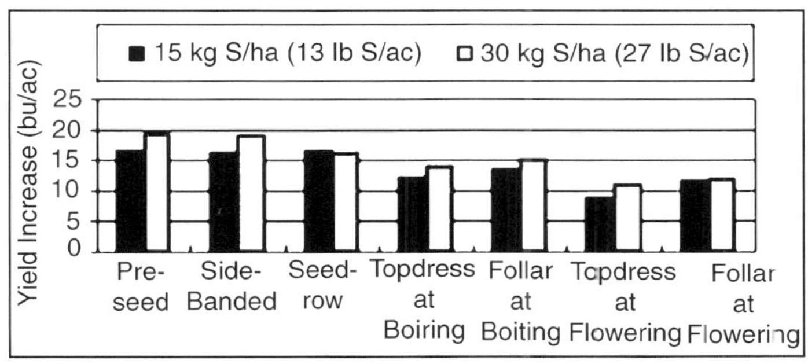

Fig. Canola Yield Increase from Sulphate S Fertilizer Applied at different Growth Stages

Although sulphate fertilization just prior to or at seeding is best, post-seeding applications can be effective. Research has found that sulphate fertilizer can be soil applied up to the rosette stage with good response provided there is sufficient rain to wash the fertilizer into the root zone before bolting. Foliar sulphate fertilizer solutions can be applied up to the bolting stage with moderate effectiveness, and can serve as a rescue treatment for deficient fields that were not adequately fertilized at or before seeding. However, the best yield response to sulphate S fertilizer is at or before seeding as illustrated in Figure from six site-years of recent research at AAFC Melfort, SK Research Centre.

In the above research, seed row placement of 30 kg S/ha (27 lb S/ac) increased yield less than side banding or preseed incorporation, probably due to seedling toxicity. This agrees with other research, indicating that only limited amounts of sulphate S fertilizer can be safely applied near the seed. The safe amount will vary with the degree of seedbed utilization, moisture conditions, soil type and amounts of other fertilizer nutrients. Until research has determined the safe amount of seed row S fertilizer under various combinations of the above factors, limit seed-placed ammonium sulphate. Include N amounts in sulphate fertilizer when calculating total N fertilizer amounts to be safely seed placed.

Elemental Sulphur

Sulphur fertilizer containing elemental sulphur must be managed differently than sulphate-based fertilizer to achieve good efficacy. Elemental sulphur has advantages of a ready supply in western Canada, low production and transportation costs, and fewer drill fill operations due to high analysis. However, elemental sulphur has a significant disadvantage availability is delayed until soil bacteria oxidize it into the sulphate form. The conversion rate from elemental sulphur to sulphate depends on the particle size, the degree of dispersion in the soil, and the growing conditions for the bacteria (moisture, temperature). Common elemental sulphur fertilizers are formulated as granules or pastilles (split pea shape) for ease of shipping and handling, each consisting of thousands of individual particles. The surface area of these individual particles is the access where the soil bacteria feed converting the elemental sulphur to sulphate. Small particles have the largest surface area and the fastest oxidation rate.

Under western Canadian conditions, research indicates that particles less than 150 microns in size will convert quickly if well mixed with soil. Some elemental sulphur fertilizers have particles consistently smaller than 150 microns such as Sulpher 95. Other products consist of a mixture of particles ranging from smaller to larger than 150 microns such as Tiger 90CR In addition, some products such as iger 90CR contain bentonite clay that swells when wet and helps to break the granule into the small particles. Granules that break down readily and completely will allow quicker oxidation and sulphate availability. Research in western Canada has found that elemental sulphur granules break down the greatest when applied to the soil surface and exposed to rain/snow and frost. Subsequent tillage will then further disperse the degraded granule. In contrast, band and seed row placement, or immediate incorporation following broadcasting, will reduce the granule dispersion and the oxidation rate. This is illustrated in Figure which is data taken from a recent lab experiment by the Agronomy Unit of Alberta Agriculture, Food and Rural Development.

The strong influence of elemental sulphur placement and timing on sulphate release subsequently affects canola yield. This effect is evident in Table which is based on research by AAFC at Melfort, SK. The effect of sulphur fertilizer type, placement, and application time on canola and wheat yield was studied. All S was applied at 20 kg S/ha (18 lb S/ac) in the first year only to canola and wheat plots. To measure residual response over three years, canola was seeded in the second year onto wheat plots grown with S treatments in the first year. No S was applied in year two. Canola was seeded in the third year onto wheat plots from year two, again without S fertilizer. Ammonium sulphate and ammonium thiosulphate corrected S deficiencies in the application year and provided residual response over the next two years. Broadcast elemental S fertilizers did not relieve S deficiency in the year of application, but broadcast Tiger 90 did provide a response equivalent to ammonium sulphate two years later. Pre-seed banded or seed-placed elemental S did not correct S deficiency, even two years later. Continuing research shows that fall broadcast elemental S fertilizer left on the surface until incorporation next spring will improve the oxidation and canola yield response compared to spring broadcast. However, this practise should still be initiated two years ahead of the canola crop to ensure consistent response.

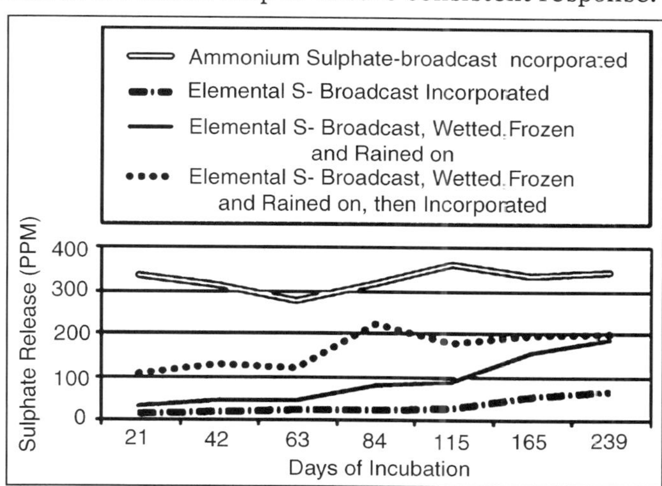

Fig. Effect of Elemental S Granule Exposure to Weathering and Sulphate Release

Another factor that influences the oxidation rate is previous use of elemental sulphur in the field. Exposure to elemental sulphur in the past has been shown to increase oxidation rates of subsequent applications, probably due to stimulation of the S-oxidizing bacterial population.

However, repeated applications of elemental S fertilizer can negatively impact agricultural soil. Saskatchewan research on Gray Wooded soil found that repeated elemental sulphur application decreased soil pH, organic C, and microbial biomass. Negative effects on soil enzymes involved with nutrient transformations were also found. Although short term crop response to elemental sulphur fertilizer is usually less than with sulphate sources, there may be residual benefits of elemental sulphur due to the slow release. While a residual benefit has been found with forages, conflicting results have been found with canola. Sulphate generated from elemental S after the period of annual crop uptake (May to August) may not benefit the soil S status in subsequent years since it is susceptible to leaching during snowmelt. Conditions most likely to exhibit residual benefit from elemental S fertilizer would be combinations of sandy, low organic matter soil, sloping topography, and high rainfall episodes after fertilizer application but before major uptake has occurred.

Effective use of elemental S fertilizer requires:

- Careful consideration of the specific product's particle size
- Application method and timing
- Severity of S deficiency
- Soil leaching risk
- Field history of elemental S use

Table. Effect of S Fertilizer Form and Placement on Canola Yield

Treatment	Yield (bu/ac)		
	1996	1997*	1998*
Check	21	16	11
Spring broadcast Tiger 90	18	12	25
Spring broadcast elemental S (99%)	23	6	11
Spring broadcast ammonium sulphate	37	31	24
Spring dribbled ammonium thiosulphate	34	31	26
Spring pre-plant banded Tiger 90	16	7	12
Spring pre-plant banded ammonium sulphate	36	31	28
Seed-placed Tiger 90	21	10	18
Seed-placed ammonium sulphate	37	10	26

The most consistent response to elemental sulphur fertilizer will be achieved by surface broadcasting the granules, allowing time for granule breakdown by rain/frost/snow, then mixing the particles with soil by tillage. Therefore, apply elemental S fertilizer at least the fall before seeding canola. In some cases, elemental sulphur fertilizer application needs to be two years before seeding canola. Adding some ammonium sulphate to the seeding fertilizer blend may still be wise insurance. Elemental S fertilizer is not advisable when S is only added to the canola rotation phase.

Several other S fertilizer forms are available, but not widely used. Elemental sulphur and ammonium sulphate has been blended into a single granule and research indicates the canola response is intermediate between sulphate and elemental S. Gypsum [$CaSO_4$ 2(H_2O)] is available from mines but its low analysis (about 17 per cent S) limits it transportation and use. Gypsum is also a by-product of phosphate fertilizer production, but contains impurities that prevent it from being land applied. Gypsum is only slightly soluble and this limits its usefulness as an immediate sulphate source, being intermediate between ammonium sulphate and elemental S. Gypsum has the added advantage that the calcium helps promote good soil structure, which could be a benefit on soils prone to crusting. Potassium sulphate (K_2SO_4) is another fertilizer S source more widely used in the tobacco, fruit and vegetable industry. Potassium sulphate should perform similarly to ammonium sulphate.

NITROGEN:SULPHUR RATIO

A proper N:S balance is important for canola production. When N is in excess (high N:S ratio), there is insufficient S to combine with the N to make protein, and thus non-protein N accumulates. A useful guideline is to add N and S fertilizer in a 7:1 ratio, which is approximately the ratio needed by the canola plant. There has been research into using the N:S ratio during tissue testing to determine S status. However, the N:S ratio of canola tissue has not proved reliable for predicting S status. The N:S ratio only indicates the relative proportions of N and S in the plant, and does not indicate their actual magnitudes. Therefore, if canola tissue tests show an optimal ratio of 7:1, there are three possibilities: both N and S levels are optimal, excessive, or deficient.

At the rosette stage, tissue testing canola for S status should include several criteria to improve the reliability:

- % S greater than 0.25 per cent
- N:S ratio of 10 or less
- A sulphate:total S ratio (as indicated by hydriodic acid reducible S:total S) greater than 0.38

Fig. Sulphur Sufficient Seedlings (left) and Deficient Seedlings

Fig. Leaf Cupping due to S Deficiency

Fig. Leaf Purpling due to S Deficiency

Fig. S-sufficient Flowering Plant (left) and Deficient Plant

Fig. Stem Leaf Cupping due to S Deficiency

Fig. Pod Purpling due to S Deficiency

CALCIUM (CA)

Calcium is a macronutrient absorbed in relatively large amounts by canola. However, deficiencies in western Canada are rare due to ample soil reserves. Calcium is often referred to as a secondary nutrient, probably due to uncommon deficiencies and non-specific roles in the plant.

Role of Calcium

Calcium performs several roles in the plant. In contrast to other macronutrients, a high proportion of Ca is found as a structural component in cell walls. Calcium structural function is to provide stable but reversible molecular linkages. Pectins are calcium compounds in cell walls that strengthen the wall and contribute to tissue resistance against fungal and bacterial infections. Calcium also plays a fundamental role in membrane stability and maintains cell integrity. This membrane protection is important under low temperature or saturated soil stress. Calcium bound at membrane surfaces can be exchanged with other cations (such as K^+, Na+ and H+). Calcium exchange with sodium (Na) at membrane surfaces is a main factor in salinity stress. Also, Ca replacement with Al^{+3}(or blocking of Ca channels) is a factor in aluminium toxicity in acid soil.

Cell extension requires Carapidly growing parts are, therefore, most affected by Ca deficiency. Root extension, shoot elongation and pollen growth are dependent on adequate Ca. The secretion of mucilage by root caps (that help root tips penetrate through soil) also needs Ca. Downward root growth (gravitropic response) relies on adequate Ca in the root caps. Callose formation is another example of a process involving Ca. In response to injury, cells will produce callose instead of cellulose, which helps wounds to heal and reduce infection.

Most plant Ca is present in leaf vacuoles where it likely contributes to the cation-anion balance. Calcium also stimulates a range of enzymes, but generally is not a constituent of enzymes. Calcium plays a key role in plants as a secondary messenger in turgor regulated processes such as stomata opening and closing.

Characteristics of Calcium

Canola roots mainly absorb calcium as the Ca^{+2} cation dissolved in the soil water. Plant available Ca also exists as exchangeable Ca adsorbed on soil organic matter, silt and clay surfaces. The amount of dissolved Ca^{+2} depends on the amount of Ca containing minerals, the soil cation exchange capacity and soil pH. High pH soils (>7.5) usually contain the highest Ca due to significant amounts of precipitated Ca salts (lime and gypsum). Since Ca is absorbed out of the soil water, the dominant processes controlling the supply to roots are mass flow, diffusion and root interception. Therefore, Ca availability is dependent on adequate soil moisture.

The Ca content varies between different plant parts and ages, ranging from 0.2 per cent to 5 per cent. The highest Ca contents are found in old leaves. At maturity, only about 10 per cent of plant Ca is found in the canola seed.

Canola Response to Calcium Fertilizer

Ca deficient soils are rare in western Canada. For example, of 352 plant samples submitted to Enviro-Test Laboratories in 1997, only three were below critical levels used in the U.S. and Australia. These criteria may not be applicable to western Canada. Calcium deficiency is possible in acid, saline or solonetzic soils due to low Ca^{+2} levels relative to other competing cations (H+, Al^{+3}, Na+ and Mg^{+2}). In addition, there are conditions where Ca containing compounds increase crop growth indirectly, either through amelioration of soil acidity or improvement of soil structure. These aspects are discussed in the soil acidity and solonetzic soil sections.

In spite of the lack of scientific evidence for Ca deficiency in western Canada, some companies promote Ca fertilizer, claiming that exchangeable Ca reserves are not readily available. Recent research in Alberta and western Canada using seed treatment with Ca_5S did not find a response in canola at nine locations. Therefore, Ca fertilization is not necessary for canola production in western Canada.

ZINC (ZN)

Zinc is a metallic micronutrient that is commonly deficient in many countries, including Australia, China and India. In western Canada, sporadic Zn deficiencies have been identified in fields of alfalfa, flax and beans, but not canola.

Role of Zinc in the Canola Plant

Zinc exists only in the Zn^{+2} form in plants, and is not involved with redox reactions. Zinc has an ability to form complexes with N, O and particularly S and performs catalytic and structural roles in enzymes. Many enzymes contain Zn as a structural, catalytic or cofactor component. Protein synthesis, hormone (auxin) and carbohydrate metabolism also require Zn. Membrane stability also relies on Zn, and the most obvious Zn deficiency symptoms (such as leaf chlorosis and inhibited stem elongation) probably arise from membrane breakdown. Pot experiments with canola have reported Zn deficiency symptoms ranging from purpling on new emerging leaves, brown spots on cotyledons, interveinal chlorosis and cupping of leaves.

Uptake of Zinc by Canola

Weathering of soil minerals is the primary source of plant available Zn^{+2}. Weathering removes Zn faster than other metals except for Cu. Zn deficiency commonly occurs in acidic, highly weathered soils (typically tropical). Zinc deficiency may also occur in high pH, calcareous soils due to Zn adsorption to lime particles. Plant available Zn exists as exchangeable Zn^{+2}, dissolved Zn^{+2} in soil water, adsorbed Zn to Mn oxide and organically bound Zn. Soil test labs often use a critical level of 0.5 ppm DTPA extractable Zn, but research in western Canada has found this level too high for predicting cereal response and that DTPA is an unsuitable extractant. No further work has been conducted to find a more suitable extractant and calibration for cereals and oilseeds on the prairies. The mechanism for Zn uptake by roots is not well understood and may involve both active and passive processes. Once absorbed by the roots, Zn is likely complexed with small organic molecules similar to other metallic micronutrients.

Zinc availability increases as soil pH decreases (becomes more acidic). Copper and other cations compete for root Zn uptake. High P levels can induce Zn deficiency by inhibiting Zn translocation within the plant rather than affecting root uptake.

MICRONUTRIENTS

There are other micronutrients and beneficial elements than those discussed above. However, deficiencies of the remaining nutrients are limited to certain plant species other than canola or deficiencies are extremely rare anywhere in the world.

Chlorine (Cl) is found in abundance in nature as chloride salts. Choride (Cl^-) is highly mobile in soil and plants and is readily absorbed by roots. Chlorine is involved in photosynthesis, charge balance, enzyme activation, stomatal regulation and disease resistance. Cereal responses to Cl fertilizer has occurred on the prairies, apparently due to disease suppression or improved water relations rather than Cl nutritional needs. Canola responses to Cl have not been reported.

Nickel (Ni) has recently been established as an essential nutrient. In higher plants, urease is the only known Ni containing enzyme. Other Ni roles include Fe absorption, seed viability, N fixation and reproductive growth. The plant available form is Ni^{+2}. Root uptake likely follows similar patterns as other micronutrient metals. Nickel appears to be readily mobile in both xylem and phloem. Ni deficiency in field grown crops has not been reported.

Silicon (Si) is the second most abundant element in the earth's crust and is a beneficial nutrient for a few wetland plant species such as rice. In non-wetland species, Si can counteract Zn deficiency induced by high P. Since Si is so abundant in nature, proving its essentiality is very difficult. Silicon may affect plant stability by influencing lignin biosynthesis as well as through deposition in cell walls. Increased leaf rigidity has been reported in cereal and cucumber crops. Silicon may also contribute to disease and insect resistance. Silicon may decrease toxicity from high levels of Mn, Fe and Al.

Sodium (Na) is an essential nutrient for some plant species that use the C_4 photosynthetic pathway. Canola uses the C_3 pathway and, therefore, Na is not a beneficial nutrient for this crop.

MAGNESIUM (MG)

Of all the macronutrients, magnesium is absorbed in the least amount. Magnesium deficiencies are rare on the prairies, similar to Ca. However, Mg has more specific roles in plant function than Ca.

Role of Magnesium in the Canola Plant

Magnesium is the central atom of the chlorophyll molecule, and depending on the Mg sufficiency level, up to 25 per cent of the total plant Mg is bound to chlorophyll. Magnesium is also needed for protein synthesis, and to activate many enzymes such as glutathione synthase, carboxylases, phosphatases, and ATPases. Most of the plant Mg is contained in cell vacuoles where it serves as a reserve for the metabolic pool and contributes to cation-anion balance.

Characteristics of Magnesium Uptake by Canola

Magnesium uptake by canola is very similar to that of calcium. The plant available form is the Mg^{+2} cation that exists in the soil water and as exchangeable Mg adsorbed on soil organic matter, silt and clay surfaces. The amount of dissolved and exchangeable Mg will depend on the extent of Mg containing minerals, soil pH and cation exchange capacity. Soils slightly acidic to neutral in pH tend to have the highest available Mg levels. This is due to Mg being a weak competitor for exchange sites on soil colloids and root binding sites.

At higher pH and in soils with free lime, Ca^{+2} will dominate the exchange sites, possibly inducing Mg deficiency. In contrast, at low pH, H+, Al^{+3}, and manganese (Mn^{+2}) under flooded conditions, will dominate the exchange sites, inducing Mg (and Ca) deficiency. High K+ or NH_4+ levels in the root zone due to fertilization may also induce Mg deficiency, although it is often short term. High Mg levels relative to Ca can induce a Ca deficiency. This has been reported in barley grown on solonetzic soil in Alberta with high Mg/Ca ratios. Magnesium level is highest during early vegetative growth (about 0.5 per cent), and declines with maturity. Stems and roots contain the least Mg. At harvest, straw Mg level is less than 0.2 per cent while canola seed contains about 0.3 per cent Mg. Roughly 1/3 to 1/2 of the above ground Mg is contained in the seed [about 7 or 8 kg/ha (6 or 7 lb/ac)].

Canola Response to Magnesium Fertilizer

Little information exists on Mg requirements of canola. There are no known cases of canola response to Mg fertilizer in Canada or Europe. Therefore, no reliable soil or canola tissue test criteria exist. However, Mg deficiency appears unlikely in slightly acidic to neutral soils that do not have excessive amounts of root zone Ca, K or NH_4+.

MICRONUTRIENTS

Micronutrients are those nutrients required in extremely small quantities (less than 100 ppm in plant dry weight). Unfortunately, the basic functions of micronutrients are less understood than macronutrients. Also, there is very limited knowledge about the forms and mechanisms of micronutrient transport in the xylem and phloem.

Micronutrient deficiencies in canola are much less common than macronutrient deficiencies. However, canola yields can be severely depressed when micronutrient deficiency occurs. This section will review the various micronutrients and canola responses.

COPPER (CU)

Knowledge about copper fertility of western Canadian soils has increased over the past two decades. Previously, Cu deficiency was thought to be limited to organic soils. The recent research has identified that Black, transitional Gray-Black and Dark Brown soils may be Cu deficient for cereal production.

Although copper deficiency and fertilizer response has been documented with cereals under field conditions, canola has not been shown to display deficiency symptoms or respond to Cu fertilizer.

Role of Copper in the Canola Plant

Copper is a transition element that forms stable complexes in the plant and soil, and is capable of electron transfer (energy processes). Copper role in plant functions is mainly as a reactive constituent of enzymes that catalyze oxidation-reduction reactions.

Some examples of Cu containing enzymes include:

- Plastocyanin (needed for energy capture through photosynthesis)
- Superoxide dismutase (needed for detoxification of oxygen radicals)

- Many different types of oxidases (enzymes that degrade or change compounds together with oxygen)

One of the phenol oxidases is involved with lignin synthesis.

Due to coppers role in photosynthesis, deficiency leads to low carbohydrates levels, at least during the vegetative stage. The low carbohydrate content in Cu deficient plants contributes to impaired pollen formation and fertilization. The reduced lignification in Cu deficient plants also affects pollen fertility since lignification of anthers is needed to release pollen.

Copper Supply from the Soil and Uptake by Canola

Copper is a metallic nutrient that originates from minerals in the soil. The total Cu content in western Canada soils usually falls in the range of 5 to 50 ppm. Approximately 1/2 to 1/4 of the total Cu exists within minerals and is unavailable to plants. Copper associated with oxides and organic matter has been found to be an important source of plant available Cu, probably by replenishment of dissolved and exchangeable Cu. The oxide and organic fractions increase with the clay content, which explains why Cu deficiency is more likely on sandier textures. Exchangeable Cu ranges from 0.1 to 10 per cent of total soil Cu. Only very small amounts of Cu exist as soluble Cu^{+2} in the soil water. Research on the prairies has found that DTPA extractable Cu is highly variable across cultivated and native fields. This means that larger numbers of soil samples are needed to obtain a precise estimate of the true soil average.

Very little information exists on the mechanisms of Cu uptake by plant roots. The driving force for Cu uptake is the electrical chemical gradient across the root cell membranes. Since free Cu levels inside the cell are kept low to avoid harmful reactions, and the membranes have a large negative potential, this creates a large force for Cu uptake. Therefore, there is no need for active Cu uptake systems. It has been suggested that Ca channels likely also allow passage of other ions such as Cu^{+2}.

Copper remobilization is much higher in old leaves and is related to N remobilization. Despite the intermediate mobility of Cu, deficiency symptoms in sensitive crops during the vegetative stage first appear in new growth. However, canola does not display strong Cu deficiency symptoms.

Pot experiments with extreme Cu deficiency have reported canola symptoms of:

- Interveinal chlorosis shortly after emergence
- Larger than normal leaves
- Wilting leaves
- Delayed flowering with a shortened flowering stem

Evidence suggests that canola growth may be affected by imbalances between Cu, molybdenum (Mo) and manganese (Mn) levels. Manganese:copper ratios (DTPA extractable) greater than 15 may result in a Cu deficiency. Molybdenum also may antagonize Cu. However, the Mo antagonism is in turn affected by S levels. Sulphur additions were found to lower Mo contents in canola plants, reducing Mo antagonism with Cu, and Cu deficiency was alleviated without adding Cu fertilizer.

Canola Response to Copper Fertilization

Although numerous field experiments have shown yield responses in cereals to Cu on deficient soils in western Canada, no positive reports in canola have been published. Two growth chamber experiments using extremely deficient organic soil did find a response to Cu fertilizer. Therefore, organic soils are likely the only soils in western Canada that may show a Cu response in canola. A compilation of research data from Saskatchewan and Alberta on mineral soil suggests that 0.30 ppm diethylenetriaminepentaacetate (DTPA) extractable Cu may be the critical level for canola. Since the critical Cu level is much lower for canola than cereals, Cu fertilization programmes should focus on application to the cereal rotation phases.

IRON (FE)

Iron is one of the most abundant metallic elements in the earthâ□™s crust. Western Canadian soils have developed from parent materials rich in Fe. Therefore, there have been no reports of Fe deficiency in field crops or responses to Fe fertilizer on the prairies. Also, there has been no work to calibrate soil test values for Fe on the prairies.

Role of Iron in the Canola Plant

Iron is a component of ferrodoxin which acts as an electron transmitter in nitrate and sulphate reduction, nitrogen fixation, and energy production. Iron is needed for chlorophyll synthesis, and low chlorophyll contents of young leaves (interveinal â□œchlorosisâ□• or yellowing) is the most obvious visible symptom of Fe deficiency. In young growing leaves, about 80 per cent of the Fe is located in the chloroplasts. Iron is also thought to be involved with protein synthesis and root tip growth.

MANGANESE (MN)

Manganese is a metallic micronutrient that is occasionally deficient in western Canadian organic, high pH soils. Although oats can be affected by Mn deficiency in cold organic soils (gray speck of oats disease), there have been no documented problems with canola.

Weathering of manganese containing soil minerals is the source of plant available Mn. The main Mn form that exists in soil solution or adsorbed to soil colloids is Mn+2, which is also the form absorbed by roots. Manganese availability decreases when pH increases above 6.2 in many soils. Low temperature and high organic matter can also decrease Mn availability. Rhizosphere microbes play a role in Mn availability by either oxidation or reduction. These microbes can either increase or decrease Mn availability. The rhizosphere acidification by canola roots likely increases Mn availability and makes this crop relatively tolerant of low soil Mn. Plant Mn status is affected by Cu levels. High Mn:Cu ratios above 15 may lead to Cu deficiency while ratios below 1 may lead to Mn deficiency. Potash (KCl) has been shown to enhance Mn uptake by several crops. In some crops, seed Mn content has been shown to be important for initial Mn nutrition and plant growth, as well as disease resistance.

MOLYBDENUM (MO)

Molybdenum is a transition element needed in extremely low amounts only nickel has a lower requirement. All the Brassica species appear to be sensitive to low Mo supply and can exhibit peculiar symptoms (for example whiptail • of cauliflower).

Role of Molybdenum in the Canola Plant

Only a few enzymes are known to contain Mo as a cofactor:

- The enzyme that helps to change nitrate to organic N in the plant (nitrate reductase)
- The major enzyme involved in nitrogen fixation in legumes (nitrogenase)
- An oxidase/dehydrogenase enzyme involved in changing the products of N fixation
- Probably an enzyme involved in sulphate metabolism (sulphite reductase)

Molybdenum functions are, therefore, closely related to N metabolism and N fixation. Some interaction occurs between Mo uptake and levels of P and S. Plant Mo uptake is usually enhanced by soluble P and decreased by sulphate. MoO_4^{-2} and SO_4^{-2} compete strongly for root uptake. Once absorbed by roots, Mo is readily mobile in both the xylem and phloem transport systems, probably as MoO_4^{-2}. Deficiencies in canola have not been documented in western Canada. If a Mo deficiency were to occur, there would be several options: seed treatment with Mo, soil or foliar fertilizer, and liming to raise soil pH. Molybdenum is unique among the micronutrients since there is a wide range between deficiency and toxicity.

BORON (B)

Boron is a micronutrient that occasionally limits canola yield in certain soils of western and eastern Canada. Unfortunately, current soil test methods do not consistently predict economic responses to B fertilizer in canola.

Role of Boron in the Canola Plant

Boronâ's role in plant nutrition is the least understood of all the nutrients. Boron is not an enzyme constituent nor does it seem to directly affect enzyme activities. Most of our understanding about B arises from symptoms observed during deficiency.

Possible roles for B include:

- Sugar transport and carbohydrate metabolism
- Cell wall synthesis and structure
- RNA metabolism
- Respiration
- Hormone metabolism
- Stomatal regulation
- Membrane function

Cell walls are dramatically affected by B deficiency. This shows up as cracked, hollow or corky stems. The cell wall diameter and proportion of plant dry weight increases under B deficiency. Most plant B is complexed with organic compounds in the cell walls, apparently serving a non-specific structural role.

One of the first plant responses to induced B deficiency is decreased root elongation, however, there is a lack of understanding how this occurs. Boron deficiency also restricts pollen tube growth. This is why B demand is higher during the reproductive stage than vegetative stage. Boron also affects fertilization and seed set by increasing the pollen production by anthers and the viability.

Boron Supply from Soil and Uptake by Canola

Boron uptake depends on adequate soil moisture, pH and the B level. The plant available forms dissolved in the soil water move to the root via mass flow and diffusion. Under drought conditions, B deficiency can occur due to reduced mass flow to roots as well as polymerization of boric acid. In contrast, under high rainfall conditions B can be leached in sandy textured soils.

Soil pH is a major factor influencing B availability. Generally, B becomes less available as pH increases above 6.3 to 6.5. At higher pH, the borate anion is likely adsorbed to clay and organic particles.

As with most nutrients, B uptake rates are increased as the level in the soil water increases. However, B uptake is unique since it is linearly related to levelâ□"there does not seem to be selective B uptake pumps that become saturated at higher levels. Therefore, B uptake appears to be a passive diffusion process. Boron uptake is affected by other nutrients. High levels of calcium and potassium have been shown to increase B deficiency symptoms. These antagonisms are not well understood.

Recent research in China and Australia reported genetic variation in canola for tolerance to B deficiency. Genetic differences in B efficiency were related to differences in root uptake or plant utilization. Research from Pakistan reported that Brassica napus was more sensitive to B deficiency than mustard (Brassica juncea), but needed relatively less B fertilizer for optimum grain yield. This area merits further research with western Canadian cultivars and conditions.

Boron deficiency symptoms in canola first appear in new growth due to the intermediate mobility.

Symptoms range from:

- Deformed, curled and rough skinned leaves with torn margins
- Yellow to brown spots in the interveinal areas of leaves
- Red to brown-purple coloured new leaves
- Early leaf drop
- Shortened stems
- Cracked stems
- Prolonged flowering
- Flower sterility
- Poor pod set and yield

Canola Response to Boron Fertilization

Boron deficiency is extremely rare in western Canada. The earliest report was at the AAFC Beaverlodge, AB Research Centre. Pot studies with B. rapa rapeseed on Alberta Gray Wooded soils showed that B deficiency symptoms and poor seed set was alleviated with added B. Subsequent research on Gray Wooded soils in Saskatchewan also found a B response but this was complicated by site and cultivar interactions. The Canola Council of Canada recorded a significant yield response to B at its Crop Production Centre site near St. Claude, MB in 2000. The only other Canadian report of a B response in canola is from northern Quebec.

In contrast, there are relatively more reports of neutral or negative responses to B in canola. In the Saskatchewan study mentioned above, 11 of the 13 sites did not have a B response. At one site, B fertilization actually lowered yield and oil content. Recent research by AAFC at Melfort, SK did not find a B response on four soils testing low in B, and one soil was similar to the responsive site in a previous study. Two years of foliar and soil applied B trials on a sandy Black 937 938 soil in central Alberta testing very low in B did not find any yield response with either B. rapa or B. napus canola. The Battle River Research Group did not find a B response after 2 years of trials in central Alberta. Research in Washington and Idaho on three soils testing low in B did not find a canola yield response to B fertilizer. Irrigated canola research conducted by Alberta Agriculture Food and Rural Development near Lethbridge, AB did not find a yield response to micronutrients, including boron, over six site-years.

Fig. Boron-deficient Canola Leaves

Fig. Boron-deficient Canola Leaves

Fig. Boron-deficient Flowering Canola (left) Compared to Boron-Sufficient Flowering Canola

Fig. Boron-deficient Podding (left) Compared to Boron-sufficient Podding Canola

Boron fertilization of canola has not consistently improved seed yield, kernel weight, protein or oil content. Boron deficiency appears rare in western

Canada, and where it does occur, it probably is in small field patches. The substantial cost of B, unlikely response and the narrow window between deficiency and toxicity make B fertilization programmes a questionable practice in western Canada. Unfortunately, there is not a consistent indicator to predict profitable canola yield response to B. The soil test criteria for B deficiency is less than 0.5 ppm (hot water extraction), but most of the above research trials did not measure a yield response with canola on soils testing much lower than 0.5 ppm. Therefore, hot water soil extraction does not appear to reliably predict plant available B, perhaps due to relatively high soil pH found in many western Canadian soils. Recent research indicates that extraction with hydrogen chloride (HCl) may improve the prediction success of soil tests, but further research is needed to confirm this and to develop calibration curves. Plant analysis at early flowering may help to identify B deficiency. However, the critical deficiency level of B in the last mature leaves at early flowering has ranged from 15 to 38 ppm in the various studies. Some of the variability in results likely arises from the dramatic difference in leaf B content due to age and the leaf part. Further refinement of plant analysis is needed, perhaps by focussing on specific leaf part and ageâ"such as the last unfolded leaf at flower initiation. For example, recent research found that the youngest open leaf was the most reliable plant part for B deficiency diagnosis.

Boron status of mature leaves was not indicative of current plant B status at the sampling time. Critical B level of the youngest open leaf was 10 to 14 ppm up to stem elongation. The youngest open leaf refers to the youngest unrolled leaf with a short, just visible petiole or midrib vein at its base and is next to the small emerging folded leaves at the growing point. Whole plant analysis at flowering is less likely to reliably predict B status since high levels in early leaves could skew whole plant measurements and hide deficiencies in new growth and flowering structures. Even after observing B deficiency symptoms and conducting soil and plant tissue analyses, prediction of a profitable yield response is difficult. Therefore, in situations of suspected deficiency, apply B fertilizer to a small affected area of the field in a carefully marked test strip. Visual observations and yields from the treated and untreated areas should help determine if a measurable response occurred. If a positive response is measured, B fertilizer could be applied in future canola crops on these areas.

Boron fertilizer can be effective either soil placed or foliar applied. Fertilizer B can be either broadcast-incorporated or banded. Ensure that B fertilizers do not come into contact with the seed at planting time. Make certain soil application rates do not exceed 1.7 kg/ha (1.5 lb/ac) on soils with a pH less than 6.5 to avoid boron toxicity problems. Foliar B fertilization appears to be effective up to the early flowering stage. Ensure foliar applications do not exceed 0.3 kg/ha (0.3 lb/ac) to avoid toxicity problems. For all applications, extreme care must be taken to apply the correct amount uniformly to avoid toxicity problems.

5

Cropping Systems with Wheat

CROPPING REQUIRES LIVESTOCK FOR DRAUGHT

Extensification of cropping requires livestock for draught, or mechanisation. Extensification may encourage supplementary pasture systems outside, as well as part of, the cropping systems themselves. It is also possible to develop extensive cropping without livestock. Agriculture in northeast Argentina and parts of Brazil, for example, has shifted directly from uncultivated grasing land or unused bush to extensive, mechanised cropping. Where cropping and livestock farming develop together, there are options as to their integration. In semi-arid Australia, mechanised extensification of 60 million ha of arable land has seen little (long term) change in the ratio of crop to pasture. Pasture land has increased in area since, about 1930 to stabilise at about 28 million ha in 1970.

Though livestock are not used for draught, this extensification has maintained a close link between crops and pasture in an integrated ley farming system. In contrast, in the semi-arid Near East, extensive cropping has been developed by private landholders without livestock. Here pasturage remains largely in the public domain with nomadic herds and common grasing land. Thus, though Australia and the Near East both have crop and livestock subsystems, in Australia they are integrated and managed by one family or corporation but in the Near East they may be separate and owned by distinct social groups.

INTENSIFICATION

Intensification involves biological or environmental modification. Biological modification occurs by increasing the complexity of a local cropping system by the development of multi-storied household gardens, relay cropping or mixed cropping. It is done by individual households. It increases labour inputs, diversifies crop management (though not necessarily crop diversity which may be greatest in hunter-gatherer systems), and usually increases the range of livestock (chickens, ducks, goats, fish and sometimes cattle are common in Javanese household gardens). It may reduce insect herbivores. Importantly, it creates food and income stability. Biological intensification can also be carried out on a larger, usually corporate scale. Multi-story plantation crops, such as beans under coffee, alley cropping and agroforestry, seek high outputs per unit area and stability in ways analogous to complex household gardens. Such intensification appreciably modifies the micro-environment as a secondary outcome. For example, tree legumes were planted along fence-lines in semi-arid Ethiopia in the late 1980s to provide quality forage for cattle and income from the sale of seeds. The trees provided shade and habitats for birds, which became more prized by villagers than the forage.

Intensification can also be carried out by deliberate modification of the environment, most obviously through irrigation.

Intensification, whether by biological or environmental modification, generally gives:

- Greater diversity of land use.
- Some intensively-used cropping land, usually in areas with favourable hydrology, fertility or social attributes (*e.g.,* prior settlement, easy transport).
- Some under-used or unused land. Cropping systems in Mediterranean lands often show these features, though there under-used land sometimes reflects lack of population pressure.

It is notable that extensification of cropping systems can create relative uniformity but intensification creates diversity. Neither, however, has a predictable social or environmental outcome. Extensification, often using monocultures, can be inherently stable and sustainable, or not. The social distribution of its benefits may be egalitarian (giving either uniform prosperity, or uneconomic, socially-deprived family farms), or it may be hierarchical with economically-strong landowners and a landless class of labourers and farmhands. Intensification at the household level benefits the family directly; intensification through village cooperatives or corporations may be advantageous or create negative social and environmental outcomes.

CROPPING SYSTEM

A cropping system may be defined as a community of plants which is managed by a farm unit to achieve various human goals. The latter include food, fibre and other raw materials, wealth and satisfaction. Conway encompassed these multiple goals in the phrase 'increased social value'. Farmers are part of the system. They are able to set, or modify, their own goals, so two farms with identical climates and soils may be managed with different aims to achieve a different mix of outputs. IRRI gives a more detailed definition of a cropping system: '.the crop production activity of a farm. It comprises all cropping patterns grown on the farm and their interaction with farm resources, other household enterprises and the physical, biological, technological and sociological factors or environments". The cropping system is one of three sub-systems likely to be present within a broader farming system.

Food security is the most basic output from a cropping system. For a subsistence household, all other outputs are secondary. For corporate agriculture, under a sophisticated cash economy, food security is established and other goals, such as profit, predominate. Thus, the first property of a cropping system is its effectiveness; that is its ability to meet the needs of the goal-setter (the farmer or corporate manager, etc.). The effectiveness of a system is a subjective attribute which should be assessed by the goal-setters and their society. It can be estimated or quantified through surveys using ranking techniques or by econometrical devices such as shadow pricing and opportunity costing.

To illustrate the importance of assessing the effectiveness of a cropping system. Here, in a survey, 270 Indonesian households were classed into six types (*e.g.,* aiming to be self-sufficient in food, wage-earning or in commercial agriculture) and, depending on their diet and the possibility of their diet being inadequate at some times of year, they were assigned to four classes of 'food insecurity'. Households managing cropping systems for self-sufficiency occur in all categories of food insecurity. A successful household whose cropping system supports them in a 'food secure' state, will have different opinions on the effectiveness of their cropping system from a neighbouring 'food insecure' household that also aims at self sufficiency. Furthermore, it is likely that their attitudes to risk-taking and change are likely to be quite different. This is in part because of the difference in effectiveness of the two systems. One family may be open to innovation and the other perhaps caught in a spiral of conservative thinking and poverty.

Production is a most obvious output and measure of the activity of a cropping system. It can be measured as the biological or economic output from the system, for example as the grain or cash generated. It is implicitly an output from the activity of one or more management units (*e.g.,* families). It is also a measure of the efficiency of the management of the cropping system and can be related to productivity-measured as output per unit of input (land, labour, capital, energy). Conway (1985) defines other ecosystem properties as sustainability, stability and equitability. Of these, sustainability, the subject of this *Bulletin,* is the most difficult to define and estimate. FAO (1989b) states that the goal of sustainable agriculture is to 'maintain production at levels necessary to meet the increasing aspirations of an expanding world population without degrading the environment', and that sustainability 'implies concern for the generation of income, the promotion of appropriate policies, and the conservation of natural resources'.

Similarly, sustainable agriculture and rural development are defined as 'the management and conservation of the natural resource base, and the orientation of technological and institutional change in such a manner as to ensure the attainment and continued satisfaction of human needs for present and future generations'.

Conway defines sustainability as the ability of an ecosystem to maintain productivity when subjected to a major disturbing force. He is thus forced to draw a continuum between stability (in the face of minor, expected forces) and sustainability (in response to major forces). He also indicates that to assess sustainability, one needs a measure 'of the effectiveness of internal adjustments that agro-ecosystems make in response to stresses and shocks'. A more pragmatic definition of sustainability, with which most agronomists would agree, is that 'a sustainable cropping system... is one which maintains resources, such as soil and water, while providing an adequate and economic level of production, both now and for generations to come'.

Similarly, a sustainable system is one in which:

- Resources are kept in balance with their use through conservation, recycling or renewal.
- Practices preserve agricultural resources and prevent environmental damage to the farm and off-site land, water and air.
- Production, profits and incentives retain their importance, because not only agriculture needs to be sustained, but so do farmers and society.

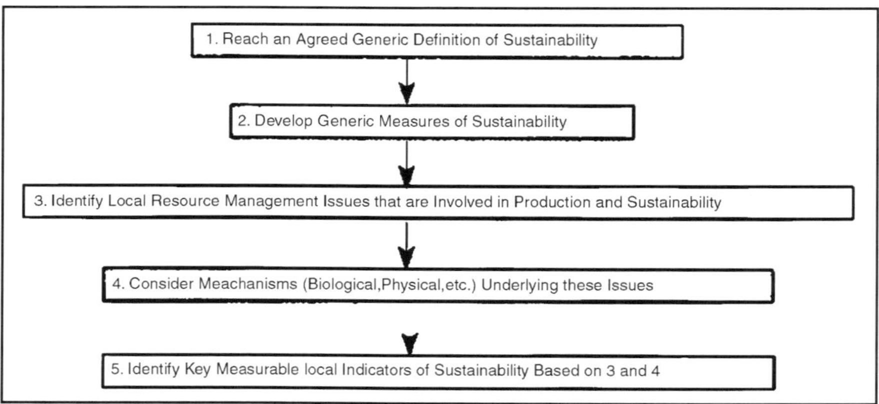

Fig. Measuring Sustainability

Conway's definition allows for levels, or degrees, of sustainability but constrains assessment of sustainability to measurements after shock. Hoare and others define a sustainable system in terms easy to relate to, but often do not accommodate degrees of sustainability, which is unhelpful for management and policy development.

Here, Conway's concept is accepted that it is useful to consider the possibility of various degrees of sustainability, and the widespread, but rarely developed, implication that sustainability should be measurable in either biological or economic terms (just as in measuring productivity). It follows that a biologically sustainable system may not be economically sustainable. This is witnessed by land in various parts of Europe and elsewhere that has gone in and out of cropping on several occasions during historic times depending on economic conditions. Here, sustainability is defined consistent with FAO, as the ability of a cropping system to maintain productivity over a long term. Since, this definition implies also maintenance of resources such as soil, it requires that all participants (farmers, researchers, extension specialists and policy-makers) agree upon appropriate measures by which the sustainability of particular cropping systems are assessed.

It is important that everyone involved in an assessment of the sustainability of a cropping system should agree on an inventory of resources which encapsulates its essential features. From this, the most critical attributes and, in some cases, those most easy to estimate, can be identified. The type of resource inventory that might be agreed necessary to define the sustainability of a semi-arid cropping system. Defined critical resources will differ for other systems. After defining the critical resources, it is necessary to identify those indices of sustainability that are easily and relatively-cheaply measured.

If a system is not sustainable, remedial action depends on agreement on the extent of non-sustainability and the changes in inputs to make it sustainable. Some research is encouraging here: for example, Rickson *et al.* find reasonable agreement between farmers' perceptions of the impact of continued soil loss on sustainability, as estimated by the extent of decline in wheat yields.

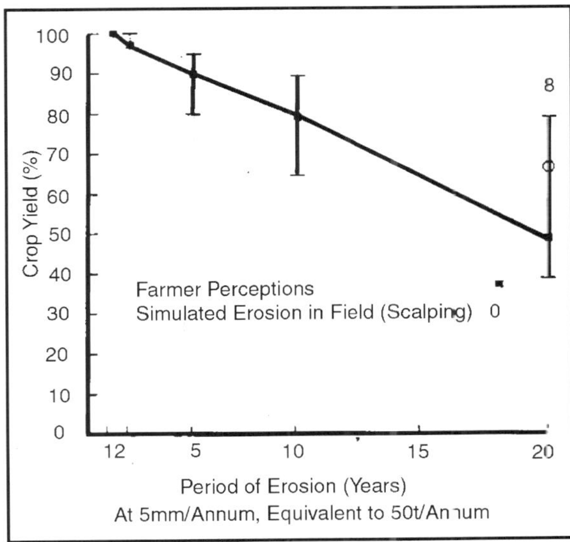

Fig. Crop Yield Declines with Continuing Soil Degradation and Erosion.

Stability (or variability) is the constancy of productivity in the face of small disturbances such as seasonal or year-to-year fluctuations in weather. It is a measure of the spatial- or year-to-year robustness of a farming practice and reflects not only variations in crop response to weather and soil type, but also variations in prices of inputs (such as labour and pesticides) and of the product.

Last, but not least, equitability is the property that reflects how the resources and outcomes of the cropping systems are shared among the population.

EVOLUTION OF FIELD CROP ECOSYSTEMS

Various terms used in the sections that follow are defined in Appendix 1. Most of the definitions are taken from IRRI and ICRAF. Cropping systems have evolved over historical time along a catena of increasing energy inputs (including mechanisation), reduced biological diversity and increased risk or instability.

Boserup (1965) developed the thesis, forwarded and embellished by others, that population growth is a precondition for agricultural change. Subsistence communities in semi-arid regions depend on herding livestock (most common in semi-arid grasslands and woodlands) or shifting cultivation. In the latter, the cultivated plot is moved to new land when fertility is depleted or the plot becomes too weedy.

Boserup's thesis, backed by worldwide experience, is that increasing population, in places partly caused by immigration (*e.g.,* settlers moving into sub-Sahelian Africa), either:

- Reduces community mobility until the villages become static and maintain permanent fields;
- Already semi-permanent villages are so deprived of hunting/foraging grounds that they too need to develop permanent cropping systems. As the fallow period between crops is shortened and the fields become more permanent, two strategies are available;
- Extensification and/or increased mechanisation. For example, use of draught animals or tractors;
- Intensification including use of shorter fallows, perhaps coupled with use of unused poorer land. This is accompanied by increased complexity of cropping, *e.g.,* household gardens, and environmental modifica-tion, *e.g.,* irrigation.

Okigbo comprehensively described these changes in cropping systems, and their implications for sustainability.

GROWING SEASON OF CLIMATES AND DRYLANDS

FAO (1987) uses the term 'drylands' to describe climates with fewer than 120 days growing season. These are divided into 'arid drylands' with less than 75 days growing season, and 'semi-arid' areas which have from 75 to 119 days growing season. Seasonal changes in indices of crop growth in tropical drylands, classed by Hutchinson *et al, as* having climate types I (hot, seasonally wet/dry: uppermost row) or E (warm, seasonally wet/dry: middle row), and in temperate drylands (lowermost row). Growth index () is a dimensionless index from zero (no growth) to 1 (maximum growth). Though most semi-arid lands border the arid or desert climates of the world, they support a diversity of cropping (mostly with annual crops). Because of this diversity their climate should be quantified and any variations within them delineated. Such drylands have either one or two wet seasons. Bimodal rainfall regimes are characteristically found in latitudes between the wet tropics and the unimodal more temperate, semi-arid zones. Their distribution, however, is governed more by pressure systems and land masses. For example, bimodal patterns are rarely found in east Asia. In the tropics, the wet season(s) coincide with a high solar angle except for a relatively small area in east Brazil. The tropical drylands are commonly called monsoon (in Asia) or savannah (in Africa) although there is confusion over these terms. Koppen uses 'monsoon' in a restricted sense, as a climate with a dry season but able to support rainforest. In temperate areas, there is usually a single wet season that coincides with a low solar angle (winter). These wet-and-dry climates are generally called Mediterranean and, on the drier margins, semi-arid.

The tropical drylands (to use FAO's term) are categorised I and E by Hutchinson *et al.,* (1992); and the temperate drylands as E, D and C on the basis of seasonably of available water and temperature. This delineation of climates is based on calculations of the relative constraints of soil water, temperature and solar radiation on growth of temperate and tropical crops. Each of the indices ranges from zero (no growth) to 1 (climate optimal for growth). The water, temperature and radiation indices are multiplied together to calculate a growth index, which is an estimate of the overall suitability of the location for crop growth in a particular week or month of the year. Hutchinson's climate classification, unlike more traditional, geographically-oriented classifications, reflects seasonal fluctuations of environmental factors related quantitatively to crop growth. Examples relevant to the tropical drylands and climatically-similar wet-and-dry temperate areas.

Tropical drylands that support cropping have mean annual water indices as low as 0.29, but have a growth index (the product of the water, temperature and radiation indices) during the highest quarter (three-month period) greater than 0.3 (commonly 0.3-0.8). This is sufficient to support good growth of a single crop. Zone I coincides with parts of the Aw, Bsh and Cwa types of Koppen (1900) and Trewartha and corresponds with the 1.3, 1.4 1.5, 1.7 and 1.9 zones of Papadakis. Hutchinson's E zone, which is not as hot coincides with parts of Koppen's Cfa, Cwa and Bsh types and Papadakis' zones 1.7, 2, 4.2 and 5 in the tropics.

PREDOMINANT SOILS OF THE SEMI-ARID CROPLANDS

The predominant soils of the semi-arid croplands are Luvisols and Calcisols in Africa and Australia, Mollisols and Luvisols in temperate areas of the Americas, Ferralsols in wet-and-dry tropical America, Luvisols and Chernozems in temperate Eurasia and Calcisols and some Acrisols in wet-and-dry tropical

Eurasia. The climatic distribution, area and chief limitations of the main major soil orders. All these soils occur in seasonally wet-and-dry climates, although the Acrisols are not extensive in drylands. Lixisols, most widespread in potential cropping areas within seasonally wet-and-dry climates, have poor water-holding capacity, hard-setting surface horizons with a risk of crusting, run-off and erosion.

The topsoil texture varies from loamy sand to loam and clay, but the hard-setting properties and relatively poor infiltration and drainage (compared with, say, Acrisols) mean that crop growth is frequently limited by water availability. Lixisols often have low levels of organic matter and generally low effective cation exchange capacity (ECEC). They are commonly deficient in phosphorus. Calcisols are extensive, particularly in north and west Africa but, lying on the dry margin of cropland, they contribute less than Luvisols to crop production in semi-arid regions.

They are high in bases, in places saline, and low in organic matter. Chemozems are relatively well structured and fertile with high levels of base saturation. Their topsoil does not harden on drying, as is common in Luvisols and Lixisols. Ferralsols and Acrisols are acid, highly weathered and more common on the wetter margins than in strictly semi-arid regions. Being highly weathered, they have low fertility and ECEC though they often have high levels of exchangeable aluminium. Applied fertilizer may be leached, because of the low ECEC, or made unavailable. They have high water infiltration capacity but a relatively low water-holding capacity and are susceptible to compaction.

Vertisols, dark-coloured cracking clays, though not extensive in all continents, are important in some seasonally wet-and-dry cropping systems. They have a high ECEC and base status and are often naturally fertile, though sometimes low in nitrogen, phosphorus and zinc. They have a large water-holding capacity. Their surface does not crust; they self-mulch as surface clods break down naturally on drying to small aggregates forming a good tilth.

The soil cracks deeply on drying, but as it re-wets the cracks close reducing infiltration and lateral movement of water. When wet the soils are so plastic and sticky that, in Australia, Vertisols were not cropped until tractors replaced horses thus enabling them to be ploughed quickly during the short period when soil moisture is optimum. It is useful to draw attention to the major soil types and broad climate and soil groupings, as above. For example, the extensive cereal-fallow-pasture system, covering 60 million ha of temperate semi-arid Australia, is found on four broad types of soil and differences in crop rotations largely reflect regional soil differences. However, the more common situation is that soils are linked through topography and moisture regime to landscape, as illustrated in West Africa.

CLASSIFICATIONS OF CROPPING SYSTEM

Depending on the resources and technology available, different types of cropping systems are adopted on farms. Mono-cropping or Single Cropping: Mono-cropping refers to growing only one crop on a particular land year after year. Or Practice of growing only one crop in a piece of land year after year *e.g.* growing only rabbi crops in dry lands or only said crops in diary lands (Lands situated in river basins which often remain flooded during rainy season). This is due to climatologically and socio economic conditions or due to specialization of a farmer in growing a particular crop.

Groundnut or cotton or sorghum is grown year due to limitation of rainfall. Flue-cured tobacco is grown in Günter (A.P.) due to specialization of a farmer in growing a particular crop. Rice crop is grown, as it is not possible to grow any other crops, in canal irrigated areas, and under water logged conditions.

Monoculture: Practice of repetitive growing only crop irrespective of its intensity as rice-rice-rice in Kerala, West Bengal and Orissa.

Sole Cropping: One crop variety grown alone in pure stand at normal density.

Multiple Cropping or Polycropping: It is a cropping system where two or three crops are gown annually on the same piece of land using high input without affecting basic fertility of the soil.

Growing two or more crops on the same piece of land in one calendar year known as multiple cropping. It is the intensification of cropping in time and space dimensions *i.e.* more number of crops within a year and more number of crops on the same piece of land at any given period. It includes inter-cropping, mixed cropping and sequence cropping.

Molested (1954) has mentioned that multiples cropping is a philosophy of maximum crop production per acre of land with minimum of soil deterioration.

- Rice-potato-green gram.
- Rice-mustard-maize.
- Rice-potato-sesame.
- Jut-rice-potato.

Cropping intensity is more that 200 per cent when the farm as a whole is considered; the Multiple Cropping Index (MCI) is determined by the number of crops and total area planted divided by the total arable area. When the value is three or more, it is said to be most promising farm. This is also called as intensive cropping.

Polyculture: Cultivation of more than two types of crops grown together on a piece of land in a crop season, *e.g.*

- Subabul + Papaya + Pigeon pea + Dinanath grass.
- Mango + Pine apple + Turmeric
- Banana + Marigold + berseem.

Relay Cropping: Growing the succeeding crop when previous crop attend its maturity stage-or-sowing of the next crop immediately after the harvest of the standing crops. Or it is a system of cropping where one crop hands over land to the crop in quick succession, *e.g.*

- Paddy-lathers
- Paddy-Lucerne.
- Cotton-Berseem.
- Rice-Cauliflower-Onion-summer gourds.

Overlapping Cropping: In this system, the succeeding crop is sown in the standing crop before harvesting. Thus, in this system, one crop is sown before the harvesting of preceding crops. Here the lucre and berseem are broadcasted in standing paddy crop just before they are ready for harvesting.

Advantages:

- Minimum tillage is needed for relay cropping and primary cost of cultivation is less.
- Weed infestation is less, as land is engaged with crops year round.
- Crop residues are added in the soil and thus more organic matter.
- Residual fertilizer of previous crops benefits succeeding crops.

ROTATION SYSTEMS FOR WHEAT PRODUCTION

In the Noble Foundation service area within 100 miles of Ardmore, Okla., a lot of wheat is grown. Is this simply because wheat is what we grew last year, and the year before and the year before that? Maybe we grow wheat because we raise cattle and need winter grazing for them. However, improvements in other crops have provided some options that we may need to examine.

Canola has received increased attention lately. It would seem to be an ideal crop to use in a rotation scenario with wheat. It has the same growing season as wheat, and we can use the same equipment for planting and harvesting. Canola is a broadleaf, so it provides the opportunity to clean up some grassy weeds that are difficult to control in a continuous wheat production system. Canola also has a deep taproot to loosen and mellow the soil. In the past, problems with canola included winterkill and a lack of marketing points/options. Varieties have now been bred with improved winterhardiness. In addition, more elevators will receive canola than in the past.

Sesame is a relatively new and unknown crop in the southern Great Plains. Like canola, it can be planted and harvested with the same equipment that we use for wheat. Plant breeding has also provided some better adapted varieties. Sesame is a summer annual crop that seems to tolerate hot, dry weather very well. As sesame acreages grow, the number of elevators handling it will also increase. Cotton was once a high input crop requiring multiple pesticide sprays to control insects and weeds. Now, with boll weevil eradication, Bollgard®, Roundup Ready® and LibertyLink®, cotton could once again be a desirable crop. Many farmers are interested in growing corn or soybeans. I would not typically recommend these crops west of I-35 without irrigation. If irrigation is available, these may be options. Another option similar to soybeans is dry or edible beans such as black-eyed peas, cow peas or snap beans.

Milo or grain sorghum is a crop that is often overlooked. If value is given to the grain production and the forage that can be baled after grain harvest, it can be a profitable crop. It may even provide income from lease hunting for dove after harvesting the grain crop.

Sunflowers share many of the same advantages as other rotation crops. They are a broadleaf crop, which provides different herbicide options than wheat alone. They also have a deep taproot and are an oil seed, which may be used for feed or biofuel. The potential for lease income from bird hunters also exists with a sunflower crop.

Benefits of rotational cropping include breaking weed, disease and insect pest cycles; diversification to spread risk; different root systems to loosen compaction; possible nitrogen benefit from including a legume; and increased yields from the "rotation effect," even if the rotation does not include legumes. Including a summer annual that is not double-cropped also provides time between wheat harvest one year and planting the following spring, thus building up or banking soil moisture. Take some time to evaluate your operation and determine if a crop rotation would benefit you.

THE BENEFITS OF WHEAT IN YOUR CROP ROTATION

There are several agronomic advantages when considering adding wheat into your crop rotation. The period when winter wheat needs the most water, the flowering and early grain-fill stage, coincides with the time of year when most areas normally receive the most rain, May and June. As a result, winter wheat usually produces a crop when irrigation is limited. Most of the groundwater-irrigated land will be subject to allocations of 14 to 16 inches of water for at least the next several years.

Winter wheat also can adapt to varying amounts of precipitation.

Wheat has a lot of ways to make grain and will provide at least half a crop even in the worst years. It also works well in rotation with other crops because its peak water need falls earlier than most summer crops now grown in irrigated rotations. During the growing season, winter wheat is very competitive with warm-season weeds. By the time these weeds come on, winter wheat already has developed enough of a canopy to block the sunlight from reaching the weeds.

Introducing wheat into more traditional irrigated crop rotations is advantageous with farmers' existing schedules and workloads. Winter wheat can spread the workload for producers of irrigated row crops. It is planted in September, following dry bean harvest. Winter wheat also will fit into double-crop and relay-crop rotation systems. Under irrigated conditions, a forage crop can be planted following wheat harvest in July, enabling a farmer to produce two crops in one year. Converting the traditional three-year crop rotation (dry beans/sugar beets/corn) into a four-year rotation by adding winter wheat also has advantages. Sugar beets are susceptible to soil-borne disease; it's good to allow an additional year between one sugar beet crop and the next.

Residue management is another advantage. Winter wheat produces a good amount of useful and resilient crop residue. Many farmers have begun harvesting with stripper headers on combines, which removes only the heads and leaves behind the remainder of the plant. This stubble has a higher silhouette factor than other crops, better protecting the soil from wind erosion. During the winter, it is effective in trapping snow, thus increasing soil moisture.

At the University of Illinois, a team of researchers found that having wheat in the crop rotation helps increase the yield of corn and soybean crops that follow. A recent three-year summary of the results showed that corn grown in a three-crop rotation (soybean/wheat/corn) yielded 4 per cent more than corn in a corn/soybean rotation. Corn in a wheat/soybean/corn rotation produced 6 per cent higher yields. Meanwhile, soybeans in both three-crop rotations yielded 4 per cent more than soybeans in the corn/soybean rotation.

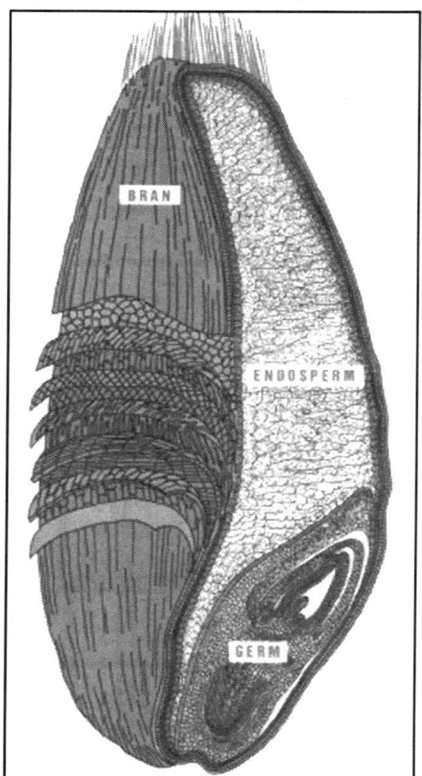

Fig. A Kernel of Wheat.

The Kernel of Wheat is sometimes called the wheat berry. The kernel is the seed from which the wheat plant grows. Each tiny seed contains three distinct parts that are separated during the milling process to produce flour.

The Endosperm makes up about 83 per cent of the kernel weight and is the source of white flour

The Bran is about 14.5 per cent of the kernel weight. Bran is included in whole wheat flour and can also be bought separately.

The Germ is about 2.5 per cent of the kernel weight. The germ is the embryo or sprouting section of the seed, often separated from flour in milling because the fat content limits flour's shelf-life.

Ash Content indicates milling performance and how well the endosperm separates from the bran. Ash content can affect flour colour. White flour has low ash content, which is often a high priority among millers, because consumers prefer white flour.

Damaged Kernels are the kernels which may be undesirable for milling because of disease, insect activity, frost or sprout damages.

Dockage is the percentage of wheat, measured by weight, easily removed from a wheat sample using the Carter Dockage Tester.

Foreign Material is any material other than wheat that remains after dockage is removed. Because foreign material may not be removed by normal cleaning equipment, it may have an adverse effect on milling quality.

Moisture Content is an indicator of grain condition and store ability. Moisture content is often standardized (12 or 14 per cent moisture basis) for other tests that are affected by moisture content. Lower moisture levels are desired to prevent spoilage in storage.

Protein Content relates to many important processing properties, such as water absorption and gluten strength, and to finished product attributes such as texture and appearance. Higher protein dough usually absorbs more water and takes longer to mix. Hard Red Winter wheat generally has a medium to high protein content, making it suitable for all-purpose flour and yeast raised flour foods.

Shrunken and Broken Kernels are the kernels which were either insufficiently filled during the growing season and as a result have shrunken and shriveled appearance or have been broken in handling. Such kernels may reduce milling yield.

Thousand-kernel weight and kernel diameter provide measurements of kernel size and density important for milling quality. Millers tend to prefer larger berries or at least berries with a consistent size.

Total Defects is the sum of damaged kernels, foreign material and shrunken and broken kernels.

Whole Wheat products are made with the whole wheat kernel. The bran (outer layer) contains the largest amount of fibre (insoluble), B vitamins, trace minerals and a small amount of protein; the endosperm (middle layer) contains mostly protein and carbohydrates along with small amounts of B vitamins, iron and soluble fibre; and the germ (inner part) is a rich source of trace minerals, unsaturated fats, B vitamins, antioxidants, phytochemicals and a minimal amount of high quality protein.

DESIGN OF AN ORGANIC FARMING CROP ROTATION EXPERIMENT

The most common organic farming system in Denmark is based on a large fraction of grass-clover and fodder crops in the rotation in combination with a stock of ruminant animals, typically for dairy production (Tersbøl and Fog, 1995). This farm type (0.9-1.4 livestock units ha-1) has with proper management proved to sustain a stable crop production with neglible problems (Askegaard et al, 1999). There is, however, a need to increase cereal production in organic farming in order to provide grain for both human consumption and non-ruminant animal feed. The design and management of organic crop rotations involves many considerations. Contrary to conventional crop production where the management factors can be optimised individually (*e.g.* fertilisation or weed control), many factors and their interactions must be included in the design and management of organic crop rotations.

The main reason is that crop management of organic crop rotations must focus on the prevention of problems like diseases, pests and weeds, rather than the curing of problems. This prevention is based on the construction of sound crop rotations, which are able to reduce the propagation of diseases, and on nitrogen self-supply through the use of N2-fixing crops and cover crops (Lampkin, 1990). Another very important prevention factor is crop establishment, where a uniform seedbed and the right time of sowing constitute the preconditions for good crop growth and development, which again will improve its competitive ability against weeds. The options for rotational crops in organic farming include: green manure crops, winter or spring cereals, pulses and oil seed crops, row crops and catch crops. The primary management options in organic crop rotations are manure application, mechanical weed control, straw removal, soil tillage and harvest time (*e.g.*, cereals for maturity or for whole-crop silage). Other factors can, however, also be used to improve the production (*e.g.*, variety choice, seed rate, plant architecture).

There have been only a limited number of studies under temperate conditions in Europe and North America, where different crop rotations have been compared under organic farming or similar production conditions. Examples of recent pure organic rotation trials are the comparison of stockless crop rotations at Elm Farm in England (Bulson et al., 1996) and the rotations with different fractions of a grass-clover ley in Scotland (Younie et al., 1996). Other factorial experiments have compared organic and integrated or conventional crop rotations. An example of this is the DOC experiment in Switzerland (Besson et al., 1992). Some factorial experiments have looked at the interaction between crop rotation and fertilisation level, *e.g.* in Norway (Uhlen et al., 1994) and in Poland (Kus and Nawrocki, 1988). This paper describes the structure and the management of an on-going crop rotation experiment, which was started at four sites in Denmark in 1997. The objective of the experiment is to explore the possibilities for both short-term and long-term increases in organic cereal production through manipulation of crop rotation design on different soil types. The performance of the crop rotations is evaluated in terms of crop production, nutrient leaching and occurrence of weeds, pests and diseases.

In addition the experiment functions as workshop facility for other projects concerned with effects of crop rotation design on soil fertility and plant growth. The representation of the crop rotation experiment in terms of soils, climate and farming systems is presented here. The experiment serves a number of purposes and the importance of these for the design of the experiment is discussed. Later papers will present and discuss results of the experiment.

MATERIAL AND METHODS

The crop rotation experiment is designed as a factorial experiment with three factors and two replicates where all fields in the rotations are represented every year. The experimental factors are: 1) Fraction of grass-clover and pulses in the rotation (crop rotation), 2) Catch crop (with/without catch crop or bi-cropped clover), and 3) Manure (with/without animal manure as slurry).

Rotations and Locations

Four different four-year crop rotations are compared. The contributions of different crop types in the rotations are shown in Table and the actual rotations are shown in Table.

Table. Percentages of each Rotation Comprised of Different Crop types. Autum Crop Cover is defined as permanent White Clover understories, grass-clover leys, or catch crops of grass or grass-clover.

Crop type	Rotation 1	Rotation 2	Rotation 3	Rotation 4
Green manure	25	25	25	0
Pulse	25	25	0	25
Cereal	50	50	50	75
Row crop (beet)	0	0	25	0
Autumn crop cover				
without catch crops	50	25	25	0
with catch crops	100	75	50	100

Table. Structure of the four different 4-course crop rotations with and without catch crops. The sign '.' indicates that a grass-clover ley, a clover or a ryegrass/clover catch crop is established in a cover crop of cereals or pulses. The sign'/' indicates a mixture of peas and spring barley or bi-cropping of winter cereals and clover.

Catch crop	Course (year)	Rotation 1	Rotation 2	Rotation 3	Rotation 4
Without	1	S. barley:ley	S. barley:ley	S. barley:ley	Spring oat
	2	Grass-clover	Grass-clover	Grass-clover	Winter wheat
	3	Spring wheat	Winter wheat	Winter wheat	Winter cereal
	4	Lupin	Peas/barley	Beet	Peas/barley
With	1	S. barley:ley	S. barley:ley	S. barley:ley	S. oat:clover
	2	Grass-clover	Grass-clover	Grass-clover	W. wheat/clover
	3	S. wheat:Grass	W. wheat:Grass	W. wheat:Grass	W. cereal/clover
	4	Lupin:Grass	Peas/barley:Grass	Beet	Peas/barley:Grass

The ranking of the crop rotations indicates a decreasing input of nitrogen through nitrogen fixation: 1) 1.5 grass-clover and 1 pulse crop, 2) 1 grass-clover and 1 pulse crop, 3) 1 grass-clover crop, and 4) 1 pulse crop. The difference between the grassclover crops in rotations 1 and 2 is that spring ploughing is used in rotation 1 whereas autumn ploughing is used in rotation 2. The cereals used in the rotations include spring and winter wheat (Triticum aestivum), winter triticale (Triticosecale), spring barley (Hordeum vulgare) and spring oat (Avena sativa).

The pulses used include a mixture of pea (Pisum sativum) and barley and a pure stand of blue lupin (Lupinus angustifolius). All cereals and pulses are harvested as grain or seed crops at maturity. Care has been taken in preventing diseases being promoted by the crop rotation. Peas (pea/barley) and lupins will alternate from one rotation period to the next. In rotation 4 oat is grown prior to winter wheat in order to minimise the risk of infection with take-all disease (Gaeumannomyces graminis). Triticale has been chosen as the second-year winter cereal instead of winter wheat on the sandy soil at Foulum in order to reduce problems with take-all.

The choice of varieties is made every year on the basis of a set of prioritised criteria. Typically, high yield characteristics attract lower priority than disease resistance. Preference is given to varieties which have been tested in Denmark for at least 2-3 years. The grass-clover in rotations 1, 2 and 3 is either a stand of white clover (Trifolium repens) and five varieties of perennial ryegrass (Lolium perenne) on the lighter soils (Jyndevad and Foulum) or the same mixture combined with red clover (Trifolium pratense) on the heavier soils (Flakkebjerg and Holeby). Red clover can be an aggressive competitor to spring barley on lighter soils. The catch crop in rotations 1, 2 and 3 is either a pure stand of perennial ryegrass Lolium perenne or a mixture of perennial ryegrass and four clover species (hop medic Medicago lupulina, trefoil Lotus corniculatus, serradella Ornithopus sativus and subterranean clover *Trifolium subterraneum*).

These catch crops are undersown in cereals or pulses in spring. The catch crop treatment in rotation 4 is a bi-crop of winter wheat in a pure stand of white clover. The white clover is undersown in oat. After harvest in the autumn and a few days before sowing of winter wheat the clover is cut as short as possible, followed by rotary cultivation in 12 cm wide bands at double normal row spacing (25 cm). The winter wheat is drilled into these bands. The white clover is controlled during the growing season by cutting it separately with a row brush weeder in order to reduce competition with the wheat. After harvest of the winter wheat the clover is allowed to grow and then again a few days before sowing, the same procedure is repeated for the establishment of the second-year winter cereal. The experiment is carried out at four sites representing different soil types and climate regions in Denmark. Not all rotations and treatments are carried out at all sites, but rotation 2 is present at all sites.

Table. Experiment sites and treatments.

Location	Soil type	Irrigation	Replicates	Crop rotations	Manure	Catch crop
Jyndevad	Sand	Yes	2	1+2	With/without	With/without
Foulum	Loamy sand	No	2	2+4	With/without	With/without
Flakkebjerg	Sandy loam	No	2	2+4	With/without	With/without
			2	3	With	With
Holeby	Loam	No	1	2+3+4	With	Without

The experiment is unirrigated at all sites except Jyndevad, where the irrigation scheduling programme MarkVand (Plauborg and Olesen, 1991) is used to define the irrigation demand. All straw and grass-clover production is incorporated or left on the soil in all treatments.

Sowing and Fertilization

The sowing of spring cereals, peas and lupins is carried out as early in the spring as possible, but not before a good preparation of the soil can be performed. The ploughing and the succeeding seedbed harrowing must be very uniform to allow the seeds to be placed at the correct depth and for the weed harrowing to be performed at uniform intensity. The winter cereals are sown as soon as a proper seedbed can be prepared after 25 September. This is slightly later than recommended for conventionally grown cereals in order to reduce weed emergence in autumn. All the catch crops and the grass-clover mixtures are sown in spring.

In the spring cereals the sowing takes place on the same day as the cover crop is sown, except for Jyndevad where sowing is delayed in order to permit weed harrowing in the rotations with catch crops. Delayed sowing is possible in Jyndevad as the use of irrigation ensures the germination of the catch crops. In the winter cereals the catch crop is sown in April just after the first weed harrowing. Plots receiving manure are supplied with animal manure (slurry) at rates of ammonium nitrogen in the slurry corresponding to 40 per cent of the nitrogen demand of the specific rotation. The nitrogen demand, based on a Danish national standard (Plantedirektoratet, 1997) is 60, 60, 93 and 113 kg N ha-1 as an average of the fields in rotations 1, 2, 3 and 4, respectively. The nitrogen demands of grass-clover and of peas/barley and lupins were set at nil. The predominant type of slurry available at the site is used, either pig slurry, cattle slurry or anaerobically digested slurry. The slurry is distributed evenly between the non-fixing crops in rotations 1 to 3.

Table. Application of slurry to the crops in the four rotations in NH^4-N and in corresponding livestock density (LU ha-1) at Jyndevad (Jy), Foulum (Fo), Flakkebjerg (Fl) and Holeby (Ho). One livestock unit (LU) corresponds to 100 kg of total N in manure produced per year.

	Crop rotation			
	1	2	3	4
Crops receiving slurry	NH_4-N in slurry (kg ha^{-1} yr^{-1})			
Spring barley	50	50	50	-
Spring wheat	50	-	-	-
Spring oat	-	-	-	40
Winter cereals	-	50	50	70
Beet	-	-	50	-
Average in rotation	25	25	38	45
Slurry type	Livestock density (LU ha^{-1})			
Cattle slurry (Jy)	0.5	0.5	-	-
Pig slurry (Fo, Ho)	-	0.4	0.6	0.7
Digested slurry (Fl)	-	0.4	0.6	0.7

In rotation 4, however, the winter cereals are favoured at the expense of oat. In order to obtain a uniform distribution of the slurry to spring cereals it is applied in the seedbed after ploughing and harrowing. Harrowing is performed immediately after application in order to minimise ammonia volatilisation. In winter cereals slurry is applied mid April at the start of growth using trail hoses. Where a wider row distance is used in winter cereals the slurry is placed as close to the rows as possible in order to fertilise the cereals more than the weeds. Harrowing is always performed just before slurry application in order to loosen the soil. This accelerates the infiltration and prevents an uneven penetration of the slurry into the soil.

Plant Protection

In the spring sown cereals and pulses where there is no catch crop, weed harrowing is carried out pre and post-emergence and if necessary, once during the later growth stages (Rasmussen and Rasmussen, 1995). In winter cereals without catch crops, pre and postemergence harrowing is carried out after sowing if the weather permits and if the harrowing can be carried out without covering more than at most 10 per cent of the crop with soil. In addition, one or more weed harrowings are carried out in the spring. These measures proved not to be sufficient on the lighter soils. Therefore in order to facilitate mechanical hoeing between rows, from 1998 larger row distances were used for the winter cereals in rotation 4 at Foulum and for the spring and winter wheat at Jyndevad and from 1999 also for the lupins at Jyndevad (Rasmussen and Pedersen, 1990). In the winter wheat with catch crops, except for rotation 4, weed harrowing is carried out in the autumn (if possible) and in the spring before sowing the catch crop.

At Jyndevad, mechanical weed control is carried out pre and post-emergence in all cereal and pulse crops before the catch crop is sown. In rotation 4, the row brushing carried out to control white clover also controls weeds emerging between the wheat rows. The sugar beets are kept weed-free by a strategy of pre-emergence flaming, row and hand-hoeing. If perennial weeds such as creeping thistle (*Cirsium arvense*), mugwort (*Artemisia vulgaris*), curled dock (*Rumex crispus*) and others occur, they are removed manually from the plots.

Creeping thistle is removed by cutting the stalk as deep under ground as possible at the time the thistles are budding, which coincides with the time of the anthesis of the cereals. At this time the reserves in the root system are at a minimum (Dock Gustavsson, 1997). If couch grass (*Elymus repens)* occurs in plots without catch crops above a threshold level of 5 shoots m-2, repeated stubble cultivation is carried out after harvest. If couch grass occurs in plots with catch crops above a threshold level of 50 shoots m-2 stubble cultivation will be carried out.

Another measure to cope with couch grass is to intensify the cutting of the green manure crop. Without occurrence of couch grass the cutting is carried out when the grass-clover has a height of about 15-20 cm in mixtures without red clover and about 20-25 cm in mixtures with red clover. With occurrence of couch grass in the rotation above a threshold of 5 shoots m-2, the cutting is carried out when the grass clover has a height of about 10-15 cm or 15-20 cm, respectively.

Management of the experiment

Local conditions can affect the plots differently, even at the same experimental site. This means that one of the two replicates of each treatment may have need for a management treatment (*e.g.* for controlling couch grass), whereas the other replicate has no need. Guidelines have therefore been set up defining the conditions under which the plots can be managed individually and when both replicates should be managed identically. If a management treatment changes the principal effect of one of the three experimental treatments (rotation, catch crop and manure) then both replicates should be managed identically. All other management treatments are based on the needs of the individual plots. Each plot is sub-divided into between three and five sub-plots.

Two of the subplots are harvested for determination of crop yield. The other sub-plots are used for plant and soil sampling and for experiments. All samplings are conducted in mini-plots, which are square plots of 1 m2. The positions of the plots and of the mini-plots are fixed through the use of permanently installed iron tubes in guard rows between all plots. The iron tubes are used for reference when managing the plots. Short-cut grass borders separate all plots in order to prevent movement of soil between plots. A soil border separates the crop of each plot from the grass border. This soil border is kept bare throughout the growing season by rotary cultivation in order to prevent weeds (*e.g.* couch grass and white clover) from entering or leaving the plots and annual weeds from establishment and seeding. The experimental treatments were started in 1997. In 1996 a spring barley crop with undersown grass-clover was grown at all sites, except Holeby where a winter wheat crop was grown. No pesticides were used in 1996.

Soil characteristics

A characterisation of the soil at the experimental sites was conducted in autumn 1996 prior to the initiation of the experiment. Sixteen soil samples were taken in each plot to one meter depth. The soil horizons of all soil samples were characterised using a standard soil taxonomy (Soil Survey Staff, 1992). The samples were divided into 25 cm layers, and the samples from each layer in the plot were mixed.

The samples from all layers were analysed for pH and contents of K and P. Selected plots and soil layers were also analysed for soil texture, carbon content, cation exchange capacity (CEC) and total nitrogen. Soil pH was determined in a mixture of soil and a solution of 0.01 M $CaCl_2$. The soil pH was calculated as $pH(CaCl_2)+0.5$. The content of K was determined after extracting the soil for 30 minutes with a solution of 0.5 M ammonium acetate. The content of P was determined after extracting the soil for 30 minutes with a mixture of 0.5 M $NaHCO_3$ and active carbon. The soil texture was determined as described by Plantedirektoratet (1994). The CEC was determined by the method described by Kalra and Maynard (1991). Total nitrogen was determined by the method described by Hansen (1989).

RESULTS

Soil characteristics and climate

The results of the initial characterisation of the experimental sites are summarised in Tables. The mean clay content varies from about 4 per cent at Jyndevad to about 24 per cent at Holeby. This variation is typical for Danish soils, where 24 per cent of the agricultural area is characterised by coarse sandy soils represented by Jyndevad, 28 per cent by loamy sands represented by Foulum, 24 per cent by sandy loams represented by Flakkebjerg, and 6 per cent by loams represented by Holeby (Madsen et al., 1992).

Table. Soil texture at the four sites at different depths. Particle size fractions, organic matter and calcium carbonate in per cent of dry soil.

Horizon (cm)	Clay < 2 μm	Silt 2-20 μm	Fine sand 20-200 μm	Coarse sand 200-2000 μm	Organic matter	$CaCO_3$
Jyndevad						
0-25	4.5	2.4	18.0	73.1	2.0	-
25-50	4.4	1.3	16.6	76.7	1.1	-
50-75	3.8	0.7	14.7	80.3	0.4	-
75-100	4.3	0.8	16.1	78.6	0.3	-
Foulum						
0-25	8.8	13.3	47.0	27.2	3.8	-
25-50	11.2	12.9	46.2	27.6	2.1	-
50-75	13.5	11.9	47.0	26.8	0.7	-
75-100	14.4	11.3	46.4	27.5	0.4	-
Flakkebjerg						
0-25	15.5	12.4	47.4	22.9	1.7	0.1
25-50	17.2	12.6	46.7	21.7	1.1	0.6
50-75	19.0	12.1	45.5	20.9	0.6	1.8
75-100	19.4	11.9	44.1	20.5	0.4	3.9
Holeby						
0-25	24.0	24.0	35.2	14.7	2.2	-
25-50	23.3	26.2	32.3	16.5	1.7	-
50-75	22.9	27.1	35.5	13.5	1.1	-
75-100	17.8	24.1	37.9	19.3	0.9	-

Table. Mean depth of the A-horizon formed at the soil surface, and classification of the soils according to the Soil Taxonomy System (Nielsen and Moberg, 1985). The values in brackets are standard deviations.

Location	Soil classification	Depth of A-horizon (cm)
Jyndevad	Orthic Halplohumod	32 (6)
Foulum	Typic Hapludult	44 (14)
Flakkebjerg	Typic Agrudalf	45 (16)
Holeby	Oxyaquic Hapludalf	35 (6)

Table. Mean values of chemical analyses of the soils at the four sites at different depth for samples taken in autumn 1996. pH is taken as pH($CaCl_2$)+0.5. P and K are measured as mg per 100 g dry soil. The cation exchange capacity (CEC) is measured as meq per 100 g dry soil. Organic C and total N is measured in per cent of dry soil.

Horizon (cm)	pH	P	K	CEC	Organic C	Total N
Jyndevad						
0-25	6.1	5.2	4.9	8.0	1.17	0.085
25-50	5.9	1.0	2.6	5.6	0.62	0.041
50-75	5.6	0.4	2.6	4.3	0.25	0.017
75-100	5.3	0.3	2.9	4.6	0.14	0.016
Foulum						
0-25	6.5	5.4	13.1	12.3	2.29	0.175
25-50	5.9	2.2	6.8	10.1	1.25	0.094
50-75	5.2	1.5	7.1	7.8	0.43	0.041
75-100	4.8	1.3	7.9	7.6	0.21	0.026
Flakkebjerg						
0-25	7.4	3.0	9.8	10.6	1.01	0.107
25-50	7.5	1.7	6.9	10.5	0.67	0.074
50-75	7.5	0.7	6.6	10.5	0.34	0.042
75-100	7.8	0.4	6.8	10.3	0.21	0.032
Holeby						
0-25	8.0	1.2	10.4	17.0	1.56	0.139
25-50	8.0	0.8	8.0	13.5	1.03	0.103
50-75	8.1	0.4	6.4	9.7	0.48	0.043
75-100	8.2	0.3	5.6	7.4	0.55	0.023

The remaining 18 per cent of the Danish agricultural area is mainly soils with a high content of organic matter or fine sand and some silty or calcareous soils. The classification of the soils at the four sites is shown in Table. The depth of the A-horizon varies between sites, and also considerably within sites. The A-horizon is the upper soil horizon, which is influenced by tillage and which visibly has a larger humus content. The larger average depth of the A-horizon at Foulum and Flakkebjerg is largely caused by very deep A-horizons in parts of the experimental area. These two sites are also the only sites where there is a consistent increase in clay content with increasing soil depth. The content of organic matter in the plough layer is about twice as high at Foulum compared with the other sites. The organic matter content does not, however, decrease as rapidly with depth at Flakkebjerg and Holeby as at Jyndevad and Foulum. This indicates considerably deeper rooting on the sandy loam and loam soils at Flakkebjerg and Holeby.

The high organic matter content at Foulum causes the CEC here to equal that at Flakkebjerg in the upper soil layers, despite the differences in clay content. The soil pH is highest on the sandy loam and loam soils, but the content of plantavailable P is lowest here, especially at Holeby. The content of plantavailable K is quite low at Jyndevad, indicating that potassium deficiency may become a problem on this coarse sandy soil. The climatic differences are modest across Denmark due to the low relief, but there are some differences between the sites in temperature, precipitation and potential evapotranspiration. The mean annual temperature is almost 1°C higher at Holeby compared with Foulum. This covers roughly the span of mean normal temperatures obtained in Denmark. The average annual precipitation varies from 626 mm at Flakkebjerg to 964 mm at Jyndevad. This covers most of the spatial variation in rainfall in Denmark. The spatial variation in potential evapotranspiration is much smaller. Table shows the application of slurry to the different crops in the rotations. Only cereals and beet receive manure. Different types of slurry are used at the four experimental sites. The actual content of ammonium and total nitrogen in the applied slurry in 1997 and 1998 was used to calculate the livestock density required for production of this manure. A slightly higher livestock density was required at Jyndevad, where cattle slurry was used. This was due to the higher content of organic nitrogen in cattle slurry compared with both pig slurry and digested slurry. The estimated livestock densities are lower than on most organic dairy farms in Denmark. The livestock density, however, increases considerably from rotation 2 to rotation 4.

Adjustments of the rotations

Some adjustments in both the design of the rotations and in the management have been made since the start of the experiment in 1997. The policy is that improvements to the rotations and management are permitted as long as the changes do not interfere with the three key factors. After 1997 rotation 1 in Jyndevad was changed from barley, 1st year grass-clover, 2nd year grass-clover, winter wheat to the one presented in Table. The reason for this was that crop rotations with a high level of grass-clover already had proved their sustainability (Askegaard et al., 1999). The change made it possible to compare spring wheat in rotation 1 with winter wheat in rotation 2, both following grass-clover and also to compare the pulse crops, lupins in rotation 1 with pea/barley in rotation 2. In 1998 more clover was introduced into the rotations in order to increase both the N2- fixation and the diversity.

Red clover was added to the grass-clover fields of the heavier soils and four clover species were mixed with the ryegrass catch crop and used in selected crops. From 1999 slurry was applied to winter cereals in mid-April instead of, as earlier, at the start of May. From the colour and growth rate of the winter wheat in the spring it was clear that even following a grass-clover crop, the wheat suffered from Ndeficiency at the beginning of the growing season. The row distance was increased at Jyndevad and Foulum in selected crops without catch crops in order to improve the weed control by mechanical hoeing. A successful establishment of the white clover is a prerequisite for the success of rotation 4 (with catch crop). Oat undersown with pure white clover showed a vigorous growth and high competition against the undersown clover at Foulum in 1997 and 1998. The plant density was therefore reduced in 1998 (from 400 to 300 plants m-2). In order to increase yields and reduce problems with takeall, triticale was introduced as the second winter cereal in rotation 4 in 1999 instead of winter wheat.

Design considerations

There were two main considerations in the design of the experiment. The first consideration was related to the wish to continue all or some of the treatments for a long time. In order to investigate the effects of the systems on soil fertility, the experiment should probably be run for at least three rotations, *i.e.* twelve years (Drinkwater et al., 1995). The second consideration was the requirement to perform other experiments and investigations within the framework of the experimental design, thus investigating the effects of treatments on the dynamics of both soils and plants and related effects on management. The requirement for a long-term experiment called for measures to eliminate soil and substance movement between plots, which can otherwise have considerable influence on the treatment effects (Sibbesen, 1986). The plots were therefore separated by both continuous vegetation and continuous bare soil. This layout ensures that neither soil nor weeds move between plots.

All management treatments and all measurements are performed with reference to fixed positions placed in the permanent vegetation between the plots. Ploughing is performed starting at the opposite side of the plots compared with the last ploughing operation in order to prevent permanent movement of soil in the plots. There is a side effect to this fixed position for field operations. The traffic by tractors and other vehicles always takes place on exactly the same parts of the plots. This may over time cause soil compaction in these strips. Measures are therefore taken to loosen the soil in these strips in conjunction with some of the tillage operations. The sub-division of plots into sub-plots and mini-plots enables experiments and studies to be performed within the systems. The basic requirement for experimental treatments to be carried out in either sub-plots or mini-plots is that they do not have long-term effects on the functioning of the systems. The broad definition of the three main factors in the experiment does, however, allow a large range of management treatments to be applied in sub-plots, including different cereal species, varieties, catch crops, mechanical weed control, strategies for manure application and soil tillage. Mini-plots can probably most efficiently be used for intensive measurements, sampling of plants and soil and for small experiments concerning effects of adding extra nutrients. The advantages of using the crop rotation experiment for such investigations is that it is possible to examine the interactions of such management effects with those of the main cultural factors: rotation, catch crop and manure application.

6

Crop Rotations: Food Security and Nutrition

KEY RESOURCE

In regions where single crops are the most realistic expectation, the seven aspects outlined below may be considered ‘key local resource issues which are involved in production and sustainability. They are not listed here for prescriptive purposes. Any prescriptive list should be compiled by people working and living in the relevant regions. They can, however, be used as realistic examples of the resource issues/problems/opportunities related to sustainable rotations for the semi-arid tropics. It is useful to link these issues schematically, and attempt to make their root causes explicit, perhaps through a flow diagram as in Figure. Similar diagrams could provide a basis of discussion and, perhaps, action which may lead to changes in cropping practice and agreement with farmers as to how to monitor the sustainability of their system.

- Too large a proportion and area of land is used for livestock or fallow which is marginally beneficial for livestock and fuel production. Reducing livestock numbers and/or keeping them penned in a cut-and-carry forage system, rather than allowing them to graze freely for most of each year, frees land for food cropping. Fallow as such confers no positive benefits to the sustainability of cropping. Indeed, contrary to the implication in Kassam*et al.* poorly-managed short fallows grazed by livestock can be associated with degradation of soil structure and chemical depletion.
- There are too few local markets to generate cash so that farmers aiming for food security could purchase livestock-products and fuel. There is a need for more markets to overcome the constraint described in 1 above. There are opportunities to add value, for example small-scale processing of soybean meal. Such steps are immediate and create local employment, and are presumably more realistic than suggesting that semi-arid Africa tries to compete internationally with, say, the USA for markets in soybeans. Subsistence farmers competing with mechanised agribusiness may be likened to a runner competing against a fast car.

- The health and strength of human (and animal) labour varies seasonally. Labour is needed for tillage when debilitating disease is most likely and when seasonal deterioration in diet starts. Sowing and weeding coincide with the seasonal minimum in diet. Thus, human labour constraints enhance traditional conservatism with respect to maintaining uncropped areas and, combined with traditional high intensity tillage, reduce the area of crops sown and make sowing untimely.
- For cropping systems to be sustainable, nutrients other than nitrogen need to be imported. This constraint cannot be overcome even if recycling of manures is made more effective and degrading practices such as mounding, burning and ploughing are stopped. It follows that cash generation (2, above) is essential to increase purchases of fertilizers.
- To sustain nitrogen levels, crop legumes are essential. Their absence near homesteads needs to be balanced by giving legumes a dominant role in rotations in outer fields.
- Parts of the cropping system with herbaceous legumes do not currently dominate cropping systems in dryland Africa nor are they likely to with present species and cultivars. Groundnuts are highly competitive and have poor water-use efficiency. They, however, have an appreciable, largely un-exploited, variation in nitrogen-fixing capacity. Cowpeas have a longer growing season and are more susceptible to pests and diseases than is desirable. For legumes to dominate part of a cropping system there is a need to move, for example, to 1:1 intercropping with cereals and 3:1 years legume/cereal rotations to maximise the net benefits of legumes to soil structure and fertility. The adoption of such rotations requires:
  - Legumes that can be used as an alternative to groundnuts;
  - Legumes which are promiscuous, that is able to be infected by many naturally-occurring rhizobia; and
  - Legumes which are more effective at nodulating and can benefit more from their rhizobia. These opportunities are developed in Chapter 5: they can be seen as part of a system that focuses on the improvement of a few selected grain legumes, and their processing. In this way the present causes of non-sustainability listed in 1 to 4 above can be addressed

Naturally-occurring tree legumes that have been sustainably managed by trimming and browsing can be augmented with the same species or introduced species, particularly as part of stable land systems, along fencelines or on permanent soil ridges.

Table. Traditional and Improved Crops and Cropping Systems for Selected Locations in the Semi-arid Tropics of India

Location	Crops		Stable cropping systems	
	Traditional	Improved	Intercrop	Sequential
Cambisols				
Jodhpur (380 mm, 11 weeks)	Pearl millet Moth bean Cluster bean Mung bean Sesame Rapeseed-mustard	Hybrid, pearl millet Improved mung bean Castor bean Cluster bean Sunflower Safflower	Green gram or cluster bean/pearl millet (BJ 104) *Cenchrus ciliaris* mung bean (T 44) (normal rain) *Cenchrus ciliaris* cluster bean (FS277) (for >500 mm rain)	Pearl millet-fallow Pearl millet (BL 104)-mustard (T 59) (for >500 mm rain)
Hisar (400 mm, 13 weeks)	Pearl millet Cluster bean Mung bean Chickpea	Hybrid pearl millet Cluster bean Improved mung bean Rapeseed-mustard	Pearl millet/mung bean Pearl millet/cowpea (fodder)	Pearl millet-chickpea Mung bean-mustard
Shallow Luvisols				
Anantapur (570 mm, 14 weeks)	Groundnut Pigeonpea Foxtail millet Sorghum	Groundnut Castor Pearl millet or sorghum Pigeonpea Mesta (rozella)	Groundnut (Kadiri-1)/pigeonpea (PDM 1) Groundnut/castor bean Pearl millet/pigeonpea	-
Medium Vertisols				
Rajkot (625 mm, 17 weeks)	Pearl millet Cotton Sorghum Groundnut	Sorghum Cotton Castor Groundnut	Groundnut (J-11)/ Groundnut (J-11)/ pigeonpea Cotton/green gram	-
Shallow Luvisols				
Hyderabad (770 mm, 17 weeks)	Sorghum Castor bean Pearl millet	Castor bean Sorghum Foxtail millet Pearl millet	Sorghum/pigeonpea Pearl millet/pigeonpea Castor bean/cluster bean Pigeonpea/mung bean	-
Deep Luvisols				
Bangalore (890 mm, 32 weeks)	Finger millet Maize Groundnut Horse gram	Finger millet Maize Groundnut	Finger millet (PR 202)/ soybean Groundnut/pigeonpea Finger millet/maize or pearl millet (fodder)	Cowpea-fingar millet

Shallow to medium Vertisols				
Solapur (722 mm, 23 weeks)	Pearl millet Sorghum Safflower Chickpea	Hybrid pearl millet Sorghum Groundnut Chickpea	Pearl millet/pigeonpea	Pearl millet-chickpea Mung bean-rabi sorghum
Bijapur (680 mm, 17 weeks)	Pearl millet Groundnut Cotton	Hybrid pearl millet Foxtail millet Sunflower Green gram Safflower	Groundnut/pigeonpea Pearl millet/pigeonpea Chickpea, safflower	Green gram-rabi sorghum Green gram-safflower
Deep Vertisols				
Hyderabad (770 mm, 25 weeks)	Sorghum Maize Safflower Chickpea	Sorghum Safflower Chickpea	Sorghum/pigeonpea	Sorghum-saff lower Sorghum-chickpea Maize-chickpea
Bellary (500 mm, 8 weeks)	Cotton Rabi sorghum Safflower Coriander	Rabi sorghum Safflower Field beans Chickpea Cotton/setaria	Sorghum/coriander Cotton/chickpea	

AREAS

The largest areas of seasonally wet-and-dry climate in south Asia are the 12 zones in India and the II and 12 zones in Laos and Cambodia. There are also much smaller, densely populated, areas, in east Java and the eastern islands of Indonesia. Venkateswarlu and others review rotations in the Indian semi-arid tropics. As in west and north-east Africa, tree and shrub legumes are sometimes present and planting of elite crops such as pigeonpea, (*Cajanus cajan*) are encouraged. Annual crops, however, are dominated by pearl millet and sorghum, as in the Sahel. Table lists traditional crops and their association with the length-of-growing season. As explained earlier this varies substantially with soil water-holding capacity in areas having the same annual rainfall. Maize is grown on favoured microsites, on soils with relatively high water-holding capacity and on the wetter margins of the region.

Animals are important in Asian farming economies, cut-and-carry systems of feeding being more common, for example, in Indonesia than in India or Africa. It is common practice at the end of the wet season, before the crops are harvested and residues available to livestock, for farmers to strip leaves from the Maize or sorghum during grain filling. As Pearson and Hall have shown, lower leaves can be removed to provide nutritious feed without harming the food crop, but removal of upper leaves, which is not uncommon, reduces grain yield.

Table. Management of two Cropping Patterns Commonly used in Upland Areas of Indonesia

Cropping pattern												Crop	Spacing (m)	Population per ha (1000 plants)
O	N	D	J	F	M	A	M	J	J	A	S			
--- maize ---												maize	2.0 × 0.4	25
-- upland rice --												upland rice	0.4 × 0.2	125
	------- cassava -------											cassava	4.0 × 0.4	6
--- maize ----												maize	2.0 × 0.4	25
--- upland rice ---												upland rice	0.4 × 0.2	167
	-------- cassava ------											cassava	4.0 × 0.4	6
				groundnuts			--- rice beans ---					groundnuts	0.2 × 0.2	250
												ricebeans	0.4 × 0.2	250

Table. Relative Yield of Transplanted and Dry-seeded Rice Obtained from on-farm trials in Iloilo and Pangasinan

Location	Seedling Establishment	No. of Fields	Grain yield (t/ha)
Iloilo	Dry-seeding	12	5.0
	Transplanting	9	5.0
Pangasinan	Dry-seeding	50	4.5
	Transplanting	26	4.0

As in Africa, trees for firewood are important in India. Shortage of fuelwood has caused other communities (*e.g.,* in east Java) to trade crop surpluses for petroleum products. Budgets of fuel and forage needs for farms and regions are published by Srivastava and Rao and others.

Where community populations are sufficiently sparse, it is realistic to recommend one-fifth of available land for silviculture and a substantial proportion for tree alley crops (16 per cent). Planting of trees for fuel and forage on 'alternative land' which has been degraded and is unstable, should be seen as a conservation measure and not as a cost against food cropping.

Tandon and Rego review crop responses in India to inorganic nutrients other than nitrogen. They conclude that responses are often large and that phosphorus and zinc are of major importance, though yield responses to potassium and sulphur are also common on some soils and iron chlorosis is an emerging problem.

Published figures giving areas of grain legumes (mostly chickpea, cowpea and gram), suggest that cereal-legume rotations and intercropping mixtures are, in practice, about 4:1 cereal and oilseeds: legumes. Thus, cropping systems in India have similarities with the Sahel though differing soils, social factors and markets should caution against extrapolating, or trying to transfer, solutions from one to the other.

Semi-arid cropping systems in India, and to a greater extent in Indo-China and east Indonesia, use dryland or upland rice when possible during the wet season. Table shows typical dryland cropping patterns in east Java, with rice the preferred wet season crop if the length of the growing season (duration of rainfall), or the availability of runon water, permit. Otherwise, monoculture Maize is grown, with little interplanting of grain legumes. The grain legumes are used as a second, relay crop, often hand-broadcast into the rice stubble or as a marginal wet-season crop, planted late and on less-favoured fields.

The advantage of semi-arid south-east Asia over, say, the Sahel, is its varied topography, so that many areas are able to capture run-on water which allows extension of the growing season. Thus, dryland rice grown with a significant period in standing water is possible. In India, ICRISAT has successfully popularised water harvesting to create local areas suitable for rice or relay cropping in less hilly regions. This increases both crop diversity and economic and ecological stability.

Traditionally, rice has been grown largely by transplanting seedlings from nurseries. Societal change, driven by women preferring to do piece-work for cash instead of maintaining transplant gardens, is encouraging direct seeding. This has several advantages. It shortens the crop life cycle allowing two crops in some situations (Gomez and Gomez 1983). It is less labour intensive and avoids transplant shock. Transplanted rice has no inherent yield advantage over directly-seeded rice.

A list of constraints on and opportunities for dryland cropping in south Asia might identify similar key issues to those listed in the previous section. Livestock is segregated from cropping in parts of the region, though notably not in India. The absence of livestock and dearth of trees for fuel elsewhere in Asia, however, illustrate that communities may move from devoting land to livestock and fuel crops to essentially food-crop only systems. Recognition of the benefits of tree legumes and their adoption seems broadly feasible. Within the food crop component of the cropping system, there remains the problem of having insufficient legume cropping to maintain a neutral nitrogen budget. The productivity of the herbaceous legumes currently grown is of the order of 0.3-0.9 t/ha only.

Thus, to achieve sustainable cropping in south Asia attention is needed to the following constraints:

- The proportion of herbaceous legumes in crop mixes, and the effectiveness of nitrogen fixation (residual value) of the legumes need to be increased.
- Visual symptoms should be used to diagnose the need for inorganic fertilizers, particularly those containing phosphorus and zinc.

Soybean and mungbean are currently grown, and may be more nitrogen-effective and less degrading than groundnuts. They are, however, less extensive than cereals, which suggests that more legumes should be grown as intercrops. Table summarises potential intercropping systems, including crop legumes, shown by ICRISAT to be agronomically viable.

Table. Promising Intercropping Systems that Include Legumes in India

System	Row Ratio	Potential Areas
1	2	3
Sorghum + pigeonpea	2:1	Semi-arid red soil regions of southern Telengana, semi-arid black soils of Vidarbha region and semi-arid black soils of Malwa plateau
Pearl millet + pigeonpea	2:1	Semi-arid black soils of Deccan region of Maharashtra as well as Karnataka and semi-arid black soils of Saurashtra
Pigeonpea + soybean	1:1	Semi-arid/sub-humid black soils of Madhya Pradesh and sub-humid red soils of Chotanagpur region
Sorghum + mung bean	1:1	Semi-arid black soils of Bidarbha and Marathwada region
Pigeonpea + sunflower	1:1	Semi-arid black soils of Deccan region of Maharashtra
Finger millet + pigeonpea Karnataka	4:2	Sub-humid red soils of Orissa and semi-arid red soils of southern
Groundnut + pigeonpea	5:1 3:1	Semi-arid red soils of Rayalaseema
Maize + mung bean	1:1	Sub-humid submontane regions of northwest Uttar Pradesh and Jammu and semi-arid black soils of Malwa plateau

In parts of Asia, for example northeast Thailand and east Indonesia, root crops, particularly cassava, are grown extensively for local consumption, local processing (*e.g.,* into packaged chips) and for export (*e.g.,* as stock-feed to Europe). The profitability of root crops for export varies widely. Their local use, however, and their good storage properties and transportability once processed, suggest that root crops will remain a component of semi-arid cropping. Inter-planting of a non-competitive legume crop shortly after planting cassava stakes could substantially increase the area sown to grain or forage legumes. The low yields of legumes suggest a need to improve both inorganic nutrition and rhizobial effectiveness.

CROPPING

Cropping in temperate dryland climates is based on cereals, mainly wheat and barley, as is cropping in the wet-and-dry regions of north east Africa above 2400 m and in the Near East. Although the climates in Australia and the Mediterranean areas under discussion are similar they have different traditions, the Mediterranean systems being aimed towards village food security until recently while the Australian systems, developed since, the

1880s, are aimed at producing bulk grain for export. Farm size differs, Australian mechanised farms averaging 2800 ha. Mediterranean systems recognise traditional practices and farmers' values, and there is more local consumption of food crops, so there is a variety of cropping systems around the theme of cereal/fallow or cereal/hay crop/fallow. Australian systems have developed more limited options around crop/legume pastures or crop/volunteer weedy species rotations. On the Australian dry margin and on fine-textured soils the systems usually include a long fallow (9-12 months), primarily to conserve water but also to assist the mineralisation of nitrogen, and control weeds and diseases. The efficiency of the system, as assessed by water use (WUE), generally ranges from 10 to 16 kg grain/ha/mm, which is very similar to that of cereal crops in the United Kingdom and Syria.

Table. Major Small-grain Production Systems of Australia Subdivided on Rainfall and Soil Texture

Winter Rainfall				Summer Rainfall
Low (<300 mm)		Medium	High	
Coarse-textured Soils	Fine-textured Soils			
C-(VP...)	F-C	C-(P...)	C-(P...)	SF-C
C-GL	C-C	C-C	C-P-P	F-SC
	F-C-(P...)	C-GL		SF-C-SF
	C-GL	C-GL-C-(P...)		-F-SC
		F-C		

Note: Key: C, cereal crop; GL, grain legume; SF, summer fallow; VP, volunteer pasture; F, fallow; P, legume pasture.

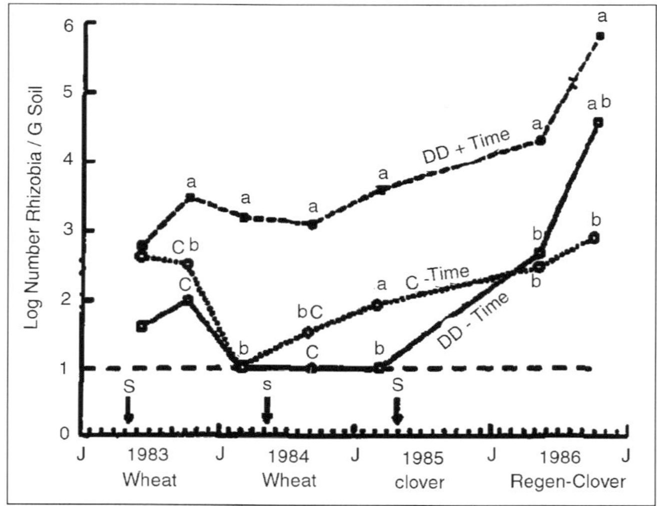

Fig. *Rhizobium Trifolii Soil* Populations over Four Seasons with Wheat and Direct-sown

The soil treatments were cultivation and lime 2.5 t/ha (•): cultivation and nil lime (O); direct drill and lime 2.5 t/ha (■); and direct drill and nil lime (□).

The dashed line at 1.0 log rhizobia/g soil denotes the limit of detection using the most probable number method. Different letters notate significant difference. Rutherglen, Australia sandy clay loam ph 4.2

Land use in Mediterranean climates presents problems for soil sustainability, particularly on hill land. These are not so much directly related to cropping practices (although contour ploughing, for example, is not uniformly practised) but to deforestation and depopulation of hill country (S. Orsi, personal communication). In Australia, land preparation methods vary greatly with location and season and may confer non-sustainability. Land to be cropped is usually heavily grazed shortly before the wet season; residual vegetation (pasture or weeds) is then killed with desiccant herbicide. The crop is then sown four hours to four days later either with minimum soil disturbance, or conventionally using 3-5 passes of a tined implement. There is some attendant loss of topsoil organic matter and structure. It is noteworthy that, in a system depending on the survival and growth of rhizobia, direct drilling creates lower bulk densities and maintains higher rhizobia populations than tillage.

Table. Estimates of Equilibrium Values of the Amount of Soil Organic Matter Nitrogen which Builds up under Various Cereal Crop-legume Pasture Rotations, and its Mineralisation in the Crop year, in Australia

	Equilibrium Soil N (kg/ha)	Mineralisation (kg/ha/yr)[1]
Continuous pasture	3350	134
2 crop: 4 pasture	1925	77
1 crop: 1 pasture	1450	58
Continuous crop	500	20

Note: [1]

A decomposition constant of 0.04 is assumed, but this would vary with season, crop and cultural condition.

Australian cereal-based cropping systems need to maintain a neutral nitrogen budget through the non-cereal phase. They require addition of phosphorus and other elements as inorganic fertilizers. Organic matter levels need to be kept up.

Dunne and Shier criticised the rotations used then as not maintaining fertility for cropping. As a result pasture legume periods were lengthened. From about 1950 to 1980, pastures were widely included in rotations, the period in pasture commonly being one and a half to three times that of the crop. The accumulation of nitrogen under pasture is illustrated in Figure.

The average annual increment of soil nitrogen from 22 estimates was 67 kg N/ha/year, the highest value being 182 kg/ha/year. Perry argues that the key to the functioning of the system is the rate of mineralisation which makes the nitrogen available to the crop. Table shows a 4-fold range in rate of mineralisation among four common crop-pasture rotations.

Nutrient elements other than nitrogen, particularly phosphorus, sulphur, zinc and molybdenum, are routinely added at sowing in quantities sufficient to compensate for their removal in the grain. As well as giving economic crop yield responses in the year of application, these nutrients have indirect benefits. Heavy applications of inorganic nutrients reduce the incidence of root diseases. The cereal-legume pasture system, so widely used from 1950 to 1980 and publicised internationally, depended on relatively favourable international prices for grains and sheep products, especially wool, and a balance between them. Successive years of drought with high grain and relatively low wool prices in the early 1980s led to intensification of the cereal crop phase. Continuous cropping became widespread and fewer legume pastures were sown.

Then, in the early 1990s, following a short period of artificially high wool prices and increased sheep numbers, wool became marginally economic so sowing and management of legume pastures were again largely abandoned. Recently, concern about erosion associated with overgrasing and the extent of bare soil that occurs under annual pasture species such as *Trifolium subterraneum* and medicago, has shifted attention towards the inclusion of grass, particularly perennial grass, in the non-cereal phase of rotations. This is supported by research which shows that the rate of soil acidification beneath perennial grass is slower than under annual ryegrass (*Lolium rigidum*) pasture in a seasonally wet-and-dry environment.

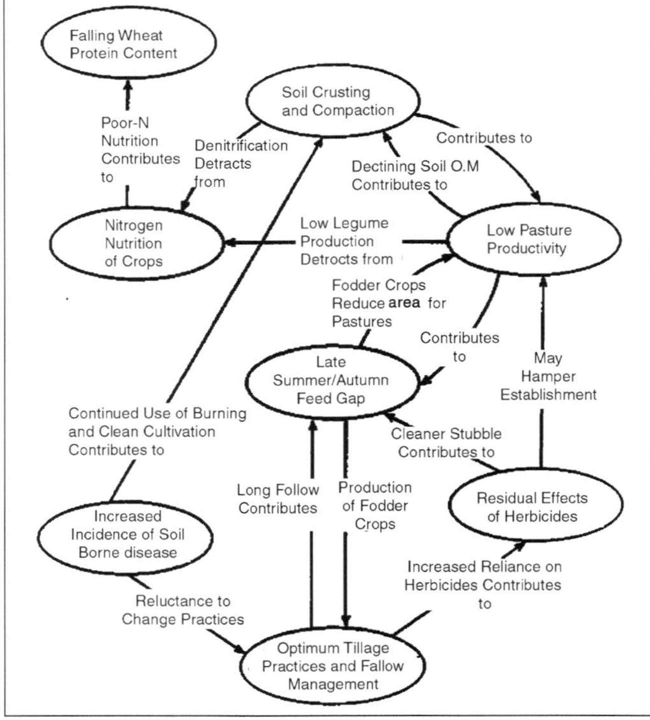

Fig. Linkages between Major Agronomic Problems or Issues Identified from the Conduct of a Topical rapid Rural Appraisal in the Forbes Shire in Australia

Table. Impact of Lupins on Soil Nitrogen under Cerealxand Cereal-Legume Rotations: Mean Soil MineralxNitrogen (NH_4+NO_3+NO_2 kg/ha, 0-20 cm) after Growing Wheat (W) and Lupins (L) in Rotation

Soil Mineral Nitrogen after							
Year 1		**Year 2**		**Year 3**			
W	71.2b[1]	WW	79.6d	WWW	65.8c	WWL	99.6b
L	105.7a	LW	102.5c	LWW	66.4c	LWL	109.2ab
		WL	127.8b	WLW	78.0c	WLL	120.5a
LL	172.3a	LLW	77.7c	LLL	126.1a		

Note: [1] Within columns, values not followed by a common letter differ significantly ($P<0.05$).

The present system is arguably not sustainable, because of the changes in markets and attitudes, depending as it does on nitrogen fixed by a pasture phase that is increasingly botanically-impoverished. This is illustrated by farmers' perceptions that the nitrogen concentration of the cereal grain is declining, and that pasture and soil factors are causing this.

The Australian system described above is a good example of how biological and soil sustainability depends on markets, in the absence of policies of land management agreed by land managers. It shows how sustainability and agronomic practices can change rapidly. Another issue raised by recent changes in Australia is the need for adapted grain legumes. The system is likely to return to sustainability only by introduction of legume crops, particularly field peas and lupins (*Lupinus angustifolius* and *L. albus*).

These fix more nitrogen than legume-based pastures and, depending on the proportion of nitrogen removed from the field as grain, usually give a residual benefit of about 20-50 kg N/ha (field peas) and 50-90 kg N/ha (lupins). The latter value exceeds that for many tropical grain legumes (Table). As Table shows, the soil nitrogen responds positively, quickly and substantially to the inclusion of a temperate grain legume (in this case, *Lupinus angustifolius* in NE Victoria).

Grain legumes currently occupy only 12 per cent in west and 4 per cent in the east of the cereal areas. These areas need to be increased to help maintain general levels of soil nitrogen, organic matter and good soil structure.

One aspect, perhaps better documented for semi-arid Australia than elsewhere is that of timeliness. As Loss *et al.* (1990), Perry (1992) and others have described, there is a window of opportunity for sowing at the beginning of each wet season. Any delay in sowing causes substantial reductions in grain yield.

This is a universal phenomenon that emphasises the need for:

- Knowledge of the probabilities of an effective beginning to the growing season, so that sowing will be neither premature and unreasonably risky or, as is common in developing countries, delayed so much by late, excessive cultivation that yield potential is routinely lost;
- Sufficient knowledge by the farmer of the best timing and type of soil

treatment at sowing for plant establishment. This requires ready availability of machinery and the use where possible of minimum tillage. Where cultivation is necessary because of local conditions, it should be coupled with sowing, so that at least some fields are sown in a timely manner and are not delayed until cultivation of the last field is complete.

Sowing is only one of several recurring operations in both crop and fallow phases. For each, timeliness may be critical. In some cases it is doubtful whether the operation should be undertaken at all. For example, ploughing and grasing at unsuitable soil-moisture contents can be detrimental to soil structure. Grasing is most beneficial if timed so that the livestock eat immature weed seeds, but it is detrimental if delayed until after seed fall and causes uprooting of plants when the topsoil is friable and dry. Within both Australia and Mediterranean regions there are examples of family farmers and large corporations with an ethic of land care who manage their soils for sustainability, and others who do not. Table summarises some of the characteristics of these extremes in Australia. Unsustainable management is almost always oriented to short-term market prices without knowledge or interest in the biological mechanisms which underlie soil and crop behaviour, and the patterns which characterise sustainable crop rotations. Table, although necessarily general, illustrates the fifth premise of this *Bulletin*, namely that tailoring an understanding of generic mechanisms to local situations is a powerful way of creating sustainable cropping systems.

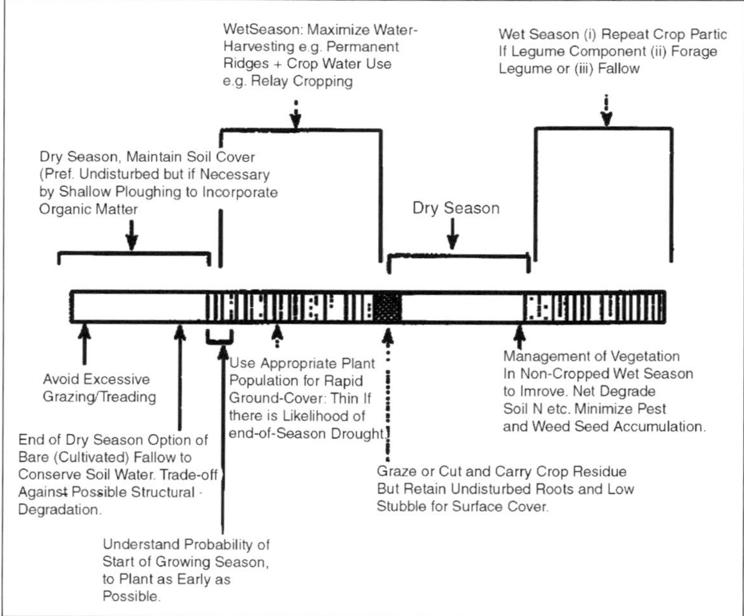

Fig. Scheme to draw Attention to the Need for clear time-lines, and Critical Timeliness of Operations, in Cropping in the Semi-arid Tropics

Figure illustrates a non-sustainable situation in east Australia. The key indication that the system is not sustainable (declining protein concentration in grain) is linked to its root causes within the system. Similar schemas, together with local knowledge of seasonal work demands and social constraints, can form the basis for debate about any action to be taken.

ROTATIONS AND SOIL PRODUCTIVITY

On the balance of experimental evidence, those cropping elements that help to maintain soil productivity in parts of Africa, Asia and Australia have been emphasised.

These include:

- The need to retain organic matter and soil structure.
- The need for appropriate tillage which, in many situations, involves minimal soil disturbance.
- The need to replenish the nutrients removed in grain, straw and livestock products.
- The essential need to maximize ground cover.
- The desirability of mixing crop species, particularly to include legumes with effective nodulation (grain crops, forage crops, trees and shrubs).
- The need for any fallow period to be managed rather than to be left bare to be exploited by livestock, or for weed seeds and pests to accumulate in the soil.

The question arises: how are these conceptual elements best exploited in crop rotations? Table provides a framework, and Figure illustrates the dynamic, and recurring, nature of decisions involving the choice of rotations and techniques for the maintenance of soil productivity.

Figure illustrates the value of effectively-nodulated legumes (in this case, a pasture ley) in increasing soil nitrogen, as long as the nitrogen is not removed as a product. Figures show associated increases in soil rhizobia and vesicular arbuscular miccorhizae under undisturbed pasture legumes. Tables show the impact of plant architecture and ground cover on soil loss and mineral leaching.

All the above give estimates of the benefits, to soil biology and chemistry, of the crop elements in a rotation. The aspects they illustrate are reflected by some Australian farmers' practices, under which the rotations chosen appear to maintain soil productivity though soil nitrogen levels at equilibrium may be only half those under continuous legume pasture. However, there are very few long-term studies which quantify the effects of different crops on soil conditions.

Indeed, such studies seem to be a priority to enccurage changes in cropping systems. For example, there are few data on nitrogen fixation by legumes on farms in Africa and Asia, and no known studies, that compare the effects of groundnuts on soil structure and fertility (which are possibly degrading) with other legumes such as soybeans. There is little information on the effects of fallow management on soil biology, particularly on weed seeds, pests and diseases.

On many farms, weeds spread uncontrolled during the fallow period, while experiments often unrealistically contrast weed-free extremes. Caporali and Onnis, in re-affirming the importance of crop/legume rotations, interestingly also show that weed density increased more during ten years of sunflower-based cropping under sunflower monoculture than under a sunflower-lucerne rotation. They show too that there are more weeds after mechanical- than chemical-control.

There is much information available on the effects of tillage and litter/stubble treatments on soil properties. The data, however, are not always consistent. This is not surprising as the timeliness of an operation is relevant to its effects on the soil and there are rapid changes in soil properties in the days following sowing. Moreover, several years of treatment are often needed before there are lasting effects upon soil characteristics that influence crop growth, for example, root length density.

Notwithstanding the variation, and at times the ephemeral effects of management on soil properties, it is clear that there are benefits from those crop rotations that enable crop residue retention. Three examples follow. Gill and Aulakh show that zero tillage for three years gave the lowest soil bulk density and highest crop yields. Radford *et al.* found that zero tillage plus stubble removal gave the least soil water storage but that zero tillage plus stubble retention gave the greatest soil water capacity.

Chan *et al.*record that ten years, conventional (three-pass) cultivation and stubble burning gave least soil organic matter whilst direct drilling plus stubble retention gave most organic matter.

In this study, tillage had a cumulative effect, causing 31 per cent differences in organic matter (2.42 per cent as against 1.68 per cent organic carbon); a loss of 1 per cent organic carbon was calculated to result in a loss of almost 3 cmole of negative charge per kilogram of soil. As a working hypothesis, it seems that stubble and litter retention (and incorporation if necessary to avoid removal by grasing or wind) is the key to sustainable rotations. Bearing in mind soil organisms, however, it may be necessary occasionally to till and mix the soil layers.

Mineral nutrition is, together with crop choice, stubble/litter treatment, and soil management, a key to sustainable rotations. For example, the removal of a 600 kg grain crop requires 1.8 kg phosphorus to be replaced using manure or mineral fertilizer. Nine kilograms of nitrogen are also needed (most realistically, from the breakdown of a previous legume crop or pasture, or by application of manure). A long fallow, in which deep-rooted perennials draw nutrients from depth, may help to balance the loss of nutrients. However, nitrate may move down the profile during the fallow, commonly giving a net annual decline in soil nitrogen in the absence of legumes, however, found that there was an increase (of 30 kg N/ha) during a short (7-8 months) fallow which balanced the loss during the crop phase. If weeds or pasture grow during the fallow and are grazed, then nutrient losses of about 20 per cent are common. These losses are by removal of animal products and volatilisation of urine. Grasing can also redistribute nutrients as dung from one field to another.

IN THE END OF THIS SATGE, WE CAN SAY

Much has been written on the theoretical and practical benefits of mixed cropping and on its problems (*e.g.,* competition). This work is not reviewed here but the aim is to describe briefly some of the cropping patterns currently used in the semi-arid tropics and contrast them with systems used in semi-arid temperate regions. The aims are to draw attention to practical constraints and to identify opportunities for more-sustainable dryland crop rotations in wet-and-dry environments.

The systems chosen are in Africa (the Sahel, east Africa and Ethiopian highlands), south Asia (India and east Indonesia) and in temperate Australian and Mediterranean regions. All include livestock. The climate varies within all these areas. Rainfall increases north to south in the Sahel, and changes with relief in the Rift Valley, and in the mountains and rain-shadow areas in E and NE Africa.

It increases from east to west in Indonesia and is lower inland than on the coast in India and Australia. In mountainous country in Ethiopia, tropical crops are replaced by temperate species as altitude increases. In contrast, temperate species only can be grown in the wet, cool season climates of much of Australia and the Mediterranean.

In labour-intensive Africa and Asia, villages are the units of production. They can be either traditional or, as in Ethiopia, reformed government-directed settlements. In Australia, the farm family is the decision-making and labour unit, whether they own or manage the properties. These differences in organisation and ownership are relevant to sustainability. Where a collectively-determined cropping strategy is not sustainable, migration from the village is accompanied by necessary (often slow) changes in land use, or the village is abandoned. In extensive, mechanised, family-managed systems, economic or biological non-sustainability causes changes in ownership which are quickly followed by changes in cropping practices.

Within labour-intensive systems food security has high priority. This and family under-employment encourage the focusing of labour near the house. The importance of the household garden thus varies with the size of the property. This is more apposite in Asia and the wet tropics than in the wet-and-dry tropics, where sequential cropping of vegetables under a household tree crop (using, if necessary, hand-carried water), gives way to a seasonal vegetable crop followed by fallow in the dry season.

When villages are moved and land use is changed to commercial agriculture, recreation or forestry, as has occurred to a marked extent in Thai highlands, the intensively-used household gardens are abandoned and the overall intensity of management and the fertility of the system is reduced.

Table. Components of Cropping Systems Illustrated in this Chapter

Location	Major crops	Major animals	Feed source
W. Africa Sahel	Millet/sorghum, maize, groundnuts, cowpea, sesame, cassava, yams, tree legumes	Cattle, goats, sheep, poultry	Fallow weeds, crop residue, vines, tree crop cuttings, oil cake, cull tubers
E. Africa	Maize/barley, sorghum, millet, tef (Ethiopia)	Donkeys, cattle, goats	Crop residues, fallow weeds, tree cuttings
S. Asia	Sorghum/millet (India), maize/rice (other); cassava; kenaf, wheat, groundnut, soybean, chickpea	Cattle, buffalo, goats, sheep	Bran, oil cake, crop residues
S. Australia	Wheat/barley; lupins, peas (subterranean clover, annual pasture medics)	Sheep, cattle	Pasture ley, crop residues

	Fatick	Foundiougné	Gossas	Kaffriné	Kaolack	Nioro	Sine-Saloum
Total area (km^2) (ST)	2646	2 959	2 330	11 847	1 880	2 277	23 939
Rural population	186 973	98 629	139 775	283 868	136 602	154 829	1 000 678
Density (population/km^2)	71	33	60	24	73	68	42
No. of farms	20 393	9 781	15 351	29 635	15 486	13 969	104615
Mean pop./farm	8.71	9.58	8.65	9.10	8.38	10.53	9.10
Livestock							
Cattle	71 300	57 800	55000	157 500	62 000	114 500	528 000
Horses	36 700	8 100	22 400	48000	1 250	5 600	140 300
Mules	13 300	2 900	5800	11 900	7900	2 500	88 460
Sheep/goats	176 000	207 500	75 000	100 500	95000	122 500	751 500
Areas							
Cultivated (ha) 83/84	93 649	96 575	151 379	285 386	98000	145 965	870 956

Proportion cropped:							
Millet and sorghum	52	35	39	41	41	33	40
Maize	-	2	1	2	1	2	2
Groundnut	47	62	59	55	57	64	57
Area/inhabitant (ares)	5.00	6.35	8.54	7.47	5.34	7.79	6.86
Area needed for wood[a] (km^2)	37 395	19 725	27 955	56 774	27 324	30 966	200 139
Area needed for livestock[b] (km^2)	256 520	184 460	172 120	435 860	185160	265 700	1 489 820
Total Areas (T)	387 564	290 760	351 454	778 020	310 484	442631	2560213
Degree if exploitation (T/ST%)	146	98	151	66	165	194	107

EXAMPLES

In substantial areas of semi-arid west Africa and the highlands of Ethiopia the present land use is not sustainable. The areas needed for food crops, for livestock and fuel exceed the actual land available.

It is unrealistic to suggest that changing crop rotations will provide the quantitative answer needed in areas such as those in Table. The most obvious answer is to remove livestock from the system. According to Table they need one to two times the area currently under food crops. Animals are kept for prestige, for their products and for traction. Substitutes for their products (*e.g.*, soy milk and meal) might be provided more cheaply, in real terms, if soil sustainability is assigned a value. They could be traded for crop surpluses. Grain legumes can be substituted for some livestock products and increased planting of a wider variety of legumes would improve the sustainability of the cropping system. This has already be done elsewhere, for example in parts of east Indonesia.

The value of draught animals is also debatable. Sandford interprets conflicting data to suggest that farmers often use animal traction to increase the area cropped, rather than for more timely management of existing land to increase yields. If this is so, livestock reduce productivity in two ways: directly through their need for food and through their traction being used to increase the area, but not necessarily the management skill.

There are two obvious difficulties when recommending removal of livestock from a district. *First*, the livestock may be owned by people other than those who cultivate. In these circumstances, integration of livestock with cropping is unlikely. Livestock removal usually involves relocation or re-employment of the herdspeople, neither of which is easy. *Secondly*, it is not easy to identify and foster peasant-based industries to generate cash for fuel and livestock products such as milk. Conventional wisdom is turned upside down by the suggestion that, in densely populated regions where fallows are short (<4 years) and land use is not sustainable, land used for livestock and fuelwood should be turned to food crops. For example, recent inventories and land assessments for east Africa place equal priority on livestock, fuel and food. Such equal emphasis is surely inconsistent with the belief that the priority product of a cropping system is food security (Chapter). Similarly, and perhaps underlying the notion that fallow-for-forage has equal status with cropping, is the widely-repeated belief that soil degradation is caused by shortened fallow periods and, therefore, that sustainable cropping depends on some minimal fallow period between crops. Many current land assessments include calculated 'fallow requirements. Fallow as such, however, does not restore soil structure, biological activity or nutrient status. A fallow, long enough to allow re-establishment of deep-rooted perennial plants crops that draw nutrients from depth and maintain soil surface cover, will improve soil productivity provided that nutrients are not removed (*e.g.,* as milk or meat) from the system. This rarely holds nowadays. Most commonly, a short fallow of weedy annual species leaves the soil bare and liable to erosion in the dry season, there is structural degradation through trampling and plant uprooting, and insufficient soil water is conserved for the next crop. There is no reason why a well-managed continuous cropping system should be less sustainable than one including unmanaged short fallows from which nutrients are removed by livestock or as fuel.

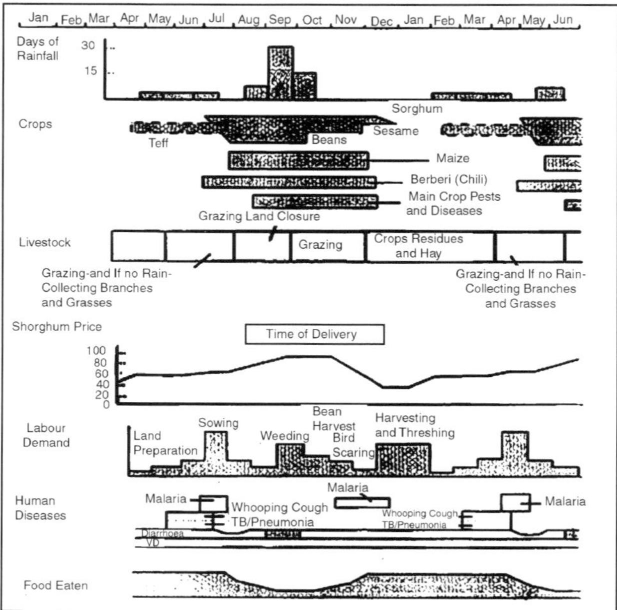

Fig. Seasonal Calendar of a Peasant Association in Wollo, Ethiopia

As well as taking up land for their feed, animals are a key constraint on crop rotations. Poulain describes a typical example, in the Serere, Senegal: 'the system comprises a typically triennial rotation and the integration of cropping with cattle rearing and fodder trees'.

The herd is enclosed in one field during the crop-growing season and immediately after harvest the cattle are released to graze on crop stubble and *Acacia albida* trees. Around the village there are permanent fields of millet, cowpeas and spices fertilised with household refuse while more distant from the settlement the rotation is typically groundnut-millet-fallow.

Seasonal cropping and grasing patterns for mid-altitude Ethiopia are shown in Figure. The specific food crops favoured in Ethiopia vary with altitude, but the seasonality of crop growth, livestock needs and labour requirements, shown in Figure, are broadly typical. This figure also illustrates the pressure on human resources in the semi-arid tropics during the period around the start of the wet season, when the highest demand for labour (for tilling and planting) occurs at a time when human diets are deteriorating and there is a high incidence of debilitating diseases. Local distributions of cropping largely radiate from the house. As shown in Figure, gradients of organic matter and nutrients from the house create semi-permanent gradients in crop rotations. Continuous cropping nearest the house is sustained by replenishment of nutrients using animal and human manures and food residues. Long-season crops (*e.g.,* Maize), and/or those favoured socially (*e.g.,* t'ef, in Ethiopia) are grown nearest the house. Examples of spatial variations in crop rotations are illustrated in Figures.

There is a wide diversity of cropping systems in dryland Africa. There are 'parkland systems' in semi-arid west Africa. On the arid margin of cropping in north and east Africa there are traditional cultivated fields among bush-fallow as well as parklands. The parklands of the Sahelian zone are predominantly *Acacia albida* with food crops, particularly pearl millet and groundnuts. These are structurally and functionally similar to the *Prosopis cineraria/pearl* millet parklands of parts of India.

One crop or, at best, two in relay, can be grown in the seasonally wet-and-dry climate of the parkland that has a perennial tree component. The cultivated ground layer carries an annual food crop, the stubble of which is grazed leaving the ground more or less bare through the dry season. The land used for food crops is traditionally cultivated, often by first cutting-back trees and bushes at the start of the wet season. The soil is ploughed and soil and organic matter piled into mounds, which are then burnt.

The ash from the mounds is then spread and the soil re-tilled as often as necessary. For t'ef in Ethiopia, there may be as many as nine tillage operations spread over three months, with consequent loss of cropping opportunity (soil water which is not used). Tillage and mounding are labour-intensive. They destroy topsoil structure and disrupt the continuity of macropores (Chapter). Nonetheless, and despite the scientifically-demonstrated benefits of maintaining soil surface cover, traditional tillage can give higher yields than minimum- or zero-tillage.

As shown in Figure, the benefits from zero tillage may take several years to accumulate so it is not surprising that one-year experiments in the Sahel failed to demonstrate the benefits of minimal soil disturbance. Ridging may also be practised. This helps to reduce wind erosion and wind-caused seedling death, improving establishment by 50 per cent in experiments. While the benefits of ridges seem intuitive, it does not follow that the ridges need to be broken and re-formed each year. Permanent ridges have the advantages of both reducing wind erosion and maintaining undisturbed topsoil.

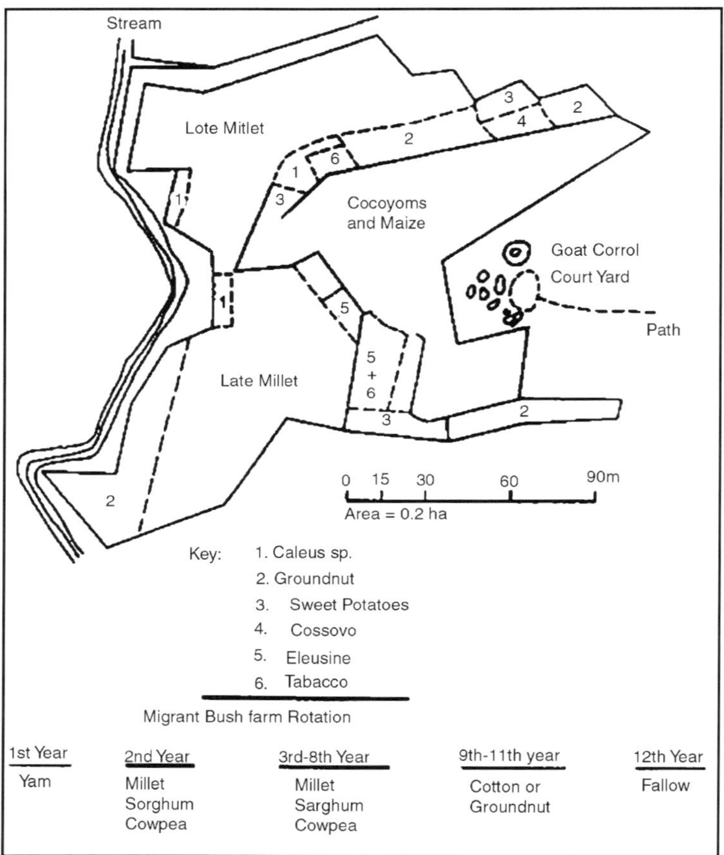

Fig. Homestead and Bush Field Land use and Cropping Pattern at Dalung, Benue Plateau

The main food crop is usually pearl millet but sorghum and Maize are grown in more favoured areas. The cereal crop is usually planted between the trees in hills or pockets of up to ten seeds. The seedlings are thinned or compete with each other so that commonly only 2-3 survive to maturity. A low plant density, commonly 5000 hills, is equivalent to between 10 000 to 15 000 plants per hectare. In these circumstances, on-farm experiments suggest that inorganic fertilizers may give only marginal economic returns. Again, the dilemma raised biological sustainability may not equate with short-term economics. As Figure illustrates, plant populations need to be increased substantially (with attendant risks of population-induced water deficits, in dry years) to get economic benefit/cost ratios from applied inorganic nutrients, at least in the year in which the fertilizers are applied.

However, this and similar analyses, highly relevant to short-term decisions of individual farmers, do not allow for:

- Residual benefits from fertilizers in the seasons following application of phosphorus and nitrogen, as found by Bationo *et al.*
- Indirect benefits from the fertilizers, such as, say, reduction of the incidence of parasitic weeds (*e.g.,* Striga) and the increased effectiveness of nodulation by rhizobia in subsequent legume crops.

Millet and sorghum yields are generally higher when grown the year following a grain legume. Though, on poor soils, legumes such as cowpea in rotation with millet have little effect on the millet yields if phosphorus fertilizer is not added. The beneficial effects of grain legumes are somewhat disputed. There is substantial evidence showing relatively small residual nitrogen benefits from groundnuts and soybeans. Australian systems are similar in this respect. The benefits, associated with relatively high nitrogen fixation, only accrue when nutrients, particularly phosphorus, are adequate for both rhizobia and crop growth. The systems require small inputs of inorganic phosphorus for sustainability.

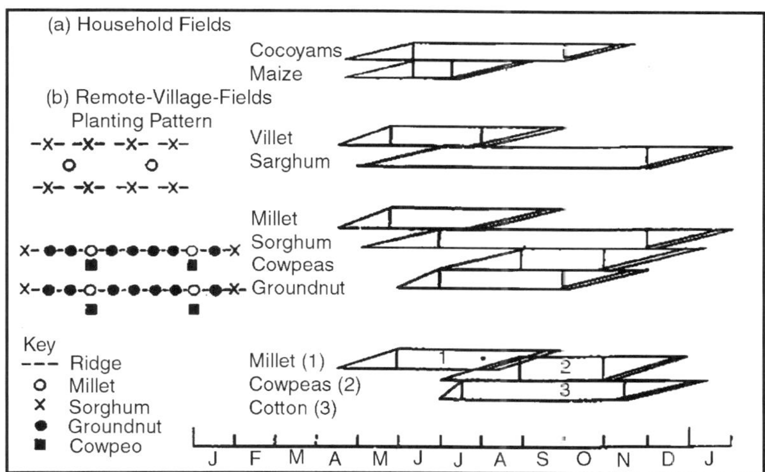

Fig. Cropping Schedules and Pattern of Planting for some Crop Combinations in Northern Nigeria

Fertility is increased under tree legumes by:

- Fixation of dinitrogen gas from the soil atmosphere.
- The gathering of livestock and humans beneath their shade so concentrating their manure.
- Recycling of nutrients from the deep subsoil by leaf fall.
- Trapping and accumulation of nutrients moved laterally by wind or water from elsewhere.

Since, soil fertility is higher beneath the trees, crop yields may also be higher (though in dry years yields can be expected to be lower because of the water used by the trees). Thus, farmers often sow sorghum under the trees to take advantage of the improved soil conditions which are not so well exploited by pearl millet, with its lower yield potential.

Figure illustrates the nitrogen benefits from the prunings of tree legumes. Juo and Payne review the use of tree legumes as windbreaks to alter the crop micro-environment. They conclude that there are beneficial effects of the trees other than nitrogen enrichment, but go on to question the use of tree legumes as alley crops in the semi-arid tropics: 'In West Africa, drought stress and acute P deficiency are major factors limiting the growth and establishment of leguminous trees and shrubs at closer (than naturally-occurring) spacings.

For example, even at row widths of 6 to 8 m, nearly double those recommended for humid and sub-humid conditions, marked yield depression was observed when alleys are pruned up to 3 times during the short cropping season in semi-arid India.

Thus, the adoption of alley cropping systems in the semi-arid tropics would largely depend upon the identification of specific locations where soil texture and fertility level, groundwater table and rainfall distribution are more favourable for alley establishment'. A similar view led to targeting existing fencelines as the most realistic sites for the introduction of tree legumes in Ethiopia.

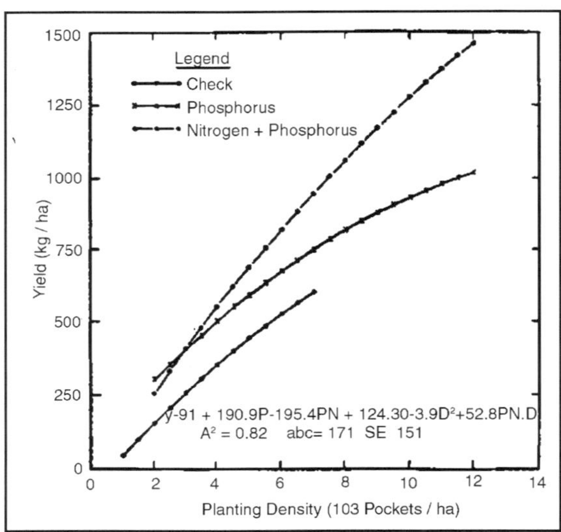

Fig. Effect of Plant Density and Fertilizer use Options on Millet Yield in farmers* fields where Y = grain yield kg/ha, D = Plant Density (10^3 Pockets/ha) and PN represent P Fertilizer only and P + N fertilizer Codes with values of 1 and 0 Represent Treatments with and without Fertilizer Respectively.

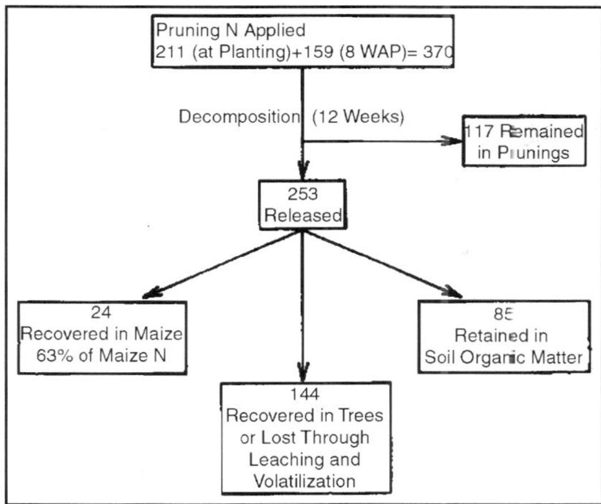

Fig. Partitioning of Nitrogen from Prunings of *Leucaena Leucocephala* to Crop, Tree and Soil in an Alley-cropping System on an Lixisol at Ibadan, Nigeria

Table. Strategies for Introducing Forage Legume into the Ethiopian Crop-livestock Systems

Strategy		**Low altitude <2000 m**	**Medium 2000-2400 m**	**High >2400 m**
	1	**2**	**3**	**4 5**
Forage strips	Trees	Leucaena *(Leucaena leucocephala)*, Sesbania (*Sesbania brachycarpa*), pigeon pea *(Cajanus cajan)*	Tagasaste (*Chamaecytisus prolifer*). sesbania Pigeon pea	Tagasaste
	Herbs	Siratro (*Macroptilium atropurpureum*), silver leaf desmodium *(Desmodium intortum)*, greenleaf desmodium *(Desmodium uncinatum)*, Verano stylo (*Stylosanthes humilis)*	Desmodium (greenleaf) Axillaris *(Axillaris* spp.) white clover, vetch	Vetch
Backyard forage	Trees	Leucaena, sesbania, pigeon pea	Tagasaste, sesbania pigeon pea	Tagasaste
	Herbs	Desmodium, alfalfa (*Madicago sativa)*	Alfalfa, vetch	Alfalfa

Undersown and relay legumes in crops	Herbs	Lablab (*Lablab purpureus*), cowpea (*Vigna unguiculata*), siratro, greenleaf desmodium,	Vetch, desmodium	Vetch
		Verano stylo, vetch (*Vicia sativa*), *Cassia rotundifolia*		
Oversown legumes in pasture and fallow	Herbs	Verano stylo, Siratro	?	?
Livestock exclusion areas	Trees	Leucaena, sesbania	Tasasaste, sesbania	Tagasaste
	Herbs	Siratro, stylos, Axillaris	Axillaris, siratro	?
Conventional pastures and special-purpose forages	Herbs	Stylo (Verano), siratro, desmodium (greenleaf and silverleaf)	Desmodium (greenleaf and silverleaf), vetch	Alfalfa, vetch

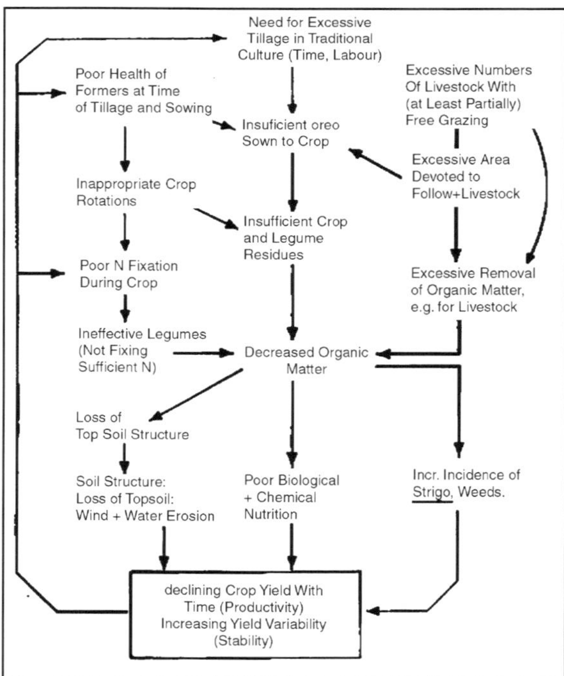

Fig. Scheme to Illustrate how Observed Deterioration in Agriculture may be linked back to Root Causes, which are Themselves Interrelated

Others rightly draw attention to the need for technical packages which incorporate various options for tree legumes: intercropping and alley cropping; planting along contour lines and farm boundaries; as wood lots; as wind-breaks and the promotion of natural regeneration of trees in pastoral areas. It is best for the farmers to decide which option fits their goals.

To conclude this overview of African dryland systems, it is appropriate to consider their constraints. Obviously, the main constraint is the rainfall amount, its distribution and reliability. These severely limit the opportunities for increasing productivity per unit area. The limitations, nonetheless, can be tackled effectively using a range of strategies. For example, in a region of Somalia, with a rainfall of 580 mm in a good year, four practices for better soil water use were tried (ridging; bunding; intercropping with a grain legume; incorporation of residue from the previous cereal crop). Even in a season of above-average rainfall, all increased the yield of sorghum, the main crop, by 30-50 per cent. Clean fallow (a crop- and weed-free season), with the aim of conserving water for the next crop, did not improve the sorghum yield. In areas with higher and more reliable rainfall, opportunities for water conservation and relay cropping increase.

7

Sustainable Management of Agricultural Land and Crop Productivity

INTERRELATIONSHIPS OF SOIL RESOURCES AND CROPPING SYSTEMS

A traditional view of the influence of soil is that it provides an opportunity, or a constraint on the type of cropping system that can be implemented and its productivity. A more responsible view is that 'the soil' combines various properties which interrelate and are directly influenced by the procedures of cropping. Figure attempts to illustrate this for a non-sustainable cropping system, in which soil resources are declining through changes in surface properties, subsoil compaction, loss of organic matter and reduced biological activity. It shows that each of these properties may affect the others and that all can directly and indirectly reduce plant performance, as well as affecting other aspects including erosion, salinisation and acidification. Some of the mechanisms underlying these interconnections are described below.

SOIL PORES AND WATER CHARACTERISTICS

Soils are made up of three parts: mineral materials, organic matter and space (called pore-or void-volume). The relative importance of these varies with soil type but pore space can occupy about half the volume of a medium-textured soil. At optimum water content for plant growth, approximately half the pore space is filled with water and half with air. The proportions of water and air can change rapidly depending on weather, evapotranspiration and other factors.

The dimensions (size, shape and arrangement) and number of pore spaces are most important in determining soil water and soil structure.

Porosity is the volume of soil voids (*pore* space). It is expressed in relation to the bulk volume of the soil. The water holding capacity of a soil depends on its porosity, and the size distribution of its pores. Small pores retain water at greater suctions than larger pores. The moisture (or water) potential is the amount of energy required to remove water from a soil; field capacity is the water-holding capacity after a free-draining soil has been allowed to drain.

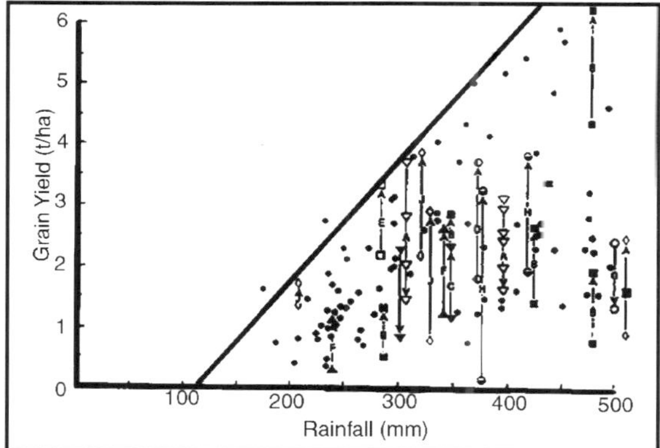

Fig. Relationship between Grain Yield of wheat and April-October Rainfall for Experimental Sites and Farmers' Paddocks in Southern Australia

The suction corresponding to this state has variously been defined as 0.33, 0.1 and 0.05 bar and so the convention used should always be checked. Wilting point, beyond which plants cannot exert sufficient suction to remove water from a soil, is generally considered to correspond to a suction of 15 bar. Available soil water (ASW) is the amount of water which is available for uptake by plants, namely that held at suctions between wilting point and field capacity. It varies with soil type and can be correlated with the clay content and structural arrangement of the soil. It varies also with soil treatment because the size and distribution of pores in the topsoil reflects surface exposure, normal seasonal wetting and drying, and management.

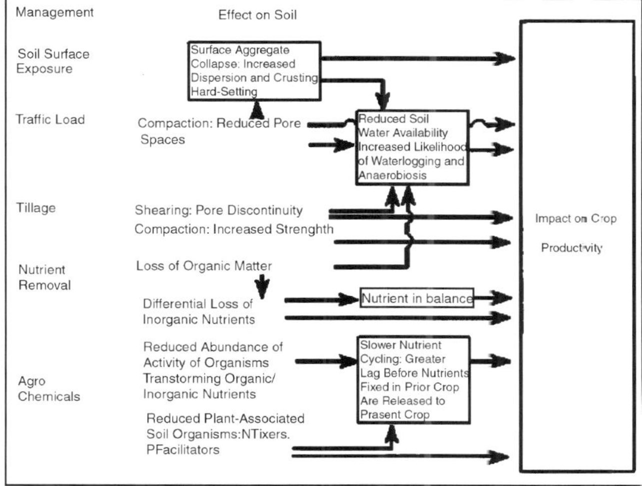

Fig. Interrelationships between Management and Soil as they Impact on Crop Productivity

Table. Generalised view of Pore Size Groups and Functions

Pore Diameter (mm)	Function	Equivalent Particle or Aggregate Size[1]	Biological Cause (if not Due to Natural Particle Arrangement)	Equivalent Soil Water[2] Tension (kPa)
>0.5	Aeration and Water Transmission	>1.6 (mostly Gravel Size, Some Coarse sand Size)	Ants, Worms	<0.6
0.50-0.05	Water Transmission (Infiltration, Permeability)	0.16-1.6 (mostly Coarse Sand Size, some Fine Sand Size)	Roots	0.6-6.0
0.0005-0.05	Water Storage	0.0016-0.16 (Mostly Silt and Fine sand Size, some Clay Size)	Lateral Roots, Root Hairs	6.0-600
< 0.0005	Residual (bound) Water, Unavailable to Plants	<0.0016 (Mostly Clay Size)	Fungal Hyphae and Bacteria	>600

Note: 1 Equivalent particle size = 3.2 × pore size (assuming spherical, uniform size particles).

2 Based on equation: Pore diameter (mm) = 0.30/soil water tension (kPa).

Table. Typical Values of Water-holding Capacity of Soils

International Textural Class	Field Capacity (10 kPa) (Gravimetric)	Wilting Point (Gravimetric) per cent	Available mm Water/m Soil
Coarse sand	8	4	80
Sand	14	4	150
Loamy sand	18	7	160
Sandy loam	26	9	180
Loam	30	13	180
Silty loam	34	16	200
Sandy clay loam	26	15	150
Clay loam	34	18	180
Silty clay loam	43	20	190
Sandy clay	29	19	140
Clay	42	25	180

Williams *et al.* (1983), studying the water content of 244 soil samples, found that the ASW of well-structured soils was one-third to twice as large as that in comparable (similarly-textured) poorly structured or degraded soils. Bearing in mind that ASW varies with natural weathering and management.

Hydraulic conductivity (K) of a soil is its conductivity to movement of water down a pressure gradient. High values of K are associated with well-structured soil and contiguous pores; they allow high infiltration rates and rapid drainage. Earthworm channels, which can have populations of 500 m^{-2} in Mediterranean climates (Barley 1959), and continuous deep voids left by dead roots (5-10 000

m^{-2}) contribute greatly to hydraulic conductivity. Hydraulic conductivity varies with soil type and management (Table). K values below 10 mm/h are low and likely to cause run-off following rainfall or problems with irrigation, given that steady rain falls at about 10 mm/h. K values of 10 to 20 mm/h can give intermittent run-off (a downpour falls at about 50 mm/h) while values up to 120 mm/h are associated with occasional, increasingly rare run-off. Values above 120 mm/h may facilitate regular drainage to the groundwater, causing potential problems for heavily-fertilised soils, and those treated with effluent, herbicides or pesticides.

Both soil water content and saturated hydraulic conductivity generally relate to the number and continuity of pores, particularly the larger macro-pores. It is, however, difficult to measure these soil attributes and they are highly location-specific, so that variability is great and they sometimes have little interpretive value. Moran *et al.*, (1988), however, in a study of a soil in a wet-and-dry environment, show that a soil treated with minimum tillage had more pores, identified directly by image analysis, and higher hydraulic conductivity, measured in the field, than did a similar soil traditionally cultivated. Figure also illustrates the impact of management on soil pore soil water characteristics; this is discussed further in the section below.

Surface sealing and crusting are common in wet-and-dry climates. Sealing increases run-off and seriously reduces the amount of water infiltrating into the soil thus reducing the water held in the soil. Sealing can also increase ponding at the surface and thereby evaporation. Infiltration rates are often reduced 1000-fold by crusting. The crust can have a skin with a conductivity of only about 0.1 mm/h, able only to accommodate the lightest rate of precipitation (fine mist) and commonly overlies a layer of poorly-aggregated material which also has a conductivity substantially lower than that of the underlying soil. Chase and Boudouresque (1989) and Chase *et al.*, (1989) illustrate the impact of crusting in the Sahel on increased run-off from soils. They show how run-off may be reduced, and the depth of wetting increased, by covering the soil surface with mulch.

SOIL FABRIC, DENSITY AND STRENGTH

The solid soil material is arbitrarily divided into fractions, the proportions of which determine the texture of the soil.

The fractions in the International Method of particle-size analysis are:

- *Sand*: Grains, which feel gritty and are large enough to be seen and felt individually; coarse sand has particle sizes between 2 and 0.2 mm and fine sand 0.2 and 0.05 mm.
- *Silt*: Imparts a smooth, soapy or silky and only slightly sticky feeling, silt grains cannot be individually detected; their particle sizes range from 0.05 to 0.002 mm.
- *Clay*: Gives a sticky feel to the soil. Clay particles are less than 0.002 mm diameter.

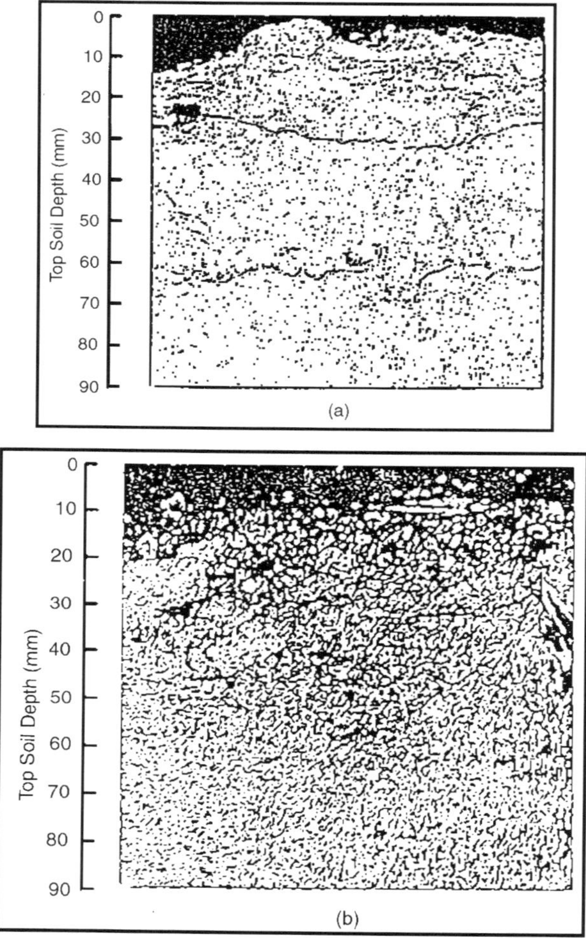

Fig. Photographic Images of Vertical Slices of Soils: The Vertical Faces of soil were taken from a Direct Drill (DD) Treatment *(a)* and a Conventionally Tilled soil *(b)* at the same Location. Hydraulic Conductivities were 42.5 and 5.0 mm/h Respectively

These solid fractions contribute to the consistence and strength of the soil, and their packing determines bulk density.

Bulk density is a measure of the packing or compression of the three constituents of soil. Just as the inherent bulk density of a soil will vary by 30 per cent according to its constituents, so the limiting values of bulk density for root penetration will range from about 1.4 g cm^3 in a soil of clay texture to 1.8 g cm^3 in a sandy one.

Soil strength is the resistance of soil to shearing or structural failure. This reflects the friction which is built up between the soil and an implement, and depends on the density, and the roughness and shape of the soil particles. The shear strength of an individual clod decreases with wetting but, more importantly, the strength of the bulk soil increases with increasing moisture to about the lower plastic limit (known to field operators as the 'sticky point'), at which each particle is surrounded by a film of water which acts as a lubricant. Soil strength drops sharply from that point to the upper plastic limit, where the soil becomes viscous. The difference between the moisture content at the upper and lower plastic limits, termed the plasticity index, is an index of the workability of the soil. A large range or high plasticity index implies a need for large amounts of energy to work the soil to a desired tilth.

SOIL STRUCTURE AND CROP GROWTH

Soil physical properties affect root and shoot growth directly and indirectly, the latter for example through poor drainage causing pores to fill with water and plants to suffer from anaerobiosis. Root growth has been described under various soil physical conditions, but relationships have only rarely been established between features such as crop yield, root growth and soil pore size distribution or conductivity, a more aggregate measure.

The difficulty in establishing simple relationships does not mean that soil has little influence upon root growth; rather it points to the complexity of the interactions and the internal homeostasis which plants maintain. Roots both elongate and proliferate and spread laterally as they grow and age. Concomitantly some roots die and others become suberised and function as conduits, but not as absorbers, of water and nutrients. Roots can elongate downwards as fast as 8 cm/d, as for example, soybean growing in a silt loam in a rhizotron. Deep-rootedness and maximum rooting depth reflect soil properties (for example, roots will not grow through pores that they cannot deform to a larger diameter than the root).

However, relationships are not often reported. Maximum rooting depth varies with species and soil type. For example, wheat roots penetrated to 0.8 m in heavy-textured soils and to 1.2 m in a loamy sand but it is often found that a variety will have a consistent rooting depth across similar soil types in a particular year or in one soil across several years. Angus *et al.* found that rice and six dryland crops (mung bean, cowpea, soybean, groundnut, Maize and sorghum) extracted different amounts of stored soil water (ranging from 100 mm for rice to 250 mm for groundnut) and that extraction was, in part, related to rooting depth.

The spread of roots with age can be related to the growth (increase in weight) of the whole plant, and to accumulated temperature or growing day-degrees (GDD); indeed, there is some evidence that temperature influences the direction of newly-appeared roots as well as the rate of appearance and extent of growth. Clearly, however, there are factors other than plant size, temperature and soil which influence root proliferation. Otherwise the plants sown at three different times of year in the same soil in Figure would align their root growth along a single growth-GDD relationship. These other factors, of which day length is probably particularly important, tend to mask underlying relationships between growth and soil structure.

Figure shows that tillage can affect root length, though in this case its effects took three years to develop. Measurable differences in soil porosity developed under two tillage treatments: in the first year both root growth and water infiltration (K approximately 5 mm/h) were the same under minimum tillage and conventional tillage. By the third year, when differences were measured between roots, infiltration rates were 84 mm/h in minimum tillage and 0.2 mm/h under conventional tillage. Despite the differences in root growth there were no substantial differences in grain yield, reflecting the overall constraint of climate in the semi-arid environment.

One of the clearer associations between soil porosity and plant growth is described by Tisdall and Cockroft and Tisdall. Soil was ameliorated by deep ploughing, inserting gypsum at depth, and incorporation of straw and green manure in the topsoil. This treatment gave 10-fold increases in the number of earthworms and a 4-fold increase in soil pores. Infiltration rates also increased 10-fold relative to untreated soil. Root growth was not measured, but fruit yields from peach trees increased from 18 to 75 t/ha.

Lateral spread of roots in three sowings of 1987, 1988 and 1989; the ratio r_L/R_L (mean of minimum and conventional tillage) increased with Days After Sowing (DAS) according to $r_L/R_L = 1 - 2.33 \times 0.97^t$ (r = 0.96) where t is DAS. are means od both cultivars and tillage treetments for 1st, 2nd and 3rd sowings (O,,, Δ, respectively) of 1987 (open), 1988 (closed and the 1989 sowing (♦).

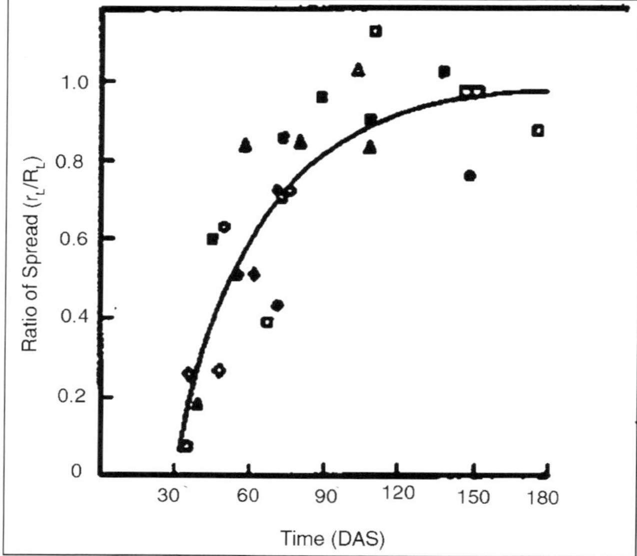

Relationships between root length at each sowing and growing day-degrees (°Cd).

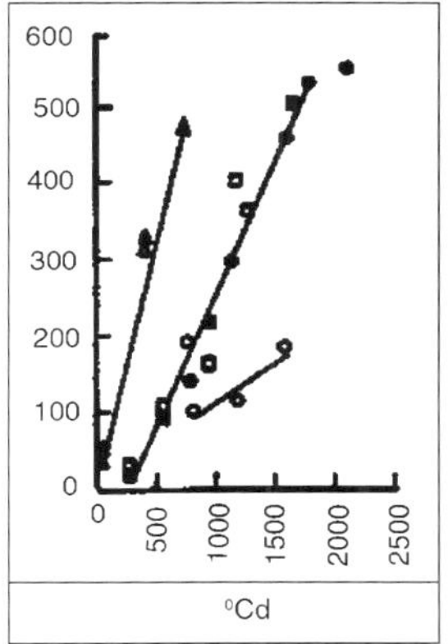

Comparison of total root length per plant in each tillage treetment (minimum tillage/conventional tillage ratio) indicated that root length were lower under minimum tillage than conventional tillage during early growth in 1987, similar in 1988 and higher under minimum than following conventional tillage in tillage in 1989; vertical bars are SE.

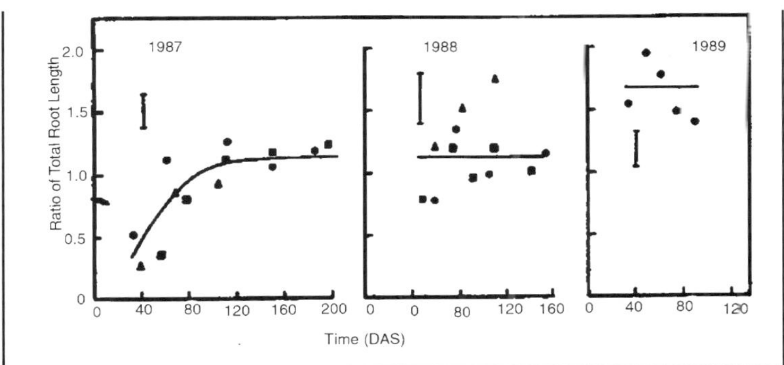

Fig. Spread of Wheat Roots with Age, and Effects on Root Growth of Minimum or Direct Drilling Compared with Conventional Tillage

Table. Values of Air-filled Porosity (Per Cent) and Bulk Density(g cm³) which are Critical and which Limit Root Growth for Various Soils

Texture class	Non-limiting	Critical[1]	Limiting
Air-filled porosity			
Fine loamy	20	10	5
Coarse silty	20	10	5
Fine silty	20	10	5
Clay:			
35-45	15	10	5
>45	15	10	5
Bulk density			
Sandy	1.60	1.69	1.85
Coarse loamy	1.50	1.63	1.80
Fine loamy	1.46	1.67	1.78

Coarse silty	1.43	1.67	1.79
Fine silty	1.34	1.54	1.65
Clayey:			
35-45 per cent	1.40	1.49	1.58
45 per cent	1.30	1.39	1.47

Note: [1] 'Critical' is defined as causing <20 per cent reduction in root growth; 'limiting' is about the value at which root growth ceases.

Increases in soil density or strength retard root penetration and thus limit the volume of soil exploited by the crop and the water available. It is difficult to quantify the relationships between these soil parameters and plant growth. In the cases of bulk density and strength, particularly, a gross measure of either for an undisturbed mass of soil can give only a remote indication of what a root encounters. A determination of gross bulk density does not assess whether a root is growing within a pore (in which case it may deform surrounding soil before its radial environment reaches the density or strength of the gross soil) or if it is growing within the soil material, in which case it has already exerted a radial force equivalent to that measured for the gross soil.

This problem of scale-what is measured in a gross estimate cannot assess the micro-environment of the root, to which it (and through hormonal signals, the plant top) responds—does not invalidate some general hypotheses about plant behaviour in response to soil compaction and structural arrangement. Roots stop growing when they are unable to deform their micro-environment. This probably occurs at suctions about 60 kPa when they cannot generate adequate internal turgor. It is thought that root elongation declines curvilinearly with increasing either bulk density or shear strength. In the case of bulk density, little effect is noted until a 'critical value' and root elongation ceases within a further 10 per cent increase in bulk density. As explained earlier, these values vary with soil type. Root elongation declines asymptotically with increasing soil strength though the actual critical values would be expected to vary with soil type, water content and the method of measurement.

Studies of the effect of bulk density or strength on other plant processes, particularly germination and shoot elongation, also suffer from the same technical problem of scale. Shoots are able to explore macropores without being subject to the gross values of the soil they are in. The actual local values which inhibit shoot elongation appear to be quite small, for example, 0.76 kPa. These values from controlled situations contrast sharply with gross field values and with the large number of inconsistent correlations which arise from attempting to correlate crop performance with grossly-measured soil structure. Perhaps, most encouragingly, studies of genetic variation in the sensitivity of crops to soil strength suggest that there is appreciable range in plant sensitivity. The relative ranking of genotypes is, however, the same when under near-critical stress as when growing with virtually no mechanical stress. Genotypes suited to stressful situations may be selected, therefore, by screening at a single soil strength.

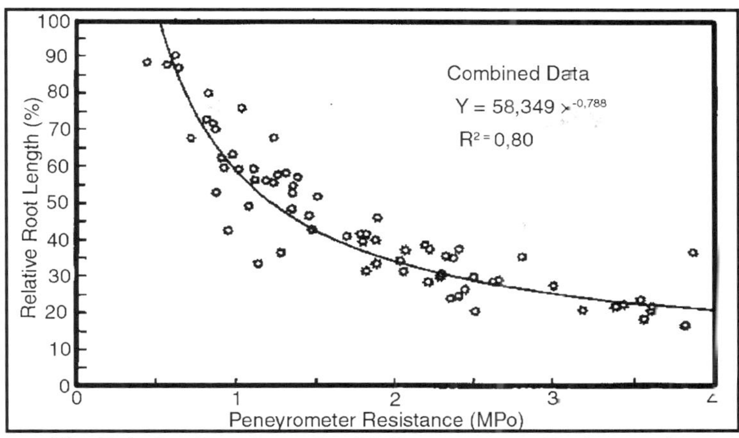

Fig. Relative Root Length with Penetrometer Resistance for Combined Maize, Wheat, Cotton and Groundnut Data

Table. Grain and Stover Yield (t/ha) of Maize and Seasonal water Run-off and Soil Loss under Maize grown with and without Alley Cropping, with two tree Legumes, and Tillage in Nigeria

Treatment	Maize		Runoff[1] (mm [% of rainfall])	Soil Loss (t/ha)
	Grain	Stover		
Without Alley Cropping				
Tilled control	2.3	3.1	66.0 (9.4)	6.18
No-tillage	2.4	3.2	5.6 (0.8)	0.43
Alley-cropped				
2 m Gliricidia	3.2	4.6	4.8 (0.7)	0.57
4 m Gliricidia	2.8	4.2	23.1 (3.3)	1.44
2 m Leucaena	3.4	4.9	2.6 (0.4)	9.17
4 m Leucaena	3.1	3.9	10.7 (1.5)	0.82

Note: [1] Seasonal rainfall (March-July 1988) = 704.2 mm.

The above comments relate to the direct effects of soil physical properties on crop growth. Two further points may be made. First, changing soil properties may not cause, or at least not immediately cause, measurable differences in plant performance.

This neither invalidates an association between the soil and the plant, nor the need to remain concerned about changes in soil properties. It is common but not often reported, that changes in a cropping system and soil characteristics first affect aspects other than the crop. It is only when a system has become significantly degraded, and there are associated environmental impacts, that reduced plant performance is noticed. This is illustrated in a study at Ibadan, Nigeria, where no-till Maize in rotation with cowpea gave similar yields to conventionally cultivated crops. The no-till fields had only one-tenth the run-off and erosion of those conventionally cultivated. Likewise, alley-cropping of the annuals between six year-old hedgerows of tree legumes (*Leucaena leucocephala, Gliricidia sepium*) caused increased crop yields, but within the alley-cropping systems the yields of Maize were very similar despite substantial variations in run-off and soil loss.

Table. Desirable Crop Attributes that Sustain Soil Productivity

Attribute	Contributing Characteristic
Rapid establishment	(Relatively) large seed for establishment and seedling vigour; abundant seed for propagation
Groundcover to reduce soil exposure, suppress weeds	Early branching; perhaps rhizomes or stolons; horizontal leaves (high canopy-extraction coefficient)
Low requirements for bacteria	• Colonisation by associative nutrients (*e.g., Brachyrhizobiun),* micorrhizae (*e.g., Glomus)* and free-living organisms (*e.g.,Azospirillum*) • Low P, K, etc., requirement per unit dry matter, *e.g.,* high 'phosphate efficiency' Efficient water use Short growth duration (to utilise residual moisture after crop); high water use efficiency
Deep rooting to reduce water table (salinity) and recover nutrients at depth and increase macropores	Vertical root distribution; roots penetrate high impedence soils
Useful products	High leaf/stem ratio; edible seeds; easily digestible; no nutritional compounds in material for livestock; leaf retention on stems for cut and carrying to livestock
Non-host for diseases and pests of main crop (to break disease cycle) or decoy (to attract diseases from concurrent crops)	Botanically unrelated to main food crop(s)
Suppress other species	Allelopathy (leachates, exudates which suppress or kill other plant species); also physical attributes (as above)

The second point about soil-crop relations is that the crop also affects the soil through ground-cover, depth of rooting and other properties. The crop attributes that most influence soil physical properties are speed of establishment and development of foliage cover. Rapid establishment and growth minimizes topsoil structural decline and soil erosion by wind and water. Thereafter, deep-rooting directly affects soil structure, particularly if deep-rooted crops, such as safflower, are grown in rotation as a 'biological plough' to create macropores and these are minimally disturbed before the next crop is sown. A general list of desirable crop characteristics is given in Table. Not all these are universally applicable (nor indeed, accepted by all scientists) because some attributes such as stolons and rhizomes, on the one hand, can provide advantageous ground cover and bind the soil, but they are undesirable if the species presents weeding or other problems. The list includes both herbaceous and tree species. Ezenwa considers the attributes important for the choice of trees in Sahelian areas as the ability *to*:

- Enrich a microsite by depositing litter;
- Fix nitrogen and not be allelopathic;
- Avoid competition;
- Withstand stress, either water or wind;
- Provide alternative produce such as fodder, fuel or fruit.

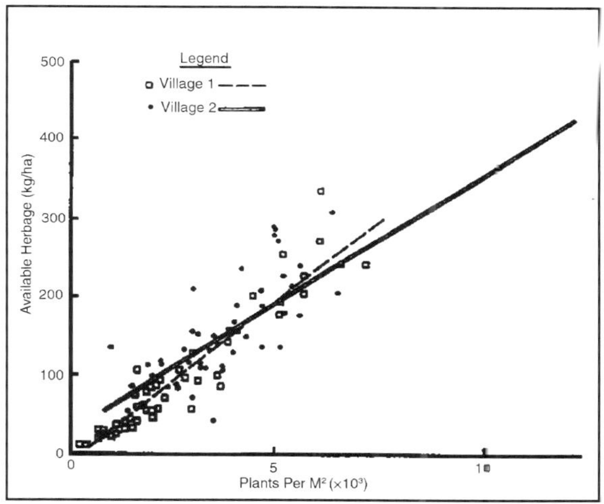

Fig. Relationship between Herbage Yield and Plant Numbers on Marginal land Near two Villages in Syria where Rainfall is about 270 mm

MANAGEMENT FOR MAINTENANCE OF SOIL PHYSICAL PROPERTIES

'In the struggle against aridisation and desertification, dry-farming is no less important or urgent than revegetation and irrigation'.

Maintenance or improvement of soil structure requires:

- Recognition of the interrelated nature of soil attributes and their interaction with cropping.
- The ability to recognise the symptoms indicating physical problems or their likely existence.
- Action.

The interrelated nature of soil and plant attributes, and measures of degradation, are discussed above. The size of the problem, regionally and internationally, is also well recognised. Likewise, the magnitude of local soil losses is established: for example, values as high as 200 t/ha/y in Niamey and measured losses of 282 t/ha/y in the highlands of Ethiopia. Lastly, and significantly, it is becoming recognised that soil physical characteristics may be less robust than chemical attributes. In Australia, the rates of degradation of structure and compaction are faster, on the same soil types, than the degradation of chemical attributes.

The issue now is to reflect in a general way, as is possible only in a bulletin like this, on appropriate action.

There are many regionally-based assessments of land degradation and various studies on how management may improve, or reduce the rate of deterioration in, soil physical structure. Regional identification of the size of the problems, and general prescriptive strategies to ameliorate them, are appropriate for planning and allocation of resources. An example from Africa (FAO 1986) is outlined in the extensive quotation that follows:

Table. Rate of Change in Soil Physical and Chemical Characteristics from Various Locations in Australia

Characteristic occur (years)	Measure of change	Time for change to
Surface structure	Loss	2-10
Subsoil compaction: traffic pan Organic matter	Formation One-third loss	1-5 5-8 sand 15-20 loam
pH	Decrease by one unit	7-12 sand 20-30 various
Salinity	Not quantified	10-100

Table. Changes in Cropping Patterns and Calculated Erosion Hazard in three Regions in Italy between 1954 and 1976

Location	Catchment and Area (ha)		Change in Cropping Systems	Change in Calculated Erosion Hazard (t/ha/year)
Tuscany	(1) (2) (3)	98 20 14	Mixed crops replaced with monoculture, particularly vines, using mechanical, non-contour tillage; minimal pastures in new catchments	6.4 to 7.7 7.8 to 8.1 1.7 to 16.2
Emilia	(1)	38	Traditionally forested; catchment logged and abandoned for regrowth and not cultivated	4.2 to 16.2
Piedmont	(1) (2)	92 51	Mixed cropping; some reduction in cropping while retaining contour tillage	40.0 to 33.5 22.5 to 21.8

The African environment is being seriously degraded as a result of the over-exploitation of cropland. The degradation of cropland can best be halted by improving soil fertility, limiting soil erosion, and introducing water harvesting and dryland farming techniques.

Options for the first two include:

- Replacing shifting cultivation with perennial tree crops such as oil palm, coconut, coffee and cacao;
- Growing crops between alleys of leguminous trees and shrubs, and supplementing the nitrogen and organic matter they supply with farmyard manure;
- Using zero or minimum tillage systems to minimize soil erosion;
- Adopting integrated plant nutrition to supply nitrogen by using animal manure alone or mixed with mineral fertilizers, or by growing leguminous crops;
- Constructing physical barriers to soil erosion, such as earth bunds, bench terraces and tied ridges.

Where population pressure has led people to cultivate hillsides with marginal soils, or to settle in semi-arid areas better suited to pastoralism, three simple measures can reduce degradation: the construction of small dams to conserve water, and tree planting on upper slopes; the use of dryland farming techniques; and, finally, the use of short-season, drought-resistant cultivars.

To tackle the problem at a local level work needs to be done in cooperation with the farmers themselves: 'start with what they know, build on what they have'. Table illustrates a locally-targeted analysis of erosion risk in the hills of Italy. It combines calculations of erosion risk with on-farm assessment of changes in land use. There is a predictable increase in erosion hazard associated with deforestation and increased cropping of monocultures, especially under vines, where there is no additional ground cover and where the land is not ploughed around the contours. Erosion risk fell where cropping declined over the 20-year study period.

It seems most likely that lasting changes to cropping practice and control of erosion will be based on farmers' recognition of trends such as those in Tables, coupled with strategies which farmers find appropriate for their goals. These strategies are listed in Table, which broadly sets out the options. Soil physical sustainability depends on activities which are explicitly aimed at its maintenance, and if necessary, any activities needed to ameliorate or control damage already done. Water harvesting can be used as a complementary strategy, the options for which are summarised by Critchley and Siegert. All the strategies in Table are described above and some have been illustrated. It is appropriate to repeat, however, that few such strategies are universally applicable in detail. For example, minimum soil disturbance may, after several years, enhance sustainability in temperate semi-arid Australia and USA whereas tillage experiments in Africa and Asia support conventional, even deep, ploughing. While the philosophy of minimal disturbance may be biologically 'right', differences in soil texture may require different approaches to sustainability. For example, Sahelian soils are commonly coarse-textured with high bulk densities and form crusts. They do not readily develop macropores. Though it is desirable to mulch such soils with straw, there is none available as it is used to feed livestock.

Table. Aspects to Consider for Maintenance or Amelioration of Soil Physical Properties

Maintenance: prevention of physical degradation

- Crop choice.
  - Rotations + sequential cropping
  - Mixed cropping
  - Relay cropping
  - Alley cropping, parkland + agroforestry
- Crop cultural practices
  - Tillage + residue management
  - Time of planting
  - Seed quality and soil organism symbioses
  - Inorganic fertilizers
  - Organic matter management
  - Cultivar: ground cover, complementarity with other crops
  - Biological pest + weed management
- Inter-crop ley and fallow
  - Cover crop
  - Pasture ley
  - Maintenance of surface litter in absence of living vegetation
- Mulches
  - *In situ* live mulch
  - Green manure crops
  - *In situ* dead residues
  - Transported residues
  - Animal wastes, composts
  - Industrial wastes
  - Inorganic covers, *e.g.*, gravel Amelioration to control damage
- Management of water erosion
  - Contour ploughing
  - Graded channels
  - Bunds
  - Grassed waterways
  - Ponds
- Management of wind erosion
  - Wind-breaks + interplanting with trees
  - Shrub + tree revegetation
  - Soil coverage
  - Ridges
- Soil surface management
  - Coverage with residues, transported waste, etc.
  - Also *Inter-crop ley and fallow* and *Mulches*

- Compaction
  - Deep tillage, subsoiling
  - Deep-rooted 'natural plough' plants

HOW MANAGEMENT DIRECTLY AFFECTS SOIL STRUCTURE

The three management practices which most affect soil physical properties (and consequent *in-situ* degradation and erosion) are groundcover, tillage and traffic load. Additionally, water harvesting techniques (FAO 1991) can be used to increase the amount of water stored within the soil and in surface catchments, though this does not necessarily affect soil structure.

Ground cover is crucial for the maintenance of soil structure in wet-and-dry climates. In marginal cropping areas in particular, plant populations (of weeds or pasture in non-crop periods) range widely and affect crop productivity. Their effects in semi-arid Syria are shown in Figure.

On resource-poor farms near the dry margins of the wet-and-dry croplands, populations of crop and weedy species tend to be small and variable during the cropping period. The effects of such sparse cover are clear, erosivity is directly related to the area exposed.

In practice, during the crop phase there is a five-fold range in erosive risk associated with the area covered by living plants, consistent with the crop factor (C) values which are used in the Universal Soil Loss Equation.

Table. Effect of Crop Cover on Crop Management Factor C in the Universal Soil Loss Equation

Crop cover	C[1]
No cover	1.0
Maize, sorghum	0.3-0.9
Groundnut	0.4-0.8
Cassava	0.2-0.8
Cotton, tobacco	0.5
Oil palm, coffee, cacao with cover crops	0.1-0.3
Rice	0.1-0.2
Rapidly growing cover crop	0.1
Savannah or pasture (without grasing)	0.01
Forest or crop with thick layer of mulch	0.001

Note: [1] C values range from almost zero, when the vegetation stops loss of soil directly related to rain, to 1 when the soil is fully exposed to rain.

The very low value for forests in Table has encouraged advisers to advocate planting of trees as part of mixed crop systems. Young points out, however, that trees are usually part of spatially-heterogeneous systems, such as hedgerows.

Thus the tree canopy is usually not extensive and does not cover and protect the cropped land. He believes that such tree canopies are not likely to reduce erosion and may indeed increase it. Under agroforestry systems, however, the use of cut leaves and litter as groundcover may directly or indirectly improve protection by providing barriers to lateral water flow and wind.

Mineral ions are lost during lateral surface erosion. Eroded sediment commonly contains a higher proportion of organic matter and nutrients than the topsoil from which it is derived. This enrichment, frequently 1.5-4-fold, but sometimes as high as 10-fold, is because the eroded soil comes from the uppermost layer with the more organic matter than the soil below and because water differentially removes light particles and soluble nutrients. Loss of plant cover also increases downward movement of ions and fine soil particles. The loss of mineral ions to groundwater is illustrated in Table. Such losses contribute to the negative nutrient balance of cropping systems and may also pollute groundwater. Under bare soil, downward translocation of silt and clay particles may fill subsoil pores and thus reduce infiltration. Vertical loss of organic material and fine particles may create a less-well-structured surface layer with less water-holding capacity, poorer infiltration and a tendency to crust.

The best ground covers are living crops and pastures. Soil stability declines in the order: living mulch; dead mulch on surface and incorporated dead mulch; and lastly, no mulch. The benefits of mulch incorporation depend on slope, soil erodibility, and on the type of organic matter.

These and the type of tillage used determine whether the dead mulch will remain on the surface. Costs and availability of both mulch and labour (human, animal or tractor) are also important. Though farmers may appreciate the benefits of surface mulching, it is unlikely that it will become practice in those tropical and temperate wet-and-dry climates where there is a shortage of feed for livestock in the dry period(s).

Type of catchment	**Catchment**	**Percent of area**			**Outflow of elements during 2 years (g m²)**				
		Arable	**Forests**	**Grass-lands**	**N-NO_3**	**P-PO_4^3**	**K^+**	**Ca^{2+}**	**Mg^{2+}**
Larger contribution of cultivated fields	1	53	12	35	1.63	0.03	3.32	37.66	2.82
	2	62	8	28	1.43	0.04	2.99	35.42	3.30
	3	53	32	13	1.13	0.04	2.57	30.95	2.28
	4	51	21	27	0.71	0.02	2.91	32.40	2.70
Mean		55	18	26	1.22	0.03	2.36	34.13	2.78
Smaller contribution of cultivated fields	1	38	44	17	0.60	0.02	1.58	23.33	1.80
	2	29	45	26	0.48	0.01	1.00	25.63	1.44
	3	21	65	14	0.39	0.01	0.91	22.54	1.56
	4	32	47	21	0.44	0.01	0.91	15.26	1.20
Mean		30	50	19	0.48	0.01	1.10	21.69	1.50

Objective 1	Type of Tillage 2	Advantage 3	Disadvantage 4
Soil conditioning	Cutting, loosening, granulating	Weed control, water conservation, structure improvement, seedbed preparation, better drying of wet soils	Greater erosion potential, high energy input[1], increased evaporation
Eradication or control of plants or plant materials	Cutting, inverting, mixing	Weed control, volunteer plant control, water conservation, establish desirable plant populations, pest control[2], better drying of wet soils, mineralization of soil nutrients	Greater erosion potential, may cause compaction, high energy input, decreased soil organic matter, increased evaporation
Establishing soil boundaries and surface configurations	Cutting as with coulters to improve ploughing, operation, land forming	Weed control, soil conservation, water conservation, residue incorporation, seedbed preparation, better drying of wet soils, warmer soil temperatures	Greater erosion potential, may cause compaction, high energy input, increased evaporation
Incorporating, covering, or handling foreign materials	Cutting, inverting, mixing	Weed control, residue incorporation, mineralization of soil nutrients, fertilizer and pesticide incorporation, pest control, better drying of wet soils, warmer soil temperatures	Greater erosion potential, may cause compaction, high energy input, decreased soil organic matter, increased evaporation
Segregation	Move soil materials from one layer to another	Wind erosion control, better drying of wet soils	High energy input, increased evaporation
Mixing	Mixing	Better drying of wet soils, improved soil amendment distribution, fertilizer and pesticide incorporation, soil texture improvement (mixing of two or more layers), soil structure improvement, mineralization of soil nutrients	Great erosion potential, high energy input, decreased soil organic matter, increased evaporation
Compaction or firming	Rolling or pressing	Improved seed-soil contact	May cause compaction

The objectives, advantages and disadvantages of tillage are summarised in Table. Had this been written 8-10 years ago, tillage would have been emphasised. Currently, tillage is seen as appropriate for some objectives, for example to break impervious layers of subsoil and to incorporate organic matter topsoil, but generalisations about the value of tillage or the 'best' type of tillage are not regarded as transportable from one system to another. Tillage may be defined as the mechanical manipulation of soil to enhance the outcomes from a cropping system. So, given the broad range of system outcomes, it is likely that tillage will continue for reasons such as tradition and satisfaction long after its beneficial benefits on soil structure and weed control have been discounted.

When advocating particular tillage methods, or interpreting experimental results and their transferability to other environments, it should be noted that tillage is usually affected by the type of ground cover. The variables of tillage are the degree of contact with the soil, the type of manipulation, and the number of implement passes through the soil. Thus, minimum tillage (one-pass) and conventional tillage (perhaps 3-5 passes, in the Australian farming system), might have the same effects on the soil, differing only in degree. Conservation tillage, which uses different implements and ground covers, is defined as 'any tillage or cultivation system that maintains at least 30 per cent of the soil surface covered by residue after planting to reduce soil erosion by water; or where soil erosion by wind is the primary concern, maintains at least 450 kg/ha of flattened small grain residue equivalent on the surface during critical erosion periods'.

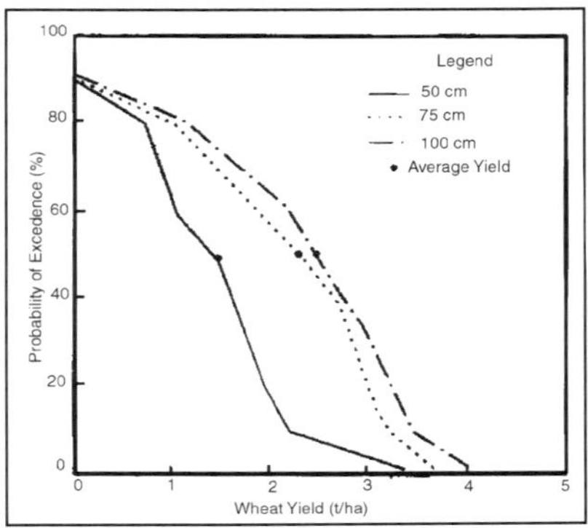

Fig. Variation in Crop Yield for Different Soil Depths of a Vertisol in the Central Highlands of Queensland, Australia

In seasonally wet-and-dry climates, the most apparent benefits from tillage are weed control and reduced evapotra-nspiration from weeds preceding the crop. This increases the water available at the beginning of the crop period and the volume of soil into which roots may penetrate (and thus, the available soil water). The impact of tillage on soil volume is modelled in Figure. This shows that rooting depth (or soil depth in this example) has a marked effect on the probability of achieving certain crop yields. Thus, exploitation of larger soil volumes may both increase yield and increase the likelihood of attaining a particular yield, that is, reduce the year-to-year risk.

The variable effects of type of tillage on crop growth (Figure) are interpreted by recognising that some effects of tillage, such as increasing the potential root volume, occur within the season of tillage, while other effects are longer term. Tillage will, over several years, reduce the outer permeability of clods, restricting water-holding capacity and increasing the power required for further effective tillage.

In some soils it dis-aggregates clods, causes downward movement of silt and clay which fills soil pores and reduces infiltration. It can also accelerate the breakdown of organic matter. Some of these effects on plant growth are discussed in the previous section and Unger (1984) presents further data on the impact of different types of tillage implement on soil and erosion.

Compaction by machinery or by treading by livestock is common and is reviewed elsewhere. The ultimate example of soil degradation through tillage and compaction is *the* formation of impermeable beds for rice production.

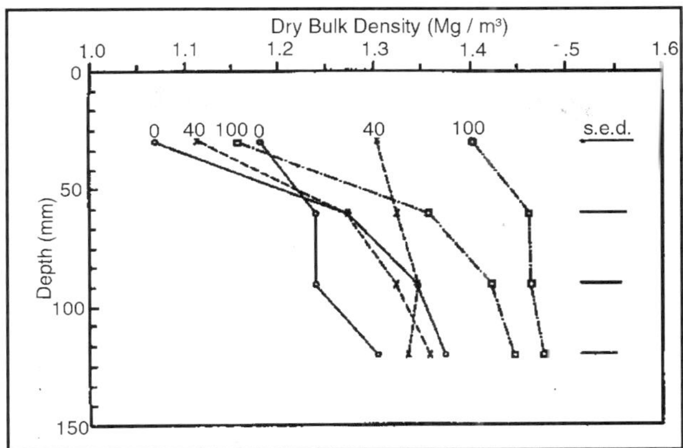

Fig. Bulk Density Profiles for each Treatment for 1987 and 1990

Figure illustrates the effects of compaction and plant activity on bulk density. Three levels of compaction were created (0, 40 and 100 kPa) under grassland. At depths below 15 cm the porosity and pore length reflected these pressures. In some places the volume of pores was halved by the heaviest compaction treatment. The subsoil bulk densities also reflect the deleterious effect of pressure. From 1987 to 1990, however, all bulk densities near the surface decreased. Grass roots were apparently able to offset the effects of machinery pressure.

FIELD INDICATORS OF PHYSICAL PROBLEMS

General field indicators of poor soil physical conditions include:

1. *Patchiness or absence of vegetation*: This can be an obvious sign of degraded structure or other factors. When structural, it may reflect surface structure degradation (see previous sections) or non-wetting characteristics which give rise to poor infiltration, or subsoil impermeability.
2. *Weedy vegetation*: *Cyperaceae* or *Juncaceae* may indicate soil structural decline because they flourish where water has been ponded on the surface, suggesting poor infiltration or an impermeable subsurface horizon.
3. *Rill and sheet erosion*: Erosive run-off may be symptomatic of poor surface structure. The turbidity of water in ponds and lakes after rain may be a good indicator of erosion.
4. Surface crusts.
5. Hard-setting surfaces.
6. *Poor infiltration and ponding*: This may be indicated by puddles following rain in an area where one would expect rapid infiltration, or by wetting to only a shallow depth (as seen when dug with a spade).
7. *Pale surface soil colour and absence of organic matter*: The surface of

degraded soils may be brittle and pale, lacking organic matter and having lost clay either through eluviation (differential movement downwards) or by water or wind erosion.

8. *Cloddiness*: This may be apparent if after a single cultivation, large, tough clods are formed requiring further cultivation to form a reasonable seedbed.
9. *Restricted root growth*: This can be seen by digging with a narrow-faced spade and washing the roots free of soil. The root mass can be restricted to the upper soil or be constricted in particular places such as a less pervious layer, above and below which the roots may proliferate.

When assessing soil degradation on a large scale as for a catchment, Hamblin (1991a) argues that most precise (scientific) measurements are time-consuming and complex, and that robust surrogate assessments amenable to remote sensing are appropriate.

Vegetation cover is one useful surrogate. Ranges in groundcover (often easier to estimate than plant densities) provide a direct index of soil productivity and they usually, although not necessarily, correlate closely with underlying problems of soil structure or fertility. Sparse groundcover might also, of course, reflect inappropriate management of non-degraded soil.

This could suggest inefficient management which will lead to loss of sustainability though there are no current problems with soil structure. The surrogate features Hamblin appears to favour are turbidity of water, transient ponding and waterlogging.

Transient ponding is particularly attractive as it can be sensed remotely, relates directly to the field being measured, and reflects several important physical aspects. These latter include surface degradation, continuity of vertical macropores, and subsurface compaction, none of which can currently be satisfactorily measured directly to give cheap meaningful data.

At farm level, attuned observers and local farmers will probably identify, often subconsciously, more indicators than those in the above list. Farmers are often good observers of their fields and can recognise soil degradation.

For example, interviews with 55 farmers in Mindanao, Philippines, found that 90 per cent reported yield declines associated with soil erosion: 'Many farmers said their soils initially had been dark on top and reddish underneath but over time the top layer had eroded away. The infertile, red subsoils were exposed. Farmers estimated that erosion reduced the depth of the darker soil from 50 cm in 1976 to 10 cm in 1986. Table illustrates their insights to declining soil productivity. Many farmers recognise changes in soil structure (*e.g.*, compaction) through the increased power needed during tillage, or by the formation of brick-like clods rather than a fine crumby tilth. These observations, however, are made at the beginning of the growing season, when there is little opportunity for soil amelioration. Farmers should be encouraged to dig small inspection holes to examine roots during the life of the crop. This supplements observations described above. Live roots are easily exposed and reflect many soil structural factors. Such examinations allow the farmer to consider soil amelioration, using stubble and crop residues, prior to the next crop. Figure illustrates clear differences in root patterns revealed by digging in three tillage systems in India.

Table. Examples of Farmer Concepts/Statements Concerning Aspects of Sustainable Crop Production

Crops and soil nutrients

"Cassava adds soil acidity."
"Cassava gobbles up soil nutrients."
"Rice is more tolerant of acidic soils than is Maize."
"Rice is more vigourous on an area previously planted in tomato."
"Intercropping is good only if there are complete chemicals."

Nutrient depletion

"Soil fertility has been used up."
"The soil is weak."
"Fertility is spotty."
"Soils are overtrained."
"The soils are getting older."
"Poor, but not used up, in the sense of the hardest part within a log."

Fallows

"The decomposing leaves of the weeds help to enrich the soil."
"The land is resting so the soil can store some nutrients."
"Rich because it is rested."
"Fertility is added and the soil is made cool."
"The soil is slightly enriched if left a short time."

Weeds

"Rice was harmed by *cogon* (*I. cylindrical*) roots."
"Poor soil if *cogon* dominates."
"*D. longiflora* and *cogon* consume soil nutrients and destroy soil quality."
"Acidity increases where *cogon* dominates."
"Weeds are thin on infertile soils."
"*R. cochinchinensis* rapidly produces seed; thus, easily soars in population; if not weeded, it exceeds the height of rice or corn."
"Fertility is added and the soil is made cool" (re.
Calapogonium spp.).
"Soil is good where there are weeds/grasses with nodules."

Soil erosion

"Soil slides down and floats away."
"Nutrients are drawn down."
"Plants are eroded along with soil."
"Soil was drawn down and fertility was washed out."
"The land was shaven and eroded after trees were removed."
"Fertilizer is collected (on lower plots) due to rain."

Erosion control

"Banana and coconut are better because they hold the soil."

"Contour plowing reduces downslope erosion losses."
"Weedy strips can decrease erosion effects."
"Trees planted above and below fields can decrease erosion effects."
"Banana planted above and below fields can decrease erosion effects."

It is important too that farmers, whether resource-poor or advanced, understand the conditions under which soil degradation is most likely to be severe. Broadly, human-induced structural damage occurs during tilling, when heavily trampled by livestock, and by poor management of soil cover.

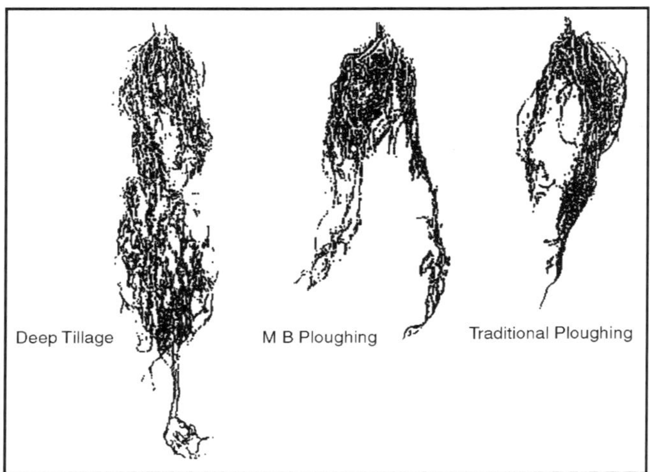

Fig. Roots of Sorghum Plants Grown on a Luvisol under three Tillage Systems

Weather-induced damage is episodic rather than continuous. Virmani (1990) explains wind and water erosion as follows:

Wind erosion is a serious problem in the arid and semi-arid tropics on sandy, loamy sand or sandy loam surface soils low in organic matter with weak topsoil structure in areas with high winds (>20 km/h). In tropical areas where wind erosion occurs the annual rainfall is generally less than 600 mm and is seasonal. The rest of the year is dry. In the tropics, 80-90 per cent of the annual wind erosion losses occurs in the hot and dry period preceding the short rainy season.

During this period the daily maximum temperature usually exceeds 35-40°C, the wind speeds exceed 20 km/h, and open pan evaporation exceeds 8 mm/d. As a rule of thumb, on uncultivated bare sandy soils, the soil displaced by wind erosion is around 1 t/ha/WED, where WED is the sum of the number of days during which the wind speed exceeds 20 km/h, the day temperature exceeds 35 °C and the wind fetch is more than 60 km. For freshly cultivated soils the amount of soil eroded may be as large as 50 t/ha/WED.

Water erosion is serious in areas with an average annual rainfall of 600-1100 mm. Using the concept of a rain erosion month (REM), that is a month in which rainfall exceeds pan evaporation, Virmani states that water-induced erosion at rates of approximately 1 t/ha/REM occurs on cropped Vertisols in areas where the rain/

pan evaporation ratio ranges between 1 and 2 and where improved soil and water conservation practices have been applied. He records rates of 4 to 8 t/ha/REM where practices are not improved.

CONCLUSION

The soil resources in this table fall into three broad groups, reflecting their underlying interrelationships: those related to soil structure (water use, structure, erosion) are dealt within this Chapter, those related to nutrition or biological factors are discussed later.

As this *Bulletin* deals with dryland crops, water availability is the major constraint on crop performance. Table suggests that water use efficiency (WUE) is a key measurement of performance. WUE reflects water availability (seasonably of rainfall) but also, importantly for management, it integrates the influences of factors such as the volume of soil exploitable by roots.

Long-term trends with time in WUE and trends between areas provide estimates of the relative sustainability of specific cropping systems. This is illustrated in Figure crops below the theoretical WUE line are less efficient or effective than they should be. One reason for such ineffectiveness is poor management during the growing season, for example weediness and nutrient deficiency. This is correctable, as shown by the vertical lines on Figure, where WUE was increased during experimental treatments. Other reasons for poor WUE are related more to fundamental issues associated with soils.

8

Single and Dual Crop Coefficient

SINGLE CROP COEFFICIENT

This chapter deals with the calculation of crop evapotranspiration (ET_c) under standard conditions. No limitations are placed on crop growth or evapotranspiration from soil water and salinity stress, crop density, pests and diseases, weed infestation or low fertility. ET_c is determined by the crop coefficient approach whereby the effect of the various weather conditions are incorporated into ET_o and the crop characteristics into the K_c coefficient:

$$ET_c = K_c\, ET_o$$

The effect of both crop transpiration and soil evaporation are integrated into a single crop coefficient. The K_c coefficient incorporates crop characteristics and averaged effects of evaporation from the soil. For normal irrigation planning and management purposes, for the development of basic irrigation schedules, and for most hydrologic water balance studies, average crop coefficients are relevant and more convenient than the K_c computed on a daily time step using a separate crop and soil coefficient. Only when values for K_c are needed on a daily basis for specific fields of crops and for specific years, must a separate transpiration and evaporation coefficient ($K_{cb} + K_e$) be considered.

The calculation procedure for crop evapotranspiration, ET_c, consists of:

- Identifying the crop growth stages, determining their lengths, and selecting the corresponding K_c coefficients;
- Adjusting the selected K_c coefficients for frequency of wetting or climatic conditions during the stage;
- Constructing the crop coefficient curve (allowing one to determine K_c values for any period during the growing period); and
- Calculating ET_c as the product of ET_o and K_c.

CROP COEFFICIENTS

Changes in vegetation and ground cover mean that the crop coefficient K_c varies during the growing period. The trends in K_c during the growing period are represented in the crop coefficient curve. Only three values for K_c are required to describe and construct the crop coefficient curve: those during the initial stage ($K_{c\ ini}$), the mid-season stage ($K_{c\ mid}$) and at the end of the late season stage ($K_{c\ end}$).

CROP COEFFICIENT FOR THE INITIAL STAGE ($K_{C\ INI}$)

Calculation Procedure

The values for $K_{c\ ini}$ are only approximations and should only be used for estimating ET_c during preliminary or planning studies. For several group types only one value for $K_{c\ ini}$ is listed and it is considered to be representative of the whole group for a typical irrigation water management. More accurate estimates of $K_{c\ ini}$ can be obtained by considering:

Time interval between wetting events

Evapotranspiration during the initial stage for annual crops is predominately in the form of evaporation. Therefore, accurate estimates for $K_{c\ ini}$ should consider the frequency with which the soil surface is wetted during the initial period. Where the soil is frequently wet from irrigation or rain, the evaporation from the soil surface can be considerable and $K_{c\ ini}$ will be large. On the other hand, where the soil surface is dry, evaporation is restricted and the $K_{c\ ini}$ will be small.

Evaporation Power of the Atmosphere

The value of $K_{c\ ini}$ is affected by the evaporating power of the atmosphere, *i.e.*, ET_o. The higher the evaporation power of the atmosphere, the quicker the soil will dry between water applications and the smaller the time-averaged K_c will be for any particular period.

Magnitude of the wetting event

As the amount of water available in the topsoil for evaporation and hence the time for the soil surface to dry is a function of the magnitude of the wetting event, $K_{c\ ini}$ will be smaller for light wetting events than for large wettings.

Depending on the time interval between wetting events, the magnitude of the wetting event, and the evaporation power of the atmosphere, $K_{c\ ini}$ can vary between 0.1 and 1.15.

Time Interval between Wetting Events

In general, the mean time interval between wetting events is estimated by counting all rainfall and irrigation events occurring during the initial period that are greater than a few millimetres. Wetting events occurring on adjacent days can be counted as one event. The mean wetting interval is estimated by dividing the length of the initial period by the number of events. Where only monthly rainfall values are available without any information on the number of rainy days, the number of events within the month can be estimated by dividing the monthly rainfall depth by the depth of a typical rain event.

The typical depth, if it exists, can vary widely from climate to climate, region to region and from season to season. Table presents some information on the range of rainfall depths. After deciding what rainfall is typical for the region and time of the year, the number of rainy days and the mean wetting interval can be estimated.

Table. Classification of Rainfall Depths

Rain Event	Depth
Very light (drizzle)	≤ 3 mm
Light (light showers)	5 mm
Medium (showers)	≥ 10 mm
Heavy (rainstorms)	≥ 40 mm

Where rainfall is insufficient, irrigation is needed to keep the crop well watered. Even where irrigation is not yet developed, the mean interval between the future irrigations should be estimated to obtain the required frequency of wetting necessary to keep the crop stress free.

The interval might be as small as a few days for small vegetables, but up to a week or longer for cereals depending on the climatic conditions. Where no estimate of the interval can be made, the user may refer to the values for $K_{c\ ini}$ of Table.

Example. Estimation of interval between wetting events

Estimate, from mean monthly rainfall data, the interval between rains during the rainy season for a station in a temperate climate (Paris, France: 50 mm/month), dry climate (Gafsa, Tunisia: 20 mm/month) and tropical climate (Calcutta, India: 300 mm/month).

Station	monthly rain (mm/month)	typical rainfall (mm)	number of rainy days	interval between rains
Paris	50	3	17	~ 2 days
Gafsa	20	5	4	weekly
Calcutta	300	20	15	~ 2 days

Determination of Kc ini

The crop coefficient for the initial growth stage can be derived which provide estimates for $K_{c\ ini}$ as a function of the average interval between wetting events, the evaporation power ET_o, and the importance of the wetting event.

Light wetting events (infiltration depths of 10 mm or less): rainfall and high frequency irrigation systems

When wetting during the initial period is only by precipitation, one will usually use to determine $K_{c\ ini}$. The graph can also be used when irrigation is by high frequency systems such as microirrigation and centre pivot and light applications of about 10 mm or less per wetting event are applied.

Graphical Determination of Kc ini

A silt loam soil receives irrigation every two days during the initial growth stage via a centre pivot irrigation system. The average depth applied by the centre pivot system is about 12 mm per event and the average ETo during the initial stage is 4 mm/day. Estimate the crop evapotranspiration during that stage.			
From Fig. 29 using the 2-day interval curve:	$K_{c\ ini}$ =	0.85	-
	$ET_c = K_c\ ET_o$ = 0.85 (4.0) =	3.4	mm/day
The average crop evapotranspiration during the initial growth stage is 3.4 mm/day			

Heavy wetting events (infiltration depths of 40 mm or more): surface and sprinkler irrigation For heavy wetting events when infiltration depths are greater than 40 mm, such as for when wetting is primarily by periodic irrigation such as by sprinkler or surface irrigation. Following a wetting event, the amount of water available in the topsoil for evaporation is considerable, and the time for the soil surface to dry might be significantly increased. Consequently, the average K_c factor is larger than for light wetting events. As the time for the soil surface to dry is, apart from the evaporation power and the frequency of wetting, also determined by the water storage capacity of the topsoil, a distinction is made between soil types.

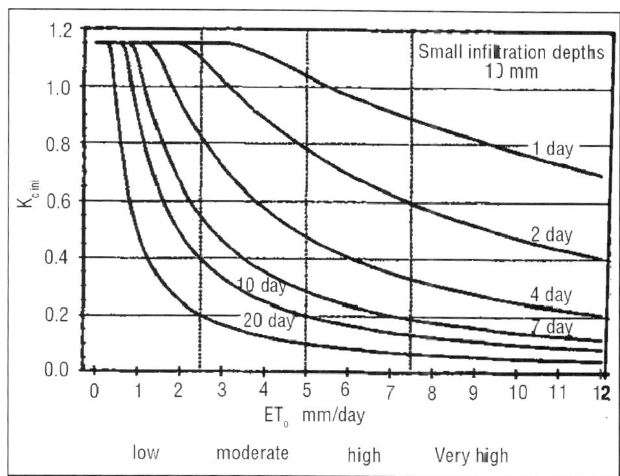

Fig. Average $K_{c\ ini}$ as related to the level of ET_o and the interval between irrigations and/ or significant rain during the initial growth stage for all soil types when wetting events are light to medium (3-10 mm per event)

Coarse textured soils include sands and loamy sand textured soils. Medium textured soils include sandy loam, loam, silt loam and silt textured soils. Pine textured soils include silty clay loam, silty clay and clay textured soils.

Average wetting events (infiltration depths between 10 and 40 mm):

Where average infiltration depths are between 10 and 40 mm, the value for $K_{c\ ini}$ can be estimated:

$$K_{cini} = K_{cin} + \frac{(1-10)}{(40-10)}[K_c \, in - K_{cin}]$$

where

$K_{c\,ini}$ value for $K_{c\,ini}$,

$K_{c\,ini}$ value for $K_{c\,ini}$,

I average infiltration depth [mm].

The values 10 and 40 are the average depths of infiltration (millimetres) upon which Figures are based.

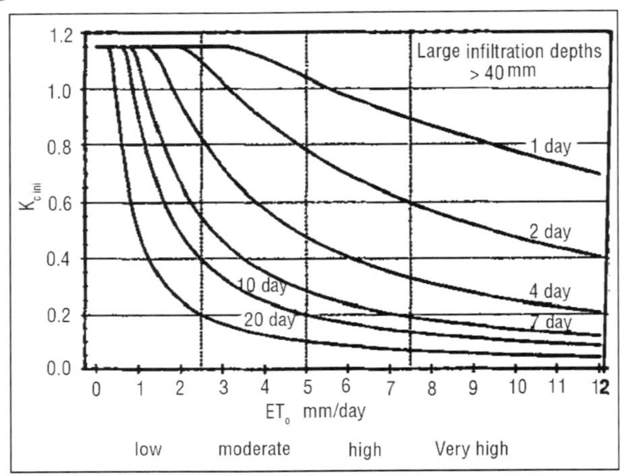

Fig. Average $K_{c\,ini}$ as Related to the Level of ET_o and the Interval between Irrigations Greater than or Equal to 40 mm per wetting event, during the initial Growth Stage for Coarse Textured Soils

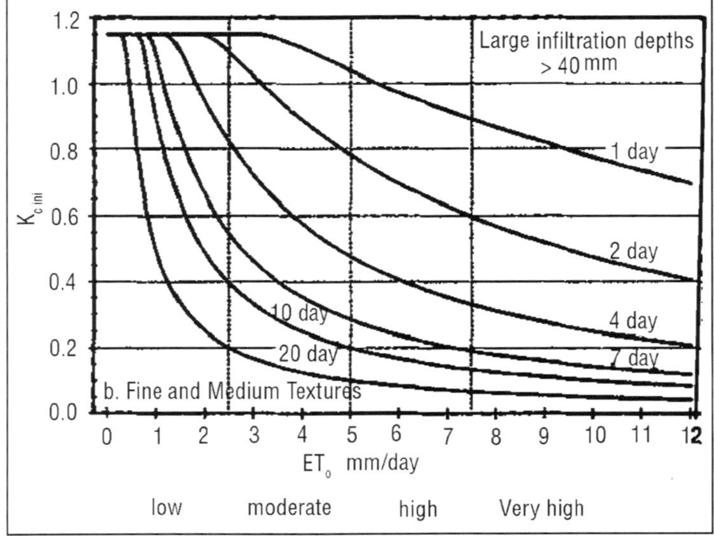

Example. Interpolation between light and heavy wetting events

Small vegetables cultivated in a dry area on a coarse textured soil receive 20 mm of water twice a week by means of a sprinkler irrigation system. The average ET_o during the initial stage is 5 mm/day. Estimate the crop evapotranspiration during that stage.			
For:	7/2=	3.5	day interval
	ET_o = and a coarse textured soil	5	mm/day
From Fig. 29:	$K_{c\ ini\ (Fig.\ 29)}$?	0.55	-
From Fig. 30. a:	$K_{c\ ini\ (Fig.\ 30a)}$?	0.7	-
For:	I =	20	mm
From Eq. 59:	$K_{c\ ini}$ = 0.55 + [(20 - 10)/(40 - 10)] (0.7 - 0.55) = 0.55 + 0.33(0.15) =	0.60	
From Eq. 58:	ET_c = 0.60 (5) =	3.0	mm/day
The average crop evapotranspiration during the initial growth stage for the small vegetables is 3.0 mm/day.			

Adjustment for Partial Wetting by Irrigation

Many types of irrigation systems wet only a fraction of the soil surface. For example, for a trickle irrigation system, the fraction of the surface wetted, f_w, may be only 0.4. For furrow irrigation systems, the fraction of the surface wetted may range from 0.3 to 0.8.

Common values for the fraction of the soil surface wetted by irrigation or precipitation.

When only a fraction of the soil surface is wetted, the value for $K_{c\ ini}$ obtained should be multiplied by the fraction of the surface wetted to adjust for the partial wetting:

$$K_{c\ ini} = f_w\ K_{c\ ini}$$

where

f_w the fraction of surfaced wetted by irrigation or rain [0 - 1],

$K_{c\ ini\ (Tab\ Fig)}$ the value for $K_{c\ ini}$.

In addition, in selecting the average infiltrated depth, expressed in millimetres over the entire field surface, should be divided by f_w to represent the true infiltrated depth of water for the part of the surface that is wetted:

$$l_w = \frac{1}{f_w}$$

where

I_w irrigation depth for the part of the surface that is wetted [mm],

f_w fraction of surface wetted by irrigation,

I the irrigation depth for the field [mm].

When irrigation of part of the soil surface and precipitation over the entire soil surface both occur during the initial period, f_w should represent the average of f_w for each type of wetting, weighted according to the total infiltration depth received by each type.

Example. Determination of $K_{c\ ini}$ for partial wetting of the soil surface

Determine the evapotranspiration of the crop in Example 24 if it had been irrigated using a trickle system every two days (with 12 mm each application expressed as an equivalent depth over the field area), and where the average fraction of surface wet was 0.4, and where little or no precipitation occurred during the initial period.			
The average depth of infiltration per event in the wetted fraction of the surface:			
From Eq. 61;	$I_w = I/f_w = 12$ mm/0.4 =	30	mm
Therefore, one can interpolate between Fig. 29 representing light wetting events (~10 mm per event) and Fig. 30.b representing medium textured soil and large wetting events (~40 mm per event).			
For:	$ET_o = 4$ mm/day	4	mm/day
and:	a 2 day wetting interval:	-	-
Fig. 29 produces:	$K_{c\ ini} = 0.85$	0.85	-
Fig. 30.b produces	$K_{c\ ini} = 1.15$	1.15	-
From Eq. 59:	$K_{c\ ini} = 0.85 + [(30-10)/(40-10)]\ (1.15 - 0.85) =$	1.05	-
Because the fraction of soil surface wetted by the trickle system is 0.4, the actual $K_{c\ ini}$ for the trickle irrigation is calculated as:			
From Eq. 60:	$K_{c\ ini} = f_w\ K_{c\ ini\ Fig} = 0.4\ (1.05) =$	0.42	-
	This value (0.42) represents the $K_{c\ ini}$ as applied over the entire field area.		
-	$ET_c = K_{c\ ini}\ ET_o = 0.42(4) =$	1.7	mm/day
The average crop evapotranspiration during the initial growth stage for this trickle irrigated crop is 1.7 mm/day.			

Kc ini for Trees and Shrubs

$K_{c\ ini}$ for trees and shrubs should reflect the ground condition prior to leaf emergence or initiation in case of deciduous trees or shrubs, and the ground condition during the dormancy or low active period for evergreen trees and shrubs. The $K_{c\ ini}$ depends upon the amount of grass or weed cover, frequency of soil wetting, tree density and mulch density. For a deciduous orchard in frost-free climates, the $K_{c\ ini}$ can be as high as 0.8 or 0.9, where grass ground cover exists, and as low as 0.3 or 0.4 when the soil surface is kept bare and wetting is infrequent. The $K_{c\ ini}$ for an evergreen orchard (having no concerted leaf drop) with a dormant period has less variation from $K_{c\ mid}$.

Kc ini for Paddy Rice

For rice growing in paddy fields with a water depth of 0.10-0.20 m, the ET_c during the initial stage mainly consists of evaporation from the standing water. The $K_{c\ ini}$ in Table is 1.05 for a sub-humid climate with calm to moderate wind speeds. The $K_{c\ ini}$ should be adjusted for the local climate as indicated in Table.

Table. $K_{c\ ini}$ for rice for various climatic conditions

Humidity	Wind speed		
	light	moderate	strong
arid - semi-arid	1.10	1.15	1.20
sub-humid - humid	1.05	1.10	1.15
very humid	1.00	1.05	1.10

Crop Coefficient for the Mid-season Stage ($K_{c\ mid}$)

Illustration of the Climatic Effect

Typical values for the crop coefficient for the mid-season growth stage, $K_{c\ mid}$, are listed in Table for various agricultural crops. The effect of the difference in aerodynamic properties between the grass reference surface and agricultural crops is not only crop specific but also varies with the climatic conditions and crop height. More arid climates and conditions of greater wind speed will have higher values for $K_{c\ mid}$. More humid climates and conditions of lower wind speed will have lower values for $K_{c\ mid}$. The relative impact of climate on $K_{c\ mid}$ where the adjustments to the values from Table are shown for various types of climates, mean daily wind speeds and various crop heights. As an example, expected variations for $K_{c\ mid}$ for tomatoes in response to regional climatic conditions are presented.

Determination of Kc Mid

For specific adjustment in climates where RH_{min} differs from 45 per cent or where u_2 is larger or smaller than 2.0 m/s, the $K_{c\ mid}$ values from Table are adjusted as:

$$K_{cmid} = K_{cmid}(Tab) + [0.04(u_2 - 2) - 0.004(RH_{min} - 45)]\left(\frac{h}{3}\right)^{0.3}$$

where

$K_{c\ mid\ (Tab)}$ value for $K_{c\ mid}$,

u_2 mean value for daily wind speed at 2 m height over grass during the mid-season growth stage [m s^{-1}], for 1 m s^{-1} $\leq u_2 \leq$ 6 m s^{-1}, RH_{min} mean value for daily minimum relative humidity during the mid-season growth stage [per cent], for 20 per cent $\leq RH_{min} \leq$ 80 per cent, h mean plant height during the mid-season stage [m] for 0.1 m < h < 10 m.

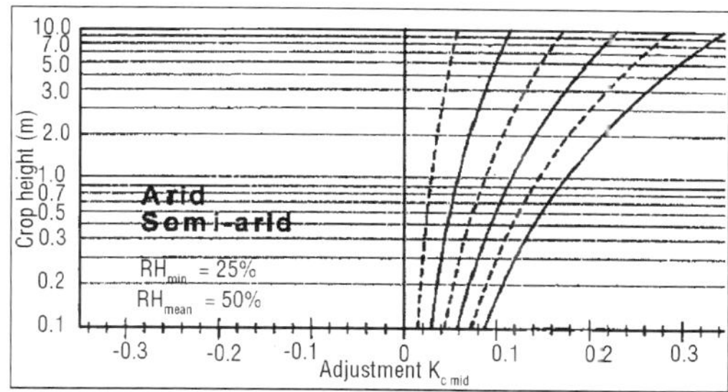

Fig. Adjustment (additive) to the $K_{c\ mid}$ values from Table for Different Crop Heights and Mean Daily Wind Speeds (u_2) for Different Humidity Conditions

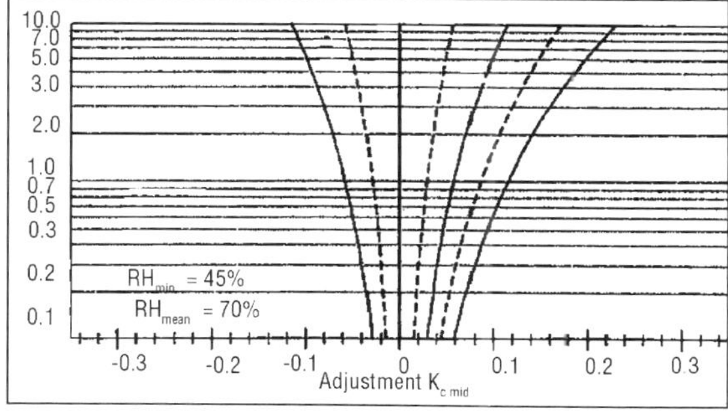

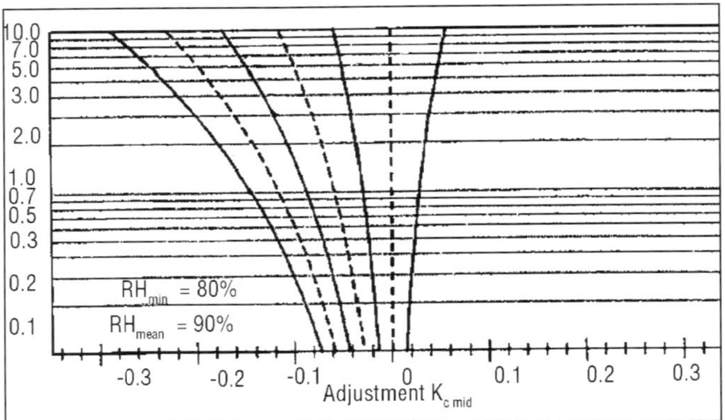

The $K_{c\ mid}$ values determined with equations are average adjustments for the midseason and late season periods. The values for parameters u_2 and RH_{min} should be accordingly taken as averages for these periods. The limits expressed for parameters u_2, RH_{min} and h should be observed.

BOX 14. Demonstration of effect of climate on $K_{c\ mid}$ for wheat crop grown under field conditions	
From Table 12 for wheat: $K_{c\ mid}$ = 1.15 and h = 1.0 m	
For semi-arid to arid conditions	
- for strong wind (4 m/s)	$K_{c\ mid}$ = 1.15 + 0.10 = 1.25
- for moderate wind (2 m/s)	$K_{c\ mid}$ = 1.15 + 0.05 = 1.20
- for calm wind (1 m/s)	$K_{c\ mid}$ = 1.15 + 0.00 = 1.17
For sub-humid conditions	
- for strong wind (4 m/s)	$K_{c\ mid}$ = 1.15 + 0.05 = 1.20
- for moderate wind (2 m/s)	$K_{c\ mid}$ = 1.15 + 0.00 = 1.15
- for calm wind (1 m/s)	$K_{c\ mid}$ = 1.15 - 0.05 = 1.12
For humid and very humid conditions	
- for strong wind (4 m/s)	$K_{c\ mid}$ = 1.15 - 0.05 = 1.10
- for moderate wind (2 m/s)	$K_{c\ mid}$ = 1.15 - 0.10 = 1.05
- for calm wind (1 m/s)	$K_{c\ mid}$ = 1.15 - 0.15 = 1.02
Depending on the aridity of the climate and the wind conditions, the crop coefficient for wheat during the mid-season stage ranges from 1.02 (humid and calm wind) to 1.25 (arid and strong wind).	

Where the user does not have access to a calculator with an exponential function, the solution of the $(h/3)^{0.3}$ expression can be approximated as $[(h/3)^{0.5}]^{0.5}$ where the square root key is used.

RH_{min} is used rather than RH_{mean} because it is easier to approximate RH_{min} from T_{max} where relative humidity data are unavailable. Moreover, under the common condition where T_{min} approaches T_{dew} (*i.e.*, $RH_{max} \approx 100$ per cent), the vapour pressure deficit ($e_s - e_a$), with e_s from Equation and e_a from Equation, becomes $[(100 - RH_{min})/200]\ e°(T_{max})$, where $e°(T_{max})$ is saturation vapour pressure at maximum daily air temperature. This indicates that RH_{min} better reflects the impact of vapour pressure deficit on K_c than does RH_{mean}.

RH_{min} is calculated on a daily or average monthly basis as:

$$RH_{min} = \frac{e^o(T_{dew})}{e^o(T_{max})}100$$

where T_{dew} is mean dewpoint temperature and T_{max} is mean daily maximum air temperature during the mid-season growth stage. Where dewpoint temperature or other hygrométrie data are not available or are of questionable quality, RH_{min} can be estimated by substituting mean daily minimum air temperature, T_{min}, for T_{dew}[1]. Then:

$$RH_{min} = \frac{e^o(T_{min})}{e^o(T_{max})}100 \tag{64}$$

In the case of arid and semi-arid climates, T_{min} in equation should be adjusted by subtracting 2°C from the average value of T_{min} to better approximate T_{dew}.

The values for u_2 and RH_{min} need only be approximate for the mid-season growth stage. This is because Equation is not strongly sensitive to these values, changing 0.04 per 1 m/s change in u_2 and per 10 per cent change in RH_{min} for a 3 m tall crop.

Measurements, calculation, and estimation of missing wind and humidity. Wind speed measured at other than 2 m height should be adjusted to reflect values for wind speed at 2 m over grass. Where no data on u_2 or RH_{min} are available, the general classification for wind speed and humidity can be used.

Table. Empirical estimates of monthly wind speed data

Description	Mean Monthly Wind Speed at 2 m
light wind	...≤ 1.0 m/s
light to moderate wind	2.0 m/s
moderate to strong wind	4.0 m/s
strong wind	... ≥ 5.0 m/s
general global conditions	2 m/s

Table. Typical values for RH_{min} compared with RH_{mean} for general climatic classifications

Climatic Classification	RH_{min} (%)	RH_{mean} (%)
Arid	20	45
Semi-arid	30	55
Sub-humid	45	70
Humid	70	85
Very humid	80	90

Equation is valid for mean plant heights up to 10 m. For plant heights smaller than 0.1 m, vegetation will behave aerodynamically similar to grass should not be applied. However, the mean plant height will greatly vary with crop variety and with cultural practices. Therefore, wherever possible, h should be obtained from general field observations.

However, the presence of the 0.3 exponent in Equation makes these equations relatively insensitive to small errors in the value used for h. Generally, a single value for h is used to represent me mid-season period.

Adjustment for Frequency of Wetting

$K_{c\ mid}$ is less affected by wetting frequency than is $K_{c\ ini}$, as vegetation during this stage is generally near full ground cover so that the effects of surface evaporation on K_c are smaller.

For frequent irrigation of crops (more frequently than every 3 days) and where the $K_{c\ mid}$ of Table is less than 1.0, the value can be replaced by approximately 1.1-1.3 to account for the combined effects of continuously wet soil, evaporation due to interception (sprinkler irrigation) and roughness of the vegetation, especially where the irrigation system moistens an important fraction of the soil surface ($f_w > 0.3$).

Determination of $K_{c\ mid}$

Calculate $K_{c\ mid}$ for maize crops near Taipei, Taiwan and near Mocha, Yemen. The average mean daily wind speed (u_2) during the mid-season stage at Taipei is about 1.3 m/s and the minimum relative humidity (RH_{min}) during this stage averages 75 per cent. The average u_2 during the mid-season near Mocha is 4.6 m/s and the RH_{min} during this stage averages 44 per cent.

From Table, the value for $K_{c\ mid}$ is 1.20 for maize. The value for h from Table is 2 m. Using Equation

For Taipei (humid climate):

$$K_{cmid} = 1.20 + [0.04(1.3-2) - 0.004(75-45)]\left(\frac{2}{3}\right)^{0.3} = 1.07$$

For Mocha (arid climate):

$$K_{cmid} = 1.20 + [0.04(4.6-2) - 0.004(44-45)]\left(\frac{2}{3}\right)^{0.3} = 1.30$$

The average crop coefficient predicted during the mid-season stage is 1.07 for Taipei and 1.30 for Mocha.

Crop Coefficient for the end of the Late Season Stage ($K_{c\ end}$)

Typical values for the crop coefficient at the end of the late season growth stage, $K_{c\ end}$, and listed in Table for various agricultural crops. The values given for $K_{c\ end}$ reflect crop and water management practices particular to those crops.

If the crop is irrigated frequently until harvested fresh, the topsoil remains wet and the $K_{c\ end}$ value will be relatively high. On the other hand, crops that are allowed to senesce and dry out in the field before harvest receive less frequent irrigation or no irrigation at all during the late season stage. Consequently, both the soil surface and vegetation are dry and the value for $K_{c\ end}$ will be relatively small.

Where the local water management and harvest timing practices are known to deviate from the typical values presented, then the user should make some adjustments to the values for $K_{c\ end}$. Some guidance on adjustment of K_c values for wetting frequency is provided. For premature harvest, the user can construct a K_c curve using the $K_{c\ end}$ value provided and a late season length typical of a normal harvest date; but can then terminate the application of the constructed curve early, corresponding to the time of the early harvest.

The $K_{c\ end}$ values are typical values expected for average $K_{c\ end}$ under the standard climatic conditions. More arid climates and conditions of greater wind speed will have higher values for $K_{c\ end}$. More humid climates and conditions of lower wind speed will have lower values for $K_{c\ end}$. For specific adjustment in climates where RH_{min} differs from 45 per cent or where u_2 is larger or smaller than 2.0 m/s, Equation can be used:

$$\text{Kcend} = \text{Kcend(Tab)} + [0.04(u_2 - 2) - 0.004(RH_{min} - 45)]\left(\frac{h}{3}\right)^{0.3}$$

where

$K_{c\ end\ (Tab)}$ value for $K_{c\ end}$ taken from Table,

u_2 mean value for daily wind speed at 2 m height over grass during the late season growth stage [m s^{-1}], for 1 m s^{-1} $\leq u_2 \leq$ 6 m s^{-1},

RH_{min} mean value for daily minimum relative humidity during the late season stage [per cent], for 20 per cent $\leq RH_{min} \leq$ 80 per cent,

h mean plant height during the late season stage [m], for 0.1 m $\leq h \leq$ 10 m.

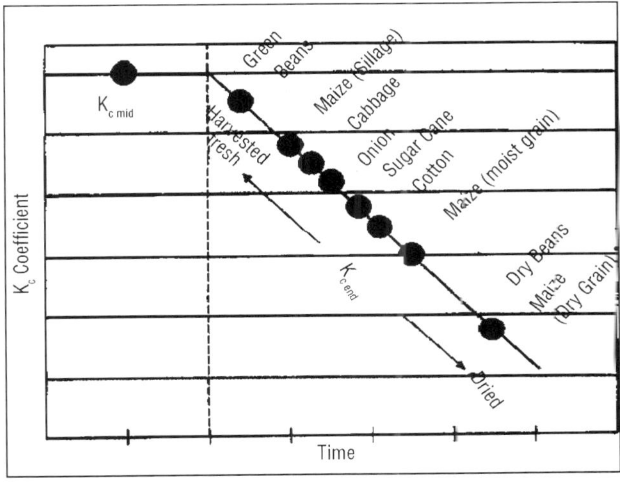

Fig. Ranges Expected for $K_{c\ end}$

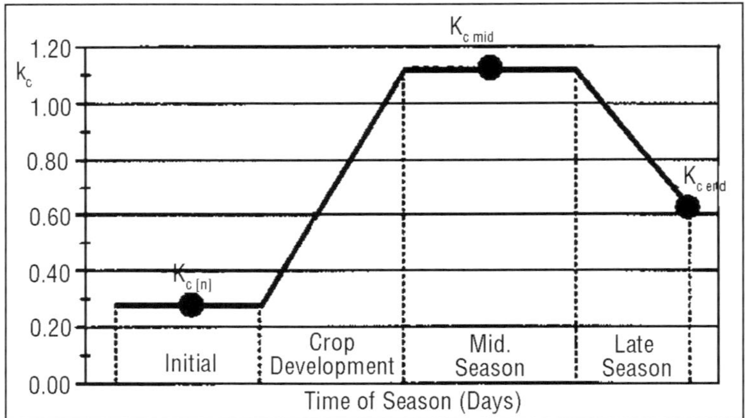

Fig. Crop Coefficient Curve

Equation is only applied when the tabulated values for $K_{c\ end}$ exceed 0.45. The equation reduces the $K_{c\ end}$ with increasing RH_{min}. This reduction in $K_{c\ end}$ is characteristic of crops that are harvested 'green' or before becoming completely dead and dry (*i.e.*, $K_{c\ end} \geq 0.45$).

No adjustment is made when $K_{c\ end\ (Table)} < 0.45$ (*i.e.*, $K_{c\ end} = K_{c\ end\ (Tab)}$). When crops are allowed to senesce and dry in the field (as evidenced by $K_{c\ end} < 0.45$), u_2 and RH_{min} have less effect on $K_{c\ end}$ and no adjustment is necessary. In fact, $K_{c\ end}$ may decrease with decreasing RH_{min} for crops that are ripe and dry at the time of harvest, as lower relative humidity accelerates the drying process.

Construction of the K_c Curve

Annual Crops

Only three point values for K_c are required to describe and to construct the K_c curve. The curve is constructed using the following three steps:

- Divide the growing period into four general growth stages that describe crop phenology or development (initial, crop development, mid-season, and late season stage), determine the lengths of the growth stages, and identify the three K_c values that correspond to $K_{c\ ini}$, $K_{c\ mid}$ and $K_{c\ end}$.
- Adjust the K_c values to the frequency of wetting and/or climatic conditions of the growth stages as outlined in the previous section.
- Construct a curve by connecting straight line segments through each of the four growth stages. Horizontal lines are drawn through $K_{c\ ini}$ in the initial stage and through $K_{c\ mid}$ in the mid-season stage. Diagonal lines are drawn from $K_{c\ ini}$ to $K_{c\ mid}$ within the course of the crop development stage and from $K_{c\ mid}$ to $K_{c\ end}$ within the course of the late season stage.

K_c Curves for Forage Crops

Many crops grown for forage or hay are harvested several times during the growing season. Each harvest essentially terminates a 'sub' growing season and associated K_c curve and initiates a new 'sub' growing season and associated K_c curve. The resulting K_c curve for the entire growing season is the aggregation of a series of K_c curves associated with each sub-cycle. **Figure** presents a K_c curve for the entire growing season constructed for alfalfa grown for hay in southern Idaho.

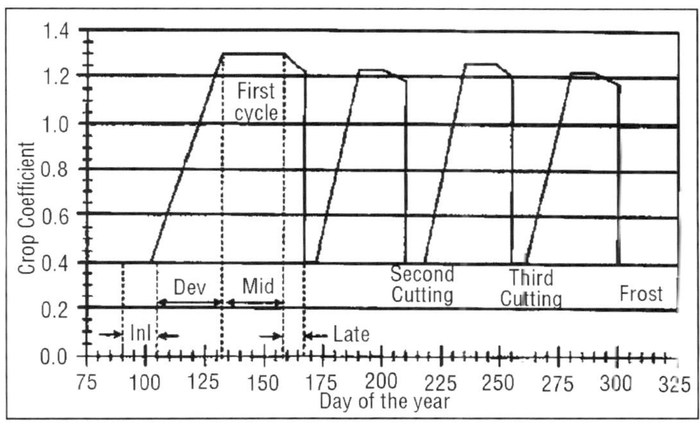

Fig. Constructed Curve for K_c for Alfalfa hay

In the southern Idaho climate, greenup (leaf initiation) begins in the spring on about day 90 of the year. The crop is usually harvested (cut) for hay three or four times during the growing season. Therefore, four K_c sub-cycles or cutting cycles: sub-cycle 1 follows greenup in the spring and the three additional K_c sub-cycles follow cuttings. Cuttings create a ground surface with less than 10 per cent vegetation cover. Cutting cycle 1 is longer in duration than cycles 2, 3 and 4 due to lower air and soil temperatures during this period that reduce crop growth rates.

The lengths for cutting cycle 1 were taken from the first entry for alfalfa (" 1st cutting cycle") in Table for Idaho, the United States (10/30/25/10). The lengths for cutting cycles 2, 3 and 4 were taken from the entry for alfalfa in Table for "individual cutting periods" for Idaho, the United States (5/20/10/10). These lengths were based on observations. In the southern Idaho climate, frosts terminate the growing season sometime in the fall, usually around day 280-290 of the year (early to mid-October).

The magnitudes of the K_c values during the mid-season periods of each cutting cycle vary from cycle to cycle due to the effects of adjusting the values for $K_{c\ mid}$ and $K_{c\ end}$ for each cutting cycle period using Equations. In applying these two adjustment equations, the u_2 and RH_{min} values were averages for the mid-season and late season stages within each cutting cycle.

Kc mid when Effects of Individual Cutting Periods are Averaged

Under some conditions, the user may wish to average the effects of cuttings for a forage crop over the course of the growing season. When cutting effects are averaged, then only a single value for $K_{c\ mid}$ and a only single K_c curve need to be employed for the whole growing season. When this is the case, a "normal" K_c curve is constructed, where only one midseason period is shown for the forage crop. The $K_{c\ mid}$ for this total midseason period must average the effects of occasional cuttings or harvesting.

The value that is used for $K_{c\ mid}$ is therefore an average of the K_c curve for the time period starting at the first attainment of full cover and ending at the beginning of the final late season period near dormancy or frost. The value used for $K_{c\ mid}$ under these averaged conditions may be only about 80 per cent of the K_c value that represents full ground cover. These averaged, full-season $K_{c\ mid}$ values are listed in Table. For example, for alfalfa hay, the averaged, full-season $K_{c\ mid}$ is 1.05, whereas, the $K_{c\ mid}$ for an individual cutting period is 1.20.

Fruit Trees

Values for the crop coefficient during the mid-season and end of late season stages. As mentioned before, the K_c values listed are typical values for standard climatic conditions and need to be adjusted by using Equations where RH_{min} or u_2 differ. As the mid and late season stages of deciduous trees are quite long, the specific adjustment of K_c to RH_{min} and u_2 should take into account the varying climatic conditions throughout the season.

Therefore, several adjustments of K_c are often required if the mid and late seasons cover several climatic seasons, *e.g.*, spring, summer and autumn or wet and dry seasons. The $K_{c\ ini}$ and $K_{c\ end}$ for evergreen non- dormant trees and shrubs are often not different, where climatic conditions do not vary much, as happens in tropical climates. Under these conditions, seasonal adjustments for climate may therefore not be required since variations in ET_c depend mostly on variations in ET_o.

Calculating ET_c

From the crop coefficient curve the K_c value for any period during the growing period can be graphically or numerically determined. Once the K_c values have been derived, the crop evapotranspiration, ET_c, can be calculated by multiplying the K_c values by the corresponding ET_o values.

Graphical Determination of Kc

Weekly, ten-day or monthly values for K_c are necessary when ET_c calculations are made on weekly, ten-day or monthly time steps. A general procedure is to construct the K_c curve, overlay the curve with the lengths of the weeks, decade or months, and to derive graphically from the curve the K_c value for the period under consideration. Assuming that all decades have a duration of 10 days facilitates the derivation of K_c and introduces little error into the calculation of ET_c.

This curve has been overlaid with the lengths of the decades. K_c values of 0.15, 1.19 and 0.35 and the actual lengths for growth stages equal to 25, 25, 30 and 20 days were used. The crop was planted at the beginning of the last decade of May and was harvested 100 days later at the end of August.

For all decades the K_c values can be derived directly from the curve. The value at the middle of the decade is considered to be the average K_c of that 10 day period. Only the second decade of June, where the K_c value changes abruptly, requires some calculation.

Case study of a dry bean crop at Kimberly, Idaho, the United States (single crop coefficient)

An example application for using the K_c procedure under average soil wetness conditions is presented for a dry bean crop planted on 23 May 1974 at Kimberly, Idaho, the United States (latitude = 42.4°N). The initial, development, mid-season and late season stage lengths are taken from Table for a continental climate as 20, 30, 40 and 20 days (the stage lengths listed for southern Idaho were not used in this example in order to demonstrate the only approximate accuracy of values provided in Table when values for the specific location are not available). Initial values for $K_{c\ ini}$, $K_{c\ mid}$ and $K_{c\ end}$ are selected from Table

The mean RH_{min} and u_2 during both the mid-season and late season growth stages were 30 per cent and 2.2 m/s. The maximum height suggested in Table for dry beans is 0.4 m. Therefore, $K_{c\ mid}$ is adjusted using Equation as:

$$K_{cmid} = 1.15 + [0.04(2.2-2) - 0.004(30-45)]\left(\frac{0.4}{3}\right)^{0.3} = 1.19$$

As $K_{c\ end}$ = 0.35 is less than 0.45, no adjustment is made to $K_{c\ end}$. The value for $K_{c\ mid}$ is not significantly different from that in Table as ≈2 m/s, RH_{min} is just 15 per cent lower than the 45 per cent represented in Table, and the height of the beans is relatively short.

The initial K_c curve for dry beans in Idaho can be drawn, for initial, planning purposes, as shown in the graph (dotted line), where $K_{c\ ini}$, $K_{c\ mid}$ and $K_{c\ end}$ are 0.4, 1.19 and 0.35 and the four lengths of growth stages are 20, 30, 40 and 20 days. Note that the $K_{c\ ini}$ = 0.4 taken from Table serves only as an initial, approximate estimate for $K_{c\ ini}$.

Constructed K_c curves using values from Tables directly (dotted line) and modified using $K_{c\ ini}$ from and L_{ini} = 25, L_{dev} = 25, L_{mid} = 30, and L_{late} = 20 days (heavy line) for dry beans at Kimberly, Idaho. Also shown are daily measured K_c.

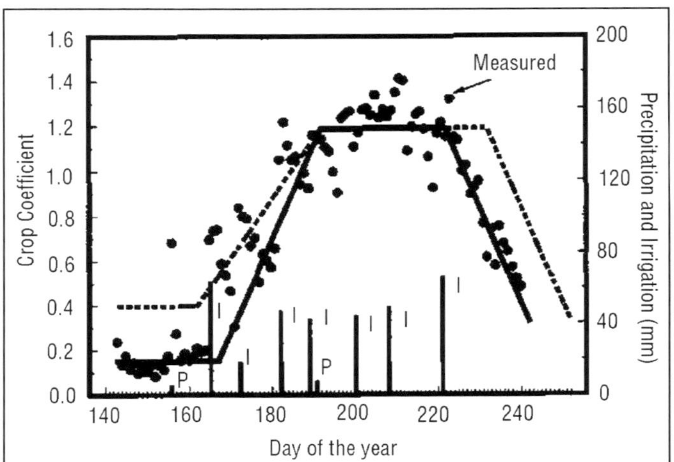

$K_{c\ ini}$ can be more accurately estimated using the approach described in this chapter. ET_0 during the initial period at Kimberly (late May - early June, 1974) averaged 5.3 mm/day, and the wetting interval during this period was approximately 14 days (2 rainfall events occurred averaging 5 mm per event).

Therefore, as the wetting events were light (< 10 mm each). The soil texture at Kimberly, Idaho is silt loam. $K_{c\ ini}$ for the 14 day wetting interval and ET_0 = 5.3 mm/day is about 0.15. This value is substantially less than the general 0.4 value suggested, and emphasizes the need to utilize local, actual precipitation and irrigation data when determining $K_{c\ ini}$.

Comparison of Constructed Curves with Measurements

Because the ET_c data for the dry bean crop at Kimberly, Idaho were measured using a precision lysimeter system during 1974 by Wright, the actual K_c measurements can be compared with the constructed K_c curves, where actual K_c was calculated by dividing lysimeter measurements of ET_c by daily ET_0 estimated using the FAO Penman-Monteith equation.

As illustrated in the graph, the mid-season length for the general, continental climate overestimated the true mid-season length for dry beans in southern Idaho, which averaged only about 30 days rather than 40 days. This illustrates the importance of using the local observation of 30 days for mid-season period length rather than the general value.

The final, best estimate for the K_c curve for the dry bean crop in southern Idaho is plotted (lower curve in graph) using K_c values of 0.15, 1.19, and 0.35 and the actual observed lengths of growth stages equal to 25, 25, 30 and 20 days. Note the impact that the error in estimating mid-season length has on the area under the K_c curve. This supports the need to obtain local observations of growth stage dates and lengths.

The value calculated for $K_{c\ mid}$ (1.19) appears to have underestimated the measured value for K_c during portions of the mid-season period at Kimberly.

Some of this effect was due to effects of increased soil water evaporation following four irrigations during the 1974 mid-season which increased the effective K_c. Where the basal $K_{cb} + K_e$ approach is introduced and demonstrated for this same example.

The 0.15 value calculated for $K_{c\,ini}$ agrees closely with measured K_c during the initial period. Measured K_c during the development period exceeded the final K_c curve during days on or following wetting events. The day to day variation in the lysimeter measured K_c is normal and is caused by day to day variations in weather, in wind direction, by errors in prediction of R_n and ET_o, and by some random errors in the lysimeter measurements and weather measurements.

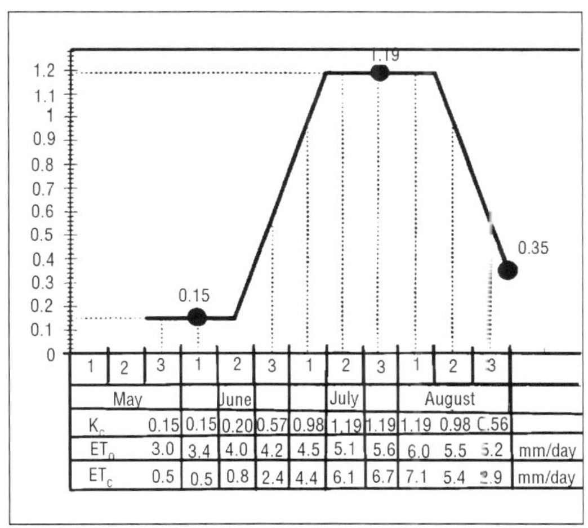

	1	2	3	1	2	3	1	2	3	1	2	3	
	May				June				July		August		
K_c			0.15	0.15	0.20	0.57	0.98	1.19	1.19	1.19	0.98	0.56	
ET_o			3.0	3.4	4.0	4.2	4.5	5.1	5.6	6.0	5.5	5.2	mm/day
ET_c			0.5	0.5	0.8	2.4	4.4	6.1	6.7	7.1	5.4	2.9	mm/day

Fig. K_c curve and Ten-day Values for K_c and ET_c derived from the Graph for the Dry Bean Crop Example

first five days of that decade, $K_c = 0.15$, while during the second part of the decade K_c varies from 0.15 to 0.36 at the end of day 10. The K_c for that decade is consequently: 5/10 (0.15) + 5/10(0.15+0.36)/2 = 0.20.

Numerical Determination of K_c

The K_c coefficient for any period of the growing season can be derived by considering that during the initial and mid-season stages K_c is constant and equal to the K_c value of the growth stage under consideration. During the crop development and late season stage, K_c varies linearly between the K_c at the end of the previous stage ($K_{c\,prev}$) and the K_c at the beginning of the next stage ($K_{c\,next}$), which is $K_{c\,end}$ in the case of the late season stage:

$$K_{ci} = K_{cprev} + \left[\frac{i - \Sigma(L_{prev})}{L_{stage}}\right](K_{cnext} - K_{cprev})$$

where

i day number within the growing season [1.. length of the growing season],
$K_{c\,i}$ crop coefficient on day i,
L_{stage} length of the stage under consideration [days],
$\Sigma (L_{prev})$ sum of the lengths of all previous stages [days].
Equation 66 applies to all four stages.
Example. Numerical determination of K_c

Determine K_c at day 20, 40, 70 and 95 for the dry bean crop (Figure 36).		
Crop growth stage	Length (days)	K_c
initial	25	$K_{c\,ini} = 0.15$
crop development	25	0.15... 1.19
mid-season	30	$K_{c\,mid} = 1.19$
late season	20	1.19 .. $K_{c\,end} = 0.35$

At i = 20:	initial stage, $K_c = K_{c\,ini}$ =	0.15	-
At i = 40	Crop development stage,		
For:	$\Sigma (L_{prev}) = L_{ini}$ =	25	days
and:	$L_{stage} = L_{dev}$ =	25	days
From Eq. 66:	$K_c = 0.15 + [(40 - 25)/25](1.19 - 0.15)$ =	0.77	-
At i = 70:	mid-season stage, $K_c = K_{c\,mid}$ =	1.19	-
At i = 95	late season stage,		
For:	$\Sigma (L_{prev}) = L_{ini} + L_{dev} + L_{mid} = (25 + 25 + 30)$ =	80	days
and:	$L_{stage} = L_{late}$ =	20	days
From Eq. 66:	$K_c = 1.19 + [(95-80)/20](0.35-1.19)$ =	0.56	-
The crop coefficients at day 20, 40, 70 and 95 for the dry bean crop are 0.15, 0.77, 1.19 and 0.56 respectively.			

Alfalfa-Based Crop Coefficients

As two reference crop definitions (grass and alfalfa) are in use in various parts of the world, two families of K_c curves for agricultural crops have been developed. These are the alfalfa-based K_c curves by Wright and grass-based curves by Pruitt and those reported in this paper. The user must exercise caution to avoid mixing grass-based K_c values with alfalfa reference ET and vice versa. Usually, a K_c based on the alfalfa reference can be 'converted' for use with a grass reference by multiplying by a factor ranging from about 1.0 to 1.3, depending on the climate (1.05 for humid, calm conditions, and 1.2 for semi-arid, moderately windy conditions, and 1.35 for arid, windy conditions):

$$K_{c\,(grass)} = K_{ratio}\, K_{c\,(alfalfa)}$$

where

$K_{c\,(grass)}$ grass-based K_c (this handbook),
$K_{c\,(alfalfa)}$ alfalfa-based K_c,
K_{ratio} conversion factor (1.0... 1.3).

A reference conversion ratio can be established for any climate by using the $K_{c\ mid}$ = 1.20 listed for alfalfa and then adjusting this $K_{c\ mid}$ for the climate using Equation. For example, at Kimberly, Idaho, the United States, where RH_{min} = 30 per cent and u_2 = 2.2 m/s are average values during the summer months, a reference conversion ratio between alfalfa and grass references using Equation is approximately:

$$K_{ratio} = 1.2 + [0.04(2.2 - 2) - 0.004(30 - 45)]\left(\frac{0.5}{3}\right)^{0.3} = 1.24$$

where

h = 0.5 m is the standard height for the alfalfa reference.

Transferability of Previous K_c Values

The values for $K_{c\ mid}$ and $K_{c\ end}$ listed in Table are for a large part based on the original values presented in FAO Irrigation and Drainage, with some adjustment and revisions to reflect recent findings. Similarly adjustments in $K_{c\ mid}$ to compensate for differences in aerodynamic roughness and leaf area, as introduced in Equation are derived from the K_c values given for different wind and RH_{min} conditions in the concerned K_c Table, with some upward adjustment to better reflect increased ET_{crop} values under high wind and low RH_{min} when applied with the FAO Penman-Monteith equation.

The K_c's from FAO-24 were based primarily on a living grass reference crop. The FAO Penman-Monteith equation presented in this publication similarly represents the same standardized grass reference. For that reason K_c values are in general not very different between these publications except under high wind and low RH_{min}.

The modified Penman was found frequently to overestimate ET_o even up to 25 per cent under high wind and low evaporative conditions and required often substantial local calibration. K_c values derived from crop water use studies which used the FAO-24 Penman equation to compute grass reference crop evapotranspiration, can therefore not be used and need to be adjusted using ET_o values estimated from the FAO Penman-Monteith equation. Similarly crop water requirement estimates based on the FAO-24 Modified Penman equation will need to be reassessed in view of the found differences between the FAO-24 Penman and the FAO Penman-Monteith reference equations.

DUAL CROP COEFFICIENT

This chapter also deals with the calculation of crop evapotranspiration (ET_c) under standard conditions where no limitations are placed on crop growth or evapotranspiration. This chapter presents the procedure for predicting the effects of specific wetting events on the value for the crop coefficient K_c. The solution consists of splitting K_c into two separate coefficients, one for crop transpiration, *i.e.*, the basal crop coefficient (K_{cb}), and one for soil evaporation (K_e):

$$ET_c = (K_{cb} + K_e)\,ET_o$$

The dual crop coefficient approach is more complicated and more computationally intensive than the single crop coefficient approach (K_c). The procedure is conducted on a daily basis and is intended for applications using computers. It is recommended that the approach be followed when improved estimates for K_c are needed, for example to schedule irrigations for individual fields on a daily basis.

The calculation procedure for crop evapotranspiration, ET_c, consists of:

- Identifying the lengths of crop growth stages, and 'selecting the corresponding K_{cb} coefficients;
- Adjusting the selected K_{cb} coefficients for climatic conditions during the stage;
- Constructing the basal crop coefficient curve (allowing one to determine K_{cb} values for any period during the growing period);
- Determining daily K_e values for surface evaporation; and
- Calculating ET_c as the product of ET_o and (K_{cb} + K_e).

DAILY CALCULATION OF K_E

Daily Water Balance

The estimation of K_e in the calculation procedure requires a daily water balance computation for the surface soil layer for the calculation of the cumulative evaporation or depletion from the wet condition. The daily soil water balance equation for the exposed and wetted soil fraction f_{ew} is:

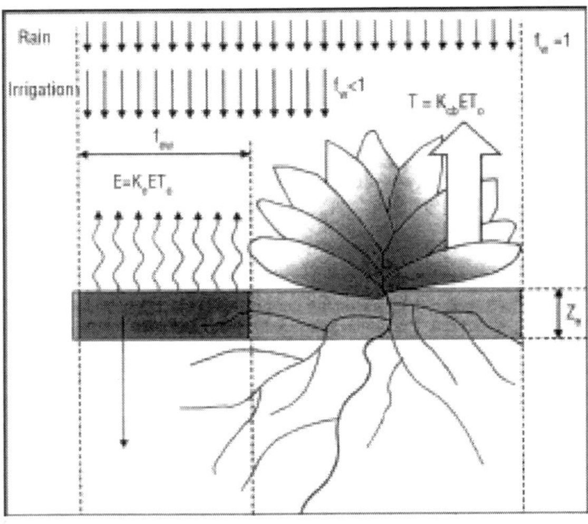

Fig. Water Balance of the Topsoil Layer

$$D_{ei} = D_{ei-1} - (P_i - RO_i) - \frac{I_i}{f_w} + \frac{E_i}{f_{ew}} + T_{ewi} + DP_{ei}$$

where

$D_{e,\, i\text{-}1}$ cumulative depth of evaporation following complete wetting from the exposed and wetted fraction of the topsoil at the end of day i-1 [mm],

$D_{e,\, i}$ cumulative depth of evaporation (depletion) following complete wetting at the end of day i [mm],

P_i precipitation on day i [mm],

RO_i precipitation run off from the soil surface on day i [mm],

I_i irrigation depth on day i that infiltrates the soil [mm],

E_i evaporation on day i (*i.e.*, $E_i = K_e\, ET_o$) [mm],

$T_{ew,\, i}$ depth of transpiration from the exposed and wetted fraction of the soil surface layer on day i [mm],

$DP_{e,i}$ deep percolation loss from the topsoil layer on day i if soil water content exceeds field capacity [mm], f_w fraction of soil surface wetted by irrigation [0.01 - 1],

f_{ew} exposed and wetted soil fraction [0.01 - 1].

Limits on De, i

By assuming that the topsoil is at field capacity following heavy rain or irrigation, the minimum value for the depletion $D_{e,\, i}$ is zero. As the soil surface dries, $D_{e,\, i}$ increases and in absence of any wetting event will steadily reach its maximum value TEW (Equation 73). At that moment no water is left for evaporation in the upper soil layer, K_r becomes zero, and the value for $D_{e,\, i}$ remains at TEW until the topsoil is wetted once again. The limits imposed on $D_{e,\, i}$ are consequently:

$0 \le D_{e,\, i} \le TEW$ (78)

Initial Depletion

To initiate the water balance for the evaporating layer, the user can assume that the topsoil is near field capacity following a heavy rain or irrigation, *i.e.*, $D_{e,\, i\text{-}1} = 0$. Where a long period of time has elapsed since the last wetting, the user can assume that all evaporable water has been depleted from the evaporation layer at the beginning of calculations, *i.e.*, $D_{e,\, i\text{-}1} = TEW = 1000\, (\theta_{FC} - 0.5\, \theta_{WP})\, Z_e$.

Precipitation and Run-off

P_i is equivalent to daily precipitation. Daily precipitation in amounts less than about 0.2 ET_o is normally entirely evaporated and can usually be ignored in the K_e and water balance calculations. The amount of rainfall lost by run-off depends on: the intensity of rainfall; the slope of land; the soil type, its hydraulic conditions and antecedent moisture content; and the land use and cover.

For general situations, RO_i can be assumed to be zero or can be accounted for by considering only a certain percentage of P_i. This is especially true for the water balance of the topsoil layer, since almost all precipitation events that would have intensities or depths large enough to cause run-off would probably replenish the water content of the topsoil layer to field capacity. Therefore, the impact of the run-off component can be ignored. Light precipitation events will generally have little or no run-off.

Irrigation

I_i is generally expressed as a depth of water that is equivalent to the mean infiltrated irrigation depth distributed over the entire field. Therefore, the value I_i/f_w is used to describe the actual concentration of the irrigation volume over the fraction of the soil that is wetted..

Evaporation

Evaporation beneath the vegetation canopy is assumed to be included in K_{cb} and is therefore not explicitly quantified. The computed evaporation is fully concentrated in the exposed, wetted topsoil. The evaporation E_i is given by K_e ET_o. The E_i/f_{ew} provides for the actual concentration of the evaporation over the fraction of the soil that is both exposed and wetted.

Transpiration

Except for shallow rooted crops (*i.e.*, where the depth of the maximum rooting zone is < 0.5 to 0.6 m), the amount of transpiration from the evaporating soil layer is small and can be ignored (*i.e.*, $T_{ew} = 0$). In addition, for row crops, most of the water extracted by the roots may be extracted from beneath the vegetation canopy. Therefore, T_{ew} from the f_{ew} fraction of soil surface can be assumed to be zero in these cases.

Example. Estimation of crop evapotranspiration with the dual crop coefficient approach

- Estimate the crop evapotranspiration, ET_c, for ten successive days. It is assumed that:
  - The soil is a sandy loam soil, characterized by $q_{FC} = 0.23$ m^3 m^{-3} and $q_{WP} = 0.10$ m^3 m^{-3},
  - The depth of the surface soil layer that is subject to drying by way of evaporation, Z_e, is 0.1 m,
  - During the period, the height of the vegetation h = 0.30 m, the average wind speed $u_2 = 1.6$ m s^{-1}, and $RH_{min} = 35$ per cent,
  - The K_{cb} on day 1 is 0.30 and increases to 0.40 by day 10,
  - The exposed soil fraction, $(1 - f_c)$, decreases from 0.92 on day 1 to 0.86 on day 10,
  - All evaporable water has been depleted from the evaporation layer at the beginning of calculations ($D_{e,\,i-1} = TEW$),

- Irrigation occurs at the beginning of day 1 (I = 40 mm), and the fraction of soil surface wetted by irrigation, $f_w = 0.8$,
- A rain of 6 mm occurred at the beginning of day 6.

From Tab. 19 REW » 8 mm

- From Equation TEW = 1000 (0.23-0.5(0.10)) 0.1 = 18 mm
- From Equation $K_{c\ max} = 1.2 + [0.04(1.6 - 2) - 0.004(35 - 45)] (0.3/3)^{0.3} = 1.21$

All evaporable water has been depleted at the beginning of calculations, $D_{e,\ i-1}$ = TEW = 18 mm

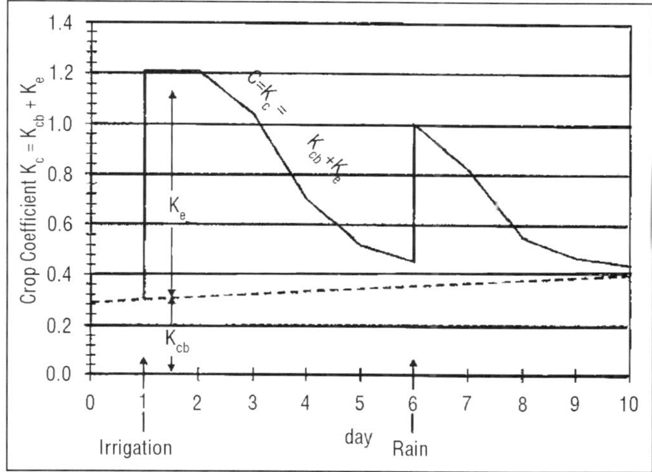

(1)	(2)	(3)	(4)	(5)	(6)	(7)	(8)	(9)	(10)	(11)	(12)	(13)	(14)	(15)	(16)	(17)
Day	ET_o	P-RO	I/f_w	$1 - f_c$	f_w	f_{ew}	K_{cb}	$D_{e,i}$ start	K_r	K_e	E/f_{ew}	DP_e	$D_{e,i}$ end	E	K_c	ET_c
	mm/d	mm	mm					mm			mm	mm	mm	mm/d		mm/d
start	-	-	-	-	-	-	-	-	-	-	-	-	18	-	-	-
1	4.5	0	50	0.92	0.8	0.80	0.30	0	1.00	0.91	5.1	32	5	4.1	1.21	5.5
2	5.0	0	0	0.91	0.8	0.80	0.31	5	1.00	0.90	5.6	0	11	4.5	1.21	6.1
3	3.9	0	0	0.91	0.8	0.80	0.32	11	0.70	0.62	3.0	0	14	2.8	1.04	4.0
4	4.2	0	0	0.90	0.8	0.80	0.33	14	0.40	0.35	1.8	0	16	1.5	0.70	2.9
5	4.8	0	0	0.89	0.8	0.80	0.34	16	0.20	0.18	1.1	0	17	0.8	0.52	2.5
6	2.7	6	0	0.89	1	0.89	0.36	11	0.75	0.64	2.0	0	13	1.7	1.00	2.7
7	5.8	0	0	0.88	1	0.88	0.37	13	0.53	0.45	3.0	0	16	2.6	0.82	4.7
8	5.1	0	0	0.87	1	0.87	0.38	16	0.20	0.17	1.0	0	17	0.9	0.55	2.8
9	4.7	0	0	0.87	1	0.87	0.39	17	0.09	0.08	0.4	0	18	0.4	0.47	2.2
10	5.2	0	0	0.86	1	0.86	0.40	18	0.05	0.04	0.2	0	18	0.2	0.44	2.3

- Day number.
- ET_o is given. Note that ET_o would be forecast values in real time irrigation scheduling but are known values after the occurrence of the day, during an update of the calculations.
- (P-RO) are known values after the occurrence of the day, during an update of the calculations.
- Net irrigation depth for the part of the soil surface wetted by irrigation.
- $(1 - f_c)$ is given (interpolated between 0.92 m on day 1 and 0.86 m on day 10).
- If significant rain: $f_{w,\ i} = 1.0$ (Tab. 20)

If irrigation: $f_{w,\,i}$ = 0.8 (given),
otherwise: $f_{w,\,i} = f_{w,\,i\text{-}1}$.

- Equation. Fraction of soil surface from which most evaporation occurs.
- K_{cb} is given (interpolated between 0.30 on day 1 and 0.40 on day 10).
- $D_{e,\,i\ start}$ (depletion at start of day)

If precipitation and irrigation occur early in the day then the status of depletion from the soil surface layer (at the start of the day) should be updated: $= \text{Max}(D_{e,\,i\text{-}1} - I_{n,\,i}/f_{wi} - (P\text{-}RO)_i$, or 0).

where $D_{e,\,i\text{-}1}$ is taken from column 14 of previous day.

If precipitation and irrigation occur late in the day, then column 6 should be set equal to $D_{e,\,i\text{-}1}$ (column 14 of previous day).

- If $D_{e,\,i} \leq$ REW $K_r = 1$
  If $D_{e,\,i} >$ REW K_r = Equation .
- Equation where $K_e = K_r\,(K_{c\ max} - K_{cb}) \leq f_{ew}\,K_{c\ max}$. (*e.g.*, K_e = min $(K_r\,(K_{c\ max} - K_{cb}), f_{ew}\,K_{c\ max})$.
- Evaporation from the wetted and exposed fraction of the soil surface $= (K_e\,ET_o)/f_{ew}$.
- Equation where $DP_e \geq 0$. (This is deep percolation from the evaporating layer).
- $D_{e,\,i}$ (depletion at end of day) is from Equation where $D_{e,\,i\text{-}1}$ is value in column 14 of previous day.
- Mean evaporation expressed as distributed over the entire field surface $= K_e\,ET_o$.
- $K_c = K_{cb} + K_e$.

The daily water balance calculation for the surface layer, even for shallow rooted crops, is not usually sensitive to T_{ew}, as T_{ew} is a minor part of the flux from the Z_e depth for the first 3-5 days following a wetting event. T_{ew} can, therefore, generally be ignored.

The effects of the reduction of the water content of the evaporating soil layer due to T_{ew} can be accounted for ulteriorly when it is assumed that T_{ew} = 0 by decreasing the value for Z_e, for example from 0.15 to 0.12 m or from 0.10 to 0.08 m.

Deep Percolation

Following heavy rain or irrigation, the soil water content in the topsoil (Z_e layer) might exceed field capacity. However, in this simple procedure it is assumed that the soil water content is at θ_{FC} nearly immediately following a complete wetting event, so that the depletion $D_{e,\,i}$ in Equation 77 is zero. Following heavy rain or irrigation, downward drainage (percolation) of water from the topsoil layer is calculated as:

$$D_{ei} = (P_i - RO_i) + \frac{l_i}{f_w} - D_{ei-1} \geq 0$$

As long as the soil water content in the evaporation layer is below field capacity (*i.e.*, $D_{e,i} > 0$), the soil will not drain and $DP_{e,i} = 0$.

Order of Calculation

In making calculations for the $K_{cb} + K_e$ procedure, for example when using a spreadsheet, the calculations should proceed in the following order: K_{cb}, h, $K_{c\,max}$, f_c, f_w, f_{ew}, K_r, K_e, E, DP_e, D_e, I, K_c, and ET_c.

CALCULATING ET_C

The calculation procedure lends itself to application by computer, either in the form of electronic spreadsheets or in the form of structured programming languages. The calculation procedure consists in determining:

A. Reference Evaporation, ET_o:

- Growth stages:
- Determine the locally adjusted lengths of the four growth stages:
  - Initial growth stage: L_{ini},
  - Crop development stage: L_{dev},
  - Mid-season stage: L_{mid},
  - Late season stage: L_{late}.

C. Basal Crop Coefficient, K_{cb}:

- Calculate basal crop coefficients for each day of the growing period:
  - Select $K_{cb\,ini}$, $K_{cb\,mid}$ and $K_{cb\,end}$;
  - Adjust $K_{cb\,mid}$ and $K_{cb\,end}$ to the local climatic conditions;
  - Determine the daily K_{cb} values
- Initial growth stage: $K_{cb} = K_{cb\,ini}$,
- Crop development stage: from $K_{cb\,ini}$ to $K_{cb\,mid}$,
- Mid-season stage: $K_{cb} = K_{cb\,mid}$,
- Late season stage: from $K_{cb\,mid}$ to $K_{cb\,end}$.

D. Evaporation Coefficient, K_e:

- Calculate the maximum value of K_c, *i.e.*, the upper limit $K_{c\,max}$, and Determine for each day of the growing period:
  - The fraction of soil covered by vegetation, f_c,
  - The fraction of soil surface wetted by irrigation or precipitation, f_w,
  - The fraction of soil surface from which most evaporation occurs, f_{ew},
  - The cumulative depletion from the evaporating soil layer, D_e, determined by means of a daily soil water balance of the topsoil,
  - The corresponding evaporation reduction coefficient, K_r, and
  - The soil evaporation coefficient, K_e.

E. Crop Evapotranspiration, ET_c:

Calculate $ET_c = (K_{cb} + K_e) ET_o$ (Equation 69).

Case Study of Dry Bean crop at Kimberly, Idaho, the United States (dual crop coefficient)

Results from applying the $K_{cb} + K_e$ procedure for a snap bean crop harvested as dry seed are shown. This example uses the same data set that was used in the case study. The measured ET_c data were measured using a precision lysimeter system at Kimberly, Idaho. Values for $K_{cb\ ini}$, $K_{cb\ mid}$, and $K_{cb\ end}$ were calculated in Example. The lengths of growth stages were 25, 25, 30, and 20 days. The K_{cb} values are plotted in Figure.

The value for $K_{c\ max}$ from Equation for the mid-season period averaged 1.24, based on u_2 = 2.2 m/s and RH_{min} = 30 per cent for Kimberly. The soil at Kimberly was a silt loam texture. Assuming that the depth of the evaporation soil layer, Z_e, was 0.1 m, values for TEW = 22 mm and REW = 9 mm, based on Equation and using soil data from Table. The occurrence and magnitudes of individual wetting events are shown in the Figure below. Nearly all wetting events were from irrigation. Because the irrigation was by furrow irrigation of alternate rows, the value for f_w was set equal to 0.5. Irrigation events occurred at about midday or during early afternoon.

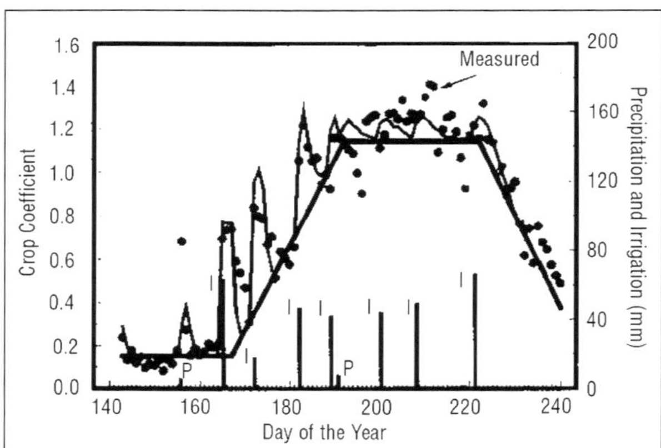

The agreement between the estimated values for daily $K_{cb} + K_e$ (thin continuous line) and actual 24-hour measurements (symbols) is relatively good. Measured and predicted $K_{cb} + K_e$ was higher following wetting by rainfall or irrigation, as expected. The two wet soil evaporation 'spikes' occurring during the late initial period and early development period (between days 160 and 180) were less than $K_{c\ max}$, because this evaporation was from wetting by furrow irrigation where f_w = 0.5. The value for f_{ew} was constrained to f_w by Equation during these two events, because during this period, $f_w < 1 - f_c$. Therefore, less than all of the 'potential energy' was converted into evaporation due to the limitation on maximum evaporation per unit surface area that was imposed by Equation. Measured (symbols) and predicted (thin line) daily coefficients (K_{cb} + K_e) and the basal crop curve (thick line) for a dry bean crop at Kimberly, Idaho. P in the Fig. denotes a precipitation event and I denotes an irrigation).

TRANSPIRATION COMPONENT (K_{CB} ET_O)

Basal Crop Coefficient (K_{cb})

The basal crop coefficient (K_{cb}) is defined as the ratio of the crop evapotranspiration over the reference evapotranspiration (ET_c/ET_o) when the soil surface is dry but transpiration is occurring at a potential rate, *i.e.*, water is not limiting transpiration. Therefore, 'K_{cb} ET_o' represents primarily the transpiration component of ET_c. The K_{cb} ET_o does include a residual diffusive evaporation component supplied by soil water below the dry surface and by soil water from beneath dense vegetation.

As the K_c values include averaged effects of evaporation from the soil surface, the K_{cb} values lie below the K_c values and a separate Table for K_{cb} is required. Recommended values for K_{cb} for the same crops. The values for K_{cb} in the Table represent K_{cb} for a sub-humid climate and with moderate wind speed. For specific adjustment in climates where RH_{min} differs from 45 per cent or where the wind speed is larger or smaller than 2 m/s, the $K_{cb\ mid}$ and $K_{cb\ end}$ values larger than 0.45 must be adjusted using the following equation:

$$K_{cb} = K_{cb(Tab)} + [0.04(u_2 - 2) - 0.004(RH_{min} - 45)]\left(\frac{h}{3}\right)^{0.3}$$

where

$K_{cb\ (Tab)}$ the value for $K_{cb\ mid}$ or $K_{cb\ end}$ (if ≥ 0.45) taken from Table ,

u_2 the mean value for daily wind speed at 2 m height over grass during the mid or late season growth stage [m s^{-1}] for 1 m s^{-1} ≤ u_2 ≤ 6 m s^{-1},

RH_{min} the mean value for daily minimum relative humidity during the mid- or late season growth stage [per cent] for 20 per cent ≤ RH_{min} ≤ 80 per cent,

h the mean plant height during the mid or late season stage [m] for 20 per cent ≤ RH_{min} ≤ 80 per cent.

For a full discussion on the impact of the climatic correction, and the numerical determination of $K_{cb\ mid}$ and $K_{cb\ end}$, the user is referred to the discussions on $K_{c\ mid}$ and $K_{c\ end}$.

The general guidelines that were used in deriving K_{cb} values from the K_c values. Where local research results are available, values for K_{cb} from Table can be modified to reflect effects of local conditions, cultural practices or crop varieties on K_{cb}. However, local values for K_{cb} should not be expected to deviate by more than 0.2 from the values. A greater deviation should signal the need to investigate or evaluate the local research technique, equipment and cultural practices. Where local K_{cb} values are used, no adjustment for climate using Equation 70 is necessary.

Determination of Daily K_{cb} values

Only three point values are required to describe and to construct the crop coefficient curve. After dividing the growing period into the four general growth stages and selecting and adjusting the K_{cb} values corresponding to the initial ($K_{cb\ ini}$), mid-season ($K_{cb\ mid}$) and end of the late season stages ($K_{cb\ end}$), the crop coefficient curve can be drawn and the K_{cb} coefficients can be derived.

Determination of Daily Values for K_{cb}

Calculate the basal crop coefficient for the dry beans at the middle of each of the four growth stages.

Initial stage (L_{ini} = 25 days), at day 12 of the growing period:

$K_{cb} = K_{cb\ ini} = 0.15$

Crop development stage (L_{dev} = 25 days), at day (25 + 25/2 =) 37 of the growing period, using Equation:

$K_{cb} = 0.15 + [(37 - 25)/25]\ (1.14 - 0.15) = 0.63$

Mid-season stage (L_{mid} = 30 days), at day (25 + 25 + 30/2 =) 65 of the growing period:

$K_{cb} = K_{cb\ mid} = 1.14$

Late season stage (L_{late} = 20 days), at day (25 + 25 + 30 + 20/2 =) 90 of the growing period, Equation:

$K_{cb} = 1.14 + [(90 - (25 + 25 + 30))/20]\ (0.25 - 1.14) = 0.70$

The basal crop coefficients, K_{cb}, at days 12, 37, 65 and 90 of the growing period are 0.15, 0.63, 1.14 and 0.70 respectively.

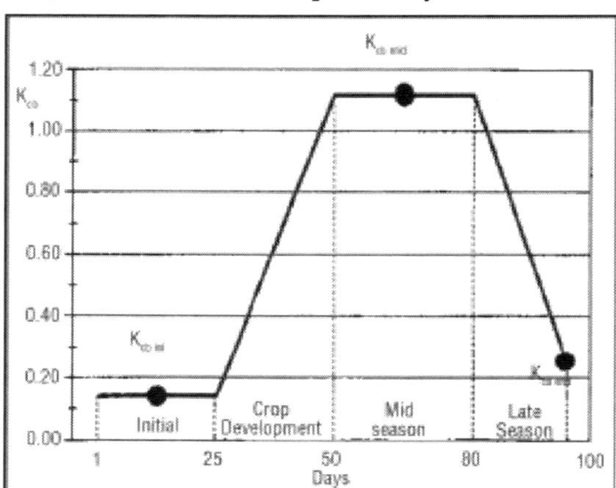

Fig. Constructed Basal Crop Coefficient (K_{cb}) Curve for a Dry Bean Crop using Growth Stage Lengths of 25, 25, 30 and 20 Days.

Evaporation Component (K_e ET_o)

The soil evaporation coefficient, K_e, describes the evaporation component of ET_c. Where the topsoil is wet, following rain or irrigation, K_e is maximal.

Where the soil surface is dry, K_e is small and even zero when no water remains near the soil surface for evaporation.

Calculation Procedure

Where the soil is wet, evaporation from the soil occurs at the maximum rate. However, the crop coefficient ($K_c = K_{cb} + K_e$) can never exceed a maximum value, $K_{c\ max}$. This value is determined by the energy available for evapotranspiration at the soil surface ($K_{cb} + K_e \leq K_{c\ max}$) or $K_e \leq (K_{c\ max} - K_{cb})$.

When the topsoil dries out, less water is available for evaporation and a reduction in evaporation begins to occur in proportion to the amount of water remaining in the surface soil layer, or:

$K_e = K_r (K_{c\ max} - K_{cb}) \leq f_{ew} K_{c\ max}$

where

K_e soil evaporation coefficient,

K_{cb} basal crop coefficient,

$K_{c\ max}$ maximum value of K_c following rain or irrigation,

K_r dimensionless evaporation reduction coefficient dependent on the cumulative depth of water depleted (evaporated) from the topsoil,

f_{ew} fraction of the soil that is both exposed and wetted, *i.e.*, the fraction of soil surface from which most evaporation occurs.

In computer programming terminology, Equation 71 is expressed as $K_e = \min (K_r (K_{c\ max} - K_{cb}), f_{ew} K_{c\ max})$.

Following rain or irrigation K_r is 1, and evaporation is only determined by the energy available for evaporation. As the soil surface dries, K_r becomes less than one and evaporation is reduced. K_r becomes zero when no water is left for evaporation in the upper soil layer.

Evaporation occurs predominantly from the exposed soil fraction. Hence, evaporation is restricted at any moment by the energy available at the exposed soil fraction, *i.e.*, K_e cannot exceed $f_{ew} K_{c\ max}$, where f_{ew} is the fraction of soil from which most evaporation occurs, *i.e.*, the fraction of the soil not covered by vegetation and that is wetted by irrigation or precipitation.

The calculation procedure consists in determining:

- The upper limit $K_{c\ max}$;
- The soil evaporation reduction coefficient K_r; and
- The exposed and wetted soil fraction f_{ew}

The estimation of K_r requires a daily water balance computation for the surface soil layer.

Upper Limit Kc max

$K_{c\ max}$ represents an upper limit on the evaporation and transpiration from any cropped surface and is imposed to reflect the natural constraints placed on available energy represented by the energy balance difference R_n - Γ- H. $K_{c\,max}$ ranges from about 1.05 to 1.30 when using the grass reference ET_o:

$$K_{c\,max} = \max\left(\left\{1.2 + [0.04(u_2 - 2) - 0.004(RH_{min} - 45)]\left(\frac{h}{3}\right)^{0.3}\right\}\{K_{cb} + 0.05\}\right)$$

where

h mean maximum plant height during the period of calculation (initial, development, mid-season, or late-season),

K_{cb} basal crop coefficient,

max ( ) maximum value of the parameters in braces {} that are separated by the comma.

Equation ensures that $K_{c\,max}$ is always greater or equal to the sum K_{cb} + 0.05. This requirement suggests that wet soil will always increase the value for K_{cb} by 0.05 following complete wetting of the soil surface, even during periods of full ground cover.

A value of 1.2 instead of 1 is used for $K_{c\,max}$ in Equation because of the effect of increased aerodynamic roughness of surrounding crops during development, mid-season and late season growth stages which can increase the turbulent transfer of vapour from the exposed soil surface. The "1.2" coefficient also reflects the impact of the reduced albedo of wet soil and the contribution of heat stored in dry soil prior to the wetting event. All of these factors can contribute to increased evaporation relative to the reference. The "1.2" coefficient in Equation represents effects of wetting intervals that are greater than 3 or 4 days. If irrigation or precipitation events are more frequent, for example daily or each two days, then the soil has less opportunity to absorb heat between wettings, and the "1.2" coefficient in Equation can be reduced to about 1.1. The time step to compute $K_{c\,max}$ may vary from daily to monthly.

Soil Evaporation Reduction Coefficient (K_r)

Soil evaporation from the exposed soil can be assumed to take place in two stages: an energy limiting stage, and a falling rate stage. When the soil surface is wet, K_r is 1. When the water content in the upper soil becomes limiting, K_r decreases and becomes zero when the total amount of water that can be evaporated from the topsoil is depleted.

Maximum Amount of Water that can be Evaporated

In the simple evaporation procedure, it is assumed that the water content of me evaporating layer of the soil is at field capacity, θ_{FC} shortly following a major wetting event and that the soil can dry to a soil water content level that is halfway between oven dry (no water left) and wilting point, θ_{WP}. The amount of water that can be depleted by evaporation during a complete drying cycle can hence be estimated as:

$$TEW = 1000\,(\theta_{FC} - 0.5\,\theta_{WP})\,Z_e$$

where

TEW total evaporable water = maximum depth of water that can be evaporated from the soil when the topsoil has been initially completely wetted [mm],

θ_{FC} soil water content at field capacity [$m^3\ m^{-3}$],

θ_{WP} soil water content at wilting point [$m^3\ m^{-3}$],

Z_e depth of the surface soil layer that is subject to drying by way of evaporation [0.10-0.15m]. Where unknown, a value for Z_e, the effective depth of the soil evaporation layer, of 0.10-0.15 m is recommended. Typical values for θ_{FC}, θ_{WP} and TEW are given in Table.

Typical Soil Water Characteristics for Different Soil Types

Soil type (USA Soil Texture Classification)	Soil water characteristics			Evaporation parameters	
	θ_{FC}	θ_{WP}	(θ_{FC} - θ_{WP})	Amount of water that can be depleted by evaporation	
				stage 1 REW	stages 1 and 2 TEW* (Z_e = 0.10m)
	m^3/m^3	m^3/m^3	m^3/m^3	mm	mm
Sand	0.07 - 0.17	0.02 - 0.07	0.05 - 0.11	2 - 7	6 - 12
Loamy sand	0.11 - 0.19	0.03 - 0.10	0.06 - 0.12	4 - 8	9 - 14
Sandy loam	0.18 - 0.28	0.06 - 0.16	0.11 - 0.15	6 - 10	15 - 20
Loam	0.20 - 0.30	0.07 - 0.17	0.13 - 0.18	8 - 10	16 - 22
Silt loam	0.22 - 0.36	0.09 - 0.21	0.13 - 0.19	8 - 11	18 - 25
Silt	0.28 - 0.36	0.12 - 0.22	0.16 - 0.20	8 - 11	22 - 26
Silt clay loam	0.30 - 0.37	0.17 - 0.24	0.13 - 0.18	8 - 11	22 - 27
Silty clay	0-30 - 0.42	0.17 - 0.29	0.13 - 0.19	8 - 12	22 - 28
Clay	0.32 - 0.40	0.20 - 0.24	0.12 - 0.20	8 - 12	22 - 29

Fig. Soil Evaporation Reduction Coefficient, K_r

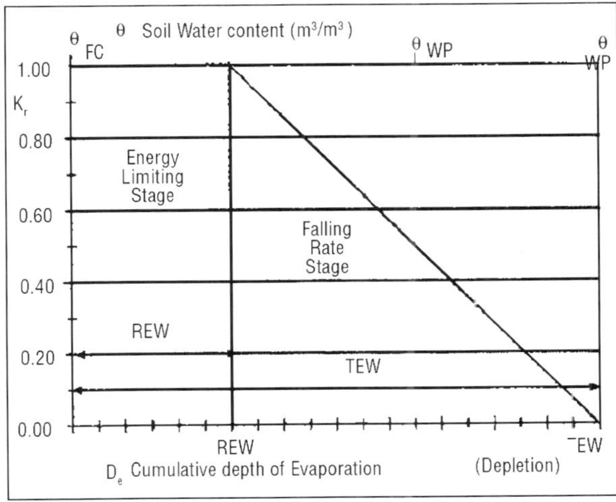

Stage 1: Energy Limiting Stage

At the start of a drying cycle, following heavy rain or irrigation, the soil water content in the topsoil is at field capacity and the amount of water depleted by evaporation, D_e, is zero. During stage 1 of the drying process, the soil surface remains wet and it is assumed that evaporation from soil exposed to the atmosphere will occur at the maximum rate limited only by energy availability at the soil surface.

This stage holds until the cumulative depth of evaporation, D_e, is such that the hydraulic properties of the upper soil become limiting and water cannot be transported to the soil surface at a rate that can supply the potential demand. During stage 1 drying, $K_r = 1$.

The cumulative depth of evaporation, De, at the end of stage 1 drying is REW (Readily evaporable water, which is the maximum depth of water that can be evaporated from the topsoil layer without restriction during stage 1). The depth normally ranges from 5 to 12 mm and is generally highest for medium and fine textured soils.

Stage 2: Falling Rate Stage

The second stage (where the evaporation rate is reducing) is termed the 'falling rate stage' evaporation and starts when D_e exceeds REW. At this point, the soil surface is visibly dry, and the evaporation from the exposed soil decreases in proportion to the amount of water remaining in the surface soil layer:

$$K_r = \frac{TEW - D_{ei-1}}{TEW - REW} \text{for} D_{ei-1} > REW$$

where

K_r dimensionless evaporation reduction coefficient dependent on the soil water depletion (cumulative depth of evaporation) from the topsoil layer (K_r = 1 when $D_{e,\,i-1} \leq$ REW),

$D_{e,\,i-1}$ cumulative depth of evaporation (depletion) from the soil surface layer at the end of day $_{i-1}$ (the previous day) [mm],

TEW maximum cumulative depth of evaporation (depletion) from the soil surface layer when $K_r = 0$ (TEW = total evaporable water) [mm],

REW cumulative depth of evaporation (depletion) at the end of stage 1 (REW = readily evaporable water) [mm].

Exposed and Wetted Soil Fraction (f_{ew})

Few: Calculation Procedure

In crops with incomplete ground cover, evaporation from the soil often does not occur uniformly over the entire surface, but is greater between plants where exposure to sunlight occurs and where more air ventilation is able to transport vapour from the soil surface to above the canopy. This is especially true where only part of the soil surface is wetted by irrigation.

It is recognized that both the location and the fraction of the soil surface exposed to sunlight change to some degree with the time of day and depending on row orientation.

The procedure presented here predicts a general averaged fraction of the soil surface from which the majority of evaporation occurs. Diffusive evaporation from the soil beneath the crop canopy is assumed to be largely included in the basal K_{cb} coefficient.

Where the complete soil surface is wetted, as by precipitation or sprinkler, then the fraction of soil surface from which most evaporation occurs, f_{ew}, is essentially defined as $(1 - f_c)$, where f_c is the average fraction of soil surface covered by vegetation and is the approximate fraction of soil surface that is exposed.

However, for irrigation systems where only a fraction of the ground surface is wetted, f_{ew} must be limited to f_w, the fraction of the soil surface wetted by irrigation. Therefore, f_{ew} is calculated as:

$$f_{ew} = \min(1 - f_c, f_w)$$

where

$1 - f_c$ average exposed soil fraction not covered (or shaded) by vegetation [0.01 - 1],

f_w average fraction of soil surface wetted by irrigation or precipitation [0.01 - 1].

The limitation imposed by Equation assumes that the fraction of soil wetted by irrigation occurs within the fraction of soil exposed to sunlight and ventilation. This is generally the case, except perhaps with drip irrigation.

In the case of drip irrigation, where the majority of soil wetted by irrigation may be beneath the canopy and may therefore be shaded, more complex models of the soil surface and wetting patterns may be required to accurately estimate total evaporation from the soil. In this case, the value for f_w may need to be reduced to about one-half to one-third of that given in Table to account for the effects of shading of emitters by the plant canopy on the evaporation rate from wetted soil. A general approach could be to multiply f_w by $[1-(2/3)f_c]$ for drip irrigation.

fw: Fraction of Soil Surface Wetted by Irrigation or Precipitation

Table presents typical values for f_w. Where a mixture of irrigation and precipitation occur within the same drying period or on the same day, the value for f_w should be based on a weighted average of the f_w for precipitation ($f_w = 1$) and the f_w for the irrigation system.

The weighting should be approximately proportional to the infiltration depths from each water source.

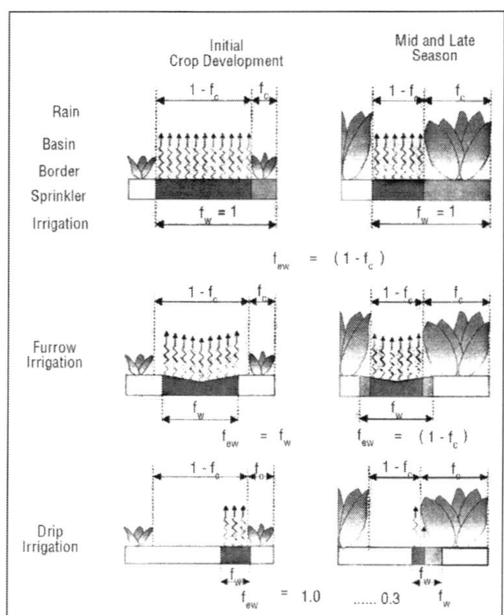

Fig. Determination of Variable f_{ew} (cross-hatched areas) as a Function of the Fraction of Ground Surface Coverage (f_c) and the Fraction of the Surface Wetted (f_w)

Table. Common values of fraction f_w of soil surface wetted by irrigation or precipitation

Wetting Event	f_w
Precipitation	1.0
Sprinkler irrigation	1.0
Basin irrigation	1.0
Border irrigation	1.0
Furrow irrigation (every furrow), narrow bed	0.6...1.0
Furrow irrigation (every furrow), wide bed	0.4... 0.6
Furrow irrigation (alternated furrows)	0.3...0.5
Trickle irrigation	0.3... 0.4

Alternatively, on each day of the application, the following rules can be applied to determine f_w for that and subsequent days in a more simplified manner:

- Surface is wetted by irrigation and rain: f_w is the f_w for the irrigation system;
- Surface is wetted by irrigation: f_w is the f_w for the irrigation system;
- Surface is wetted by significant rain (*i.e.*, > 3 to 4 mm) with no irrigation: $f_w = 1$;
- Where there is neither irrigation nor significant precipitation: f_w is the f_w of the previous day.

1 - fc: Exposed Soil Fraction

The fraction of the soil surface that is covered by vegetation is termed f_c. Therefore, $(1 - f_c)$ represents the fraction of the soil that is exposed to sunlight and air ventilation and which serves as the site for the majority of evaporation from wet soil. The value for f_c is limited to < 0.99. The user should assume appropriate values for the various growth stages. Typical values for f_c and $(1 - f_c)$ are given in Table.

Table Common values of fractions covered by vegetation (f_c) and exposed to sunlight ($1 - f_c$)

Crop growth stage	f_c	$1 - f_c$
Initial stage	0.0 - 0.1	1.0 - 0.9
Crop development stage	0.1 - 0.8	0.9 - 0.2
Mid-season stage	0.8 - 1.0	0.2 - 0.0
Late season stage	0.8 - 0.2	0.2 - 0.8

Where f_c is not measured, f_c can be estimated using the relationship:

$$f_c = \left(\frac{K_{cb} - K_{c\,min}}{K_{c\,max} - K_{c\,min}} \right)^{(1+0.5h)}$$

where

f_c the effective fraction of soil surface covered by vegetation [0 - 0.99],

K_{cb} the value for the basal crop coefficient for the particular day or period,

$K_{c\,min}$ the minimum K_c for dry bare soil with no ground cover [≈ 0.15 - 0.20],

$K_{c\,max}$ the maximum K_c immediately following wetting (Equation 72),

h mean plant height [m].

This equation should be used with caution and validated from field observations. $K_{c\,min}$ is the minimum crop coefficient for dry bare soil when transpiration and evaporation from the soil are near baseline (diffusive) levels. $K_{c\,min} \approx 0.15 - 0.20$ is recommended. The value of $K_{c\,min}$ is an integral part of all K_{cb} coefficients. $K_{c\,min}$ ordinarily has the same value as the $K_{cb\,ini}$ used for annual crops under nearly bare soil conditions (0.15 - 0.20).

Equation assumes that the value for K_{cb} is largely affected by the fraction of soil surface covered by vegetation. This is a good assumption for most vegetation and conditions.

The '1+0.5h' exponent in the equation represents the effect of plant height on shading the soil surface and in increasing the value for K_{cb} given a specific value for f_c. The user should limit the difference $K_{cb} - K_{c\,min}$ to ≥ 0.01 for numerical stability. The value for f_c will change daily as K_{cb} changes. Therefore, the above equation is applied daily.

Application of Equation predicts that f_c decreases during the late season period in proportion to K_{cb}, even though the ground may remain covered with senescing vegetation. This prediction helps to account for the local transport of sensible heat from senescing leaves to the soil surface below.

Calculation of the Crop Coefficient (K_{cb} + K_e) under Drip Irrigation

The cotton field in the previous example has been irrigated by drip irrigation, where the emitters are placed beneath the cotton canopy. The fraction of the field surface wetted by the irrigation is 0.3.

The f_{ew} in this case is calculated from Equation as $f_{ew} = \min(1 - f_c, f_w)$. Because the emitters are beneath the canopy so that less energy is available for evaporation, the value for f_w is reduced by multiplying by 1 - (2/3)f_c, so that:

$f_{ew} = \min[(1 - f_c),(1 - 0.67\, f_c)\, f_w)] = \min[(1\text{-}0.53), (1 - 0.67(0.53))(0.3)] = 0.19$

Assuming that the irrigation was sufficient to fill the f_w portion of the evaporating layer to field capacity, so that $K_r = 1$, evaporation would be in stage 1.

From Equation: $K_e = 1.00\ (1.30\text{-}0.90) = 0.40$.

The value is compared to the upper limit $f_{ew}\, K_{c\,max} = 0.19\ (1.30) = 0.25$. Because $0.25 < 0.40$, K_e from the f_w fraction of the surface area is constrained by the available energy. Therefore $K_e = 0.25$.

The total K_c for the drip irrigated field, assuming no moisture stress due to dry soil, is:

$K_c = K_{cb} + K_e = 0.9 + 0.25 = 1.15$. This K_c value is less than that for sprinkler and furrow irrigation.

Calculation of the Crop Coefficient (Kcb + Ke) under Sprinkler Irrigation

A field of cotton has just been sprinkler irrigated. The K_{cb} for the specific day (during the development period) has been computed and then interpolated from the K_{cb} curve as 0.9. The ET_o = 7 mm/day, u_2 = 3 m/s and RH_{min} = 20 per cent. Estimate the crop coefficient (K_{cb} + K_e).

Assuming h = 1 m, from Equation, $K_{c\,max}$ for this arid climate is:

$$K_{c\,max} = \max\left(\left\{1.2 + [0.04(3-2) - 0.004(20-45)]\left(\frac{1}{3}\right)^{0.3}\right\}\{0.9 + 0.05\}\right) = 1.30$$

From Equation , where $K_{c\,min} = 0.15$:

$f_c = [(K_{cb} - K_{c\,min})/(K_{c\,max} - K_{c\,min})]^{(1 + 0.5h)} = [(0.9\text{-}0.15)/(1.3\text{-}0.15)]^{(1+0.5(1))} = 0.53$.

As the field was sprinkler irrigated, $f_w = 1.0$ and from Equation:

$f_{ew} = \min(1 - f_c, f_w) = \min(1\text{-}\ 0.53, 1.0) = 0.47$.

Assuming that the irrigation was sufficient to fill the evaporating layer to field capacity, so that $K_r = 1$, evaporation would be in stage 1.

From Equation: $K_e = 1.00\ (1.30 - 0.90) = 0.40$

The value is compared against the upper limit $f_{ew}\, K_{c\,max}$ to ensure that it is less than the upper limit:

$f_{ew}\, K_{c\,max} = 0.47\ (1.30) = 0.61$, which is greater than the value for K_e. Therefore, the value for K_e can be used with no limitation.

The total K_c for the field, assuming no moisture stress due to a dry soil profile, is

$K_c = K_{cb} + K_e = 0.9 + 0.40 = 1.30$.

This value is large because of the very wet soil surface, the relatively tall rough crop as compared to the grass reference, and the arid climate ($u_2 = 3$ m/s and $RH_{min} = 20$ per cent). In this situation, K_c happens to equal $K_{c\ max}$, as the field has just been wetted by sprinkler irrigation.

Calculation of the Crop Coefficient ($K_{cb} + K_e$) under Furrow Irrigation

The cotton field in the previous example has been irrigated by furrow irrigation of alternate rows rather than by sprinkler, and the fraction of the field surface wetted by the irrigation is 0.3.

The f_{ew} in this case is calculated from Equation as:

$f_{ew} = \min(1 - f_c, f_w) = \min(1 - 0.53, 0.3) = 0.3$.

Assuming that the irrigation was sufficient to fill the f_{ew} portion of the evaporating layer to field capacity, so that $K_r = 1$, evaporation would be in stage 1.

From Equation: $K_e = 1.00\ (1.30 - 0.9) = 0.40$

The value is compared to the upper limit $f_{ew}\ K_{c\ max}$ which is $0.30\ (1.30) = 0.39$. Because $0.40 > 0.39$, K_e from the f_{ew} surface area is constrained to 0.39.

The total K_c for the furrow irrigated field, assuming no moisture stress due to dry soil, is $K_c = K_{cb} + K_e = 0.9 + 0.39 = 1.29$. This value is essentially the same as for the previous example because the procedure assumes that the soil between alternate rows is the portion that is wetted by the irrigation, so that the majority of the field surface has either vegetation cover or wet soil.

Bibliography

A. Arunachalam and M.L. Khan: *Sustainable Management of Forest India*, International Book Distributors, Delhi, 2001.

A. Rashid: *Introduction to Genetic Engineering of Crop Plants : Aims and Achievements*, I.K. International Publication, Delhi, 2009.

B.N. Pandey, Shivesh Pratap Singh and Rashmi Singh: *Sustainable Management and Conservation of Biodiversity*, Narendra Publishing House, Delhi, 2010.

B.N. Singh, Raj Kumar, Sarfaraz Alam, S.K. Singh and V.K. Kumar: *Strategies for Sustainable Management of Disaster in India*, Shree Publication, Delhi, 2011.

Bidhan Roy and Asit Kumar Basu: *Abiotic Stress Tolerance in Crop Plants : Breeding and Biotechnology*, New India Publication, Delhi, 2009.

C. Rajendran, K. Ramamoorthy and S. Juliet Hepziba: *Nutritional and Physiological Disorders in Crop Plants*, Scientific Publication, Delhi, 2009.

Franklin P. Gardner, R. Brent Pearce and Roger L. Mitchell: *Physiology of Crop Plants*, Scientific Publication, Jaipur, 2010.

G. Kalloo, Mathura Rai, Major Singh and Sanjeet Kumar: *Heterosis in Crop Plants*, Researchco Book Center, Delhi, 2006.

Herbert Kendall Hayes and Ralph John Garber: *Breeding Crop Plants*, Asiatic Publication, Delhi, 2008.

Ishi Khosla: *Is Wheat Killing You?: The Essential Cookbook and Guide to a Wheat-Free Life*, Penguin Books, Delhi, 2011.

Jyoti Parikh and Hemant Datye: *Sustainable Management of Wetlands : Biodiversity and Beyond*, Sage Publication, Delhi, 2003.

K.D. Srivastava: *Biology and Management of Wheat Pathogens*, Studium Press, Delhi, 2010.

Khundrakpam Moirangleima: *Sustainable Management of Wetlands : Central Valley of Manipur*, B.R. Publication, Delhi, 2010.

Klaus Seeland and Franz Schmithusen: *Man in the Forest : Local Knowledge and Sustainable Management of Forests and Natural Resources in Tribal Communities in India*, D K Printworld, Delhi, 2000.

M. Prakash and K. Balakrishnan: *Abiotic Stress Tolerance in Crop Plants*, Satish Serial Publishing House, Delhi, 2014.

Madan Pal Singh, Sangeeta Khetarpal, Renu Pandey and Pramod Kumar: *Climate Change : Impacts and Adaptations in Crop Plants*, Today and Tomorrow Publication, Delhi, 2011.

N R Das: *Wheat Crop Management*, Scientific Publication, Delhi, 2008.

N.C. Patel, R. Subbaiah, P.M. Chauhan, K.C. Patel and J.N. Nandasana: *Sustainable Management of Water Resources*, Himanshu Publication, Delhi, 2005.

Overy, Richard,Wheatcroft: *Road to War, The*, Andrew, Random House India, Delhi, 2009.

P. Sudhakar, P. Latha and P.V. Reddy: *Phenotyping Crop Plants for Physiological and Biochemical Traits*, BS Publications, Delhi, 2014.

Peter Wheatcroft: *World Class IT Service Delivery*, Viva Books, Delhi, 2008.

Pradeep K. Sharma: *Foodgrain Economy of India : Government Intervention in Rice & Wheat Markets*, Shipra Publication, Delhi, 1997.

Ravindra N. Chibbar and James Dexter: *Wheat Science Dynamics: Challenges and Opportunities (Proceedings International Wheat Quality Conference-IV)*, Agrobios (International), Delhi, 2011.

Rekha Sharma: *Wheat Germ : A Nutritional Capsule*, Dattsons Publication, Delhi, 2003.

S.M. Paul Khurana: *Pathological Problems of Economic Crop Plants and Their Management*, Scientific Publication, Delhi, 1998.

Sultan Singh and I S Pawar: *Trends in Wheat Breeding*, CBS Publication, Delhi, 2006.

V.L. Chopra and R.S. Paroda: *Approaches for Incorporating Drought and Salinity Resistance in Crop Plants*, New India Publishing Agency, Delhi, 2015.

Wheatcroft, Andrew: *Enemy at the Gate, The*, Random House India, Delhi, 2009.

Index